The making of British bioe

MANCHESTER
1824

Manchester University Press

6187 8716

The making of British bioethics

Duncan Wilson

Manchester University Press

Published by Manchester University Press
Altrincham Street, Manchester M1 7JA, UK
www.manchesteruniversitypress.co.uk

British Library Cataloguing-in-Publication Data
A catalogue record for this book is available from the British Library

Library of Congress Cataloging-in-Publication Data applied for

ISBN 978 0 7190 9619 8 hardback
ISBN 978 1 8477 9887 9 open access

First published 2014

Typeset in 10/12 Sabon by
Servis Filmsetting Ltd, Stockport, Cheshire
Printed in Great Britain
by TJ International Ltd, Padstow

To Nisha and my parents,
and in memory of John Pickstone

Contents

Acknowledgements

My interest in British bioethics started during my PhD research, when I noticed that philosophers, lawyers and other 'outsiders' had begun to enter debates that had traditionally been the preserve of doctors and scientists from the mid 1970s onwards. I soon realised that the growth of this outside input, or 'bioethics', was well documented for the United States, but that there was much less written for Britain, where it emerged slightly later and for different reasons. I decided then to make the history of British bioethics my next research topic.

The progression from that initial idea to this book owes a lot to the help of many people. I first want to thank the Wellcome Trust for funding my research (grant number 081493). I am also grateful to the many archivists who provided access to materials. I want to thank staff at the National Archives, London; the Wellcome Library for the History of Medicine, London; the John Rylands Library at the University of Manchester; the British Film Institute, London; the Durham Cathedral archives; and the Historical Collections at the University of Virginia. I owe particular thanks to those people who allowed me to look through private papers and uncatalogued archives. I want to thank Catherine Spanswick, from the University of Manchester's Institute for Science, Ethics and Innovation (ISEI), for letting me access papers from the Centre for Social Ethics and Policy; David Archard, from the University of Belfast, who let me look at papers from the Society for Applied Philosophy when he was working at the University of Lancaster; and Jonathan Montgomery, from University College London, who let me look through correspondence relating to the Nuffield Council on Bioethics.

I am extremely grateful to all the interviewees listed in the bibliography. My research would not have been possible had they not spared the time to meet me and patiently answer my questions. I am also grateful to the many people whom I did not formally interview, but who provided me with crucial materials and leads. I owe particular thanks to Robin Downie, for his hospitality in Glasgow and many useful readings. I also owe thanks to David Archard, Margaret Brazier, Raymond Plant, Rebecca Bennett, Richard Ashcroft, Adam Hedgecoe, David Weatherhall, Nathan Emmerich, Noortje Jacobs and Bella Starling, for providing valuable material and information on developments in sociology, bioethics, philosophy, law and the biomedical sciences.

I am also indebted to everyone who gave feedback on seminar papers at the University of Leeds, the London School for Hygiene and Tropical Medicine, University College London, Imperial College London, the University of Liverpool, the University of Manchester, the Postgraduate Conference on Bioethics and the Institute for Historical Research's 2011 conference on Anglo-American History. I am especially grateful to colleagues at the University of Manchester's Centre for the History of Science, Technology and Medicine (CHSTM), who gave me valuable advice on ideas, drafts and potential titles. These include, in no particular order, Michael Worboys, Neil Pemberton, Emma Jones, Michael Bresalier, Carsten Timmermann, Ed Ramsden, Rob Kirk, Elizabeth Toon and Jenny Goodare. I owe particular thanks to my former colleague Sam Alberti, now director of the Hunterian Museum at the Royal College of Surgeons in London, and to John Pickstone, most recently emeritus professor at CHSTM, who both gave feedback on several draft chapters.

We were all deeply saddened when John died following a short illness in February 2014. He was CHSTM's founding director and remained a huge presence in the department, and the field more generally, until his death. Aside from an incomparable body of work, John's legacy is evident in the number of historians he mentored and encouraged – myself included. I valued his help with this book as much as I did when I started my PhD under his supervision, and will miss his guidance and friendship. John's advice was peerless, as always, and improved the book a great deal. It goes without saying that any mistakes that may appear in the following chapters are mine alone.

I also want to thank friends and family who have put up with me throughout this whole process. My parents, John and Helen Wilson, continue to provide invaluable support and still don't tell me to 'get a real job'. Nisha, meanwhile, is the best proofreader, friend and wife anyone could wish for.

Abbreviations

AAMR	Association for the Advancement of Medicine by Research
BAAS	British Association for the Advancement of Science
BMA	British Medical Association
BSSRS	British Society for Social Responsibility in Science
CCH	Churches' Council of Healing
CCHMS	Central Committee for Hospital Medical Services
CEC	Central Ethical Committee
CHAI	Commission for Healthcare Audit and Inspection
CHI	Commission for Health Improvement
CMO	Chief Medical Officer
CSEP	Centre for Social Ethics and Policy
DES	Department of Education and Science
DHSS	Department for Health and Social Security
EMG	Edinburgh Medical Group
ESBAC	Emerging Science and Bioethics Advisory Committee
FMG	Frontier Medical Group
GMAG	Genetic Manipulation Advisory Group
GMC	General Medical Council
HFEA	Human Fertilisation and Embryology Authority
HGC	Human Genetics Commission
HRA	Health Research Agency
HTA	Human Tissue Authority
IME	Institute of Medical Ethics
IRM	Institute of Religion and Medicine
LMG	London Medical Group
MMG	Manchester Medical Group
MRC	Medical Research Council
NCP	National Committee for Philosophy

PHS	Public Health Service
RAE	Research Assessment Exercise
REC	Research Ethics Committee
RCP	Royal College of Physicians
SCM	Student Christian Mission
SPUC	Society for the Protection of Unborn Children
SSME	Society for the Study of Medical Ethics
UCL	University College London
UCLA	University of California, Los Angeles
UFC	Universities Funding Council
UGC	University Grants Commission

Introduction

What is bioethics?

Recent decades have witnessed profound shifts in the politics of medicine and the biological sciences, in which members of several professions now consider issues that were traditionally the preserve of doctors and scientists. In government committees and organisations such as the General Medical Council, professional conduct is determined by a diverse group of participants that includes philosophers, lawyers, theologians, social scientists, doctors, scientists, healthcare managers and representatives from patient or pressure groups. Teaching ethics, once a matter of professional etiquette, takes place on dedicated courses and in specialised departments that emphasise law and moral philosophy. A growing body of interdisciplinary journals considers topics that were once confined to the correspondence pages of the *Lancet* or the *British Medical Journal*. And public discussion of issues such as embryo research, cloning, genetic engineering or assisted dying are now as likely to be led by a lawyer or a philosopher as a doctor or a scientist.

This new approach is known as 'bioethics': a neologism derived from the Greek words *bios* (life) and *ethike* (ethics), which the *Oxford English Dictionary* defines as the discussion and management of 'the ethical issues relating to the practice of medicine and biology, or arising from advances in these subjects'.[1] The dictionary attributes the term 'bioethics' to the American biochemist and research oncologist Van Rensselaer Potter, who introduced it in a 1970 article.[2] But Potter's view of bioethics differs from the dictionary's definition, and designates an approach that is not familiar to us today. Potter characterised bioethics as a new system of ethics, or a 'science of survival', that drew on 'modern concepts of biology' in order to guide moral choices and ensure human survival in the face of environmental problems.[3] He argued that political or ethical decisions 'made in ignorance of biological knowledge, or in defiance of it, may jeopardize man's future and indeed the future of earth's biological resources for human needs'.[4] 'Bioethics', he continued, 'should develop a realistic knowledge of biological knowledge and its limitations in order to make recommendations in the field of public policy.'[5]

Quite independently of Potter, the Dutch obstetrician André Hellegers and the political activist Sargent Shriver also coined the term 'bioethics' in 1970, when they opened the Joseph and Rose Kennedy Institute for the Study of Human Reproduction and Bioethics at Georgetown University, a private Jesuit institution in Washington DC.[6] Hellegers and Shriver's definition is the one we recognise today. Amid growing discussion of the social impact of biological research, the rationing of new medical technologies such as kidney dialysis and the rights of patients and experimental subjects, they viewed bioethics as the ethical scrutiny of specific problems raised by medicine and the biological sciences.[7] These debates on the ethical aspects of science and medicine quickly became known as 'bioethics' following the opening of the Georgetown Institute. Between 1972 and 1974, the theologian Warren Reich began work on an *Encyclopedia of Bioethics*, the philosopher Daniel Callahan wrote an article on 'Bioethics as a Discipline' and the Library of Congress adopted 'bioethics' as a subject heading.[8]

The participants in this burgeoning field claimed that doctors and scientists could not solve ethical problems on their own. They differentiated bioethics from the prior tradition of medical ethics, in which doctors governed their own conduct through professional codes and bedside training, by arguing that outsiders should play a greater role in discussing ethical issues and, crucially, in determining professional conduct.[9] Figures such as Callahan argued that public disquiet and the moral problems raised by new technologies meant that 'a good training in medicine' no longer qualified doctors 'to make good ethical decisions'. Instead, philosophers, lawyers and theologians had now become vital 'for the definition of issues, methodological strategies, and procedures for decision-making'.[10] Bioethics, Callahan stated, should 'be so designed, and its practitioners so trained that it will directly – at whatever the cost to disciplinary elegance – serve those physicians and biologists whose position demands they make the practical decisions'.[11]

These arguments appealed to politicians, doctors and scientists concerned by falling public confidence, and ensured that Helleger's view of bioethics superseded Potter's more general calls for a new ethical framework. Its impact was evident in 1974, when President Richard Nixon responded to controversies surrounding human experimentation by convening a National Commission for the Protection of Human Subjects in Biomedical and Behavioral Research. The National Research Act that established the Commission stipulated that no more than five of its eleven members should be scientists or doctors – with the majority drawn from philosophy, law, sociology, theology and the general public.[12]

Senator Edward Kennedy, a critical figure in the Commission's formation, argued that policy should not emanate 'just from the medical profession, but from ethicists, the theologians, the lawyers and many other disciplines'.[13] This Commission was widely recognised as the first national bioethics committee, and in 1978 its recommendations led President Jimmy Carter to establish a permanent Presidential Commission for the Study of Ethical Problems in Medicine and Biomedical and Behavioral Research.[14] These events, coupled with the establishment of centres such as the Georgetown Institute and an Institute for Society, Ethics and the Life Sciences, later renamed the Hastings Center, led many to view bioethics as an important approach. By 1978, as David Rothman remarks, it was clear 'that the monopoly of the medical profession in medical ethics was over. The issues were now public and national – the province of an extraordinary variety of outsiders.'[15]

Contextualising bioethics

Although bioethics first emerged in the United States, the term and the approach it signifies quickly became a global phenomenon. As Sheila Jasanoff notes, today 'most Western governments, and increasingly developing states, have supplemented funding for the life sciences and technologies with public support for ethical analysis'.[16] Bioethicists now play a significant role in determining policies and guiding public debates across Europe, in Australia, Canada, Latin America, Israel, Pakistan, Japan, Singapore and South Korea, among other locations.[17]

The international growth and influence of bioethics has led

some to identify it as a decisive shift in the location and exercise of 'biopower', which Michel Foucault defined as the range of actors and strategies involved in the governance of individual and collective health.[18] In their work on modern configurations of biopower, Paul Rabinow and Nikolas Rose claim that bioethics has reshaped professional conduct in a range of settings so that the 'practices and dilemmas of life politics are not monopolized by states or even doctors'.[19] This 'bioethical complex', they argue, ensures that medical and scientific practices are 'now regulated by other authorities as never before'.[20]

In their work on the regulation of reproductive technologies and genetic research, Brian Salter, Charlotte Salter and Mavis Jones similarly identify bioethics as a 'new epistemic power' that is capable of setting agendas 'on the basis of an expert authority that can be used by governments to legitimize subsequent regulatory policy outputs'.[21] Brian Salter believes the emergence of bioethicists adds a new dimension to Foucault's work on biopower, since their ethical expertise regularly complements, or even replaces, the technical expertise of scientists and doctors. Salter argues this shift has created a new form of governance he terms 'cultural biopolitics'. Here, he claims, 'the focus for the operation of biopower is not the control of populations or bodies but the control of the values that permit or proscribe the development of health technologies … that, in turn, may subsequently act as modes of population or individual control'.[22]

As these accounts demonstrate, bioethics offers a rich subject for historical investigation. It reveals changing relations between professions, emerging notions of expertise and speaks to our perennial concerns with power – who wields it and to what ends. It appears an especially important subject in light of recent claims that we should focus on the ways in which biopower operated and was reconfigured during the twentieth century, along with the consequences for our notions of health, illness, morality and what it means to be human.[23] With this in mind, it is little wonder that writers from several fields have begun to chart the history of our 'insatiable demand for bioethics'.[24] The first accounts came from bioethicists themselves, often as a preface to books on bioethics or applied philosophy. These participant histories portrayed bioethics as a response to the unprecedented moral dilemmas and public concerns raised by new developments in medicine and the biological sciences during the late 1960s and 1970s. One bioethicist states, for example, that 'when biomedical sciences became capable of things previously unimaginable (like organ transplantation or artificially sustaining life), bioethics was invented to deal with the moral issues thus raised'.[25]

The authors of these participant histories also claim that bioethics drew on 'the climate of political radicalism and student activism' to critique professional authority and stand up for the rights of patients, research subjects and even experimental animals.[26] One early bioethicist argues the field was 'inextricably linked to public protests, teach-ins, and to civil rights, antiwar, and pro-feminist activities'.[27] These claims established a dichotomy between radical bioethicists and a conservative medical profession. And this was reinforced in the first major history of bioethics, David Rothman's *Strangers at the Bedside*, which adopted a 'twofold classification of doctors and outsiders' and claimed that outside involvement 'came over the strenuous objections of doctors, giving the whole process an adversarial quality'.[28]

But several historians, sociologists, anthropologists and political commentators have challenged these 'origin myths'.[29] In a historical volume on medical ethics, for example, Robert Baker, Dorothy Porter and Roy Porter claim that we cannot explain the emergence of bioethics by simply framing it as a response to the moral dilemmas raised by new medical technologies during the 1960s. They argue that:

> It is nothing new for physicians to be confronted with novel and agonizing problems of unexplored biotechnical possibilities and uncertain public response. Examined with care, the formulations of medical ethics over previous centuries, both theoretical and practical, are revealed to have been as complex and entangled in philosophical principle as we feel today's situation to be.[30]

Other historians criticise assumed links between bioethics and civil rights politics, claiming it does not represent a radical break from older and more paternalistic traditions in medical ethics. Roger Cooter, for one, argues that the bioethical emphasis on patient choice was a shallow appropriation of civil rights ideals, which failed to analyse how 'choice' was an ideological construct that varied across institutional, social and cultural settings.[31] Cooter also claims that instead of challenging medicine or science on behalf of patients and the public, bioethics performed the same function as medical ethics by insulating doctors and researchers from truly critical questions.[32] In her history of bioethics in the United States, Tina Stevens similarly concludes that 'bioethical impulses found their way into enduring social institutions not because they represented the social challenges of the 1960s but because they successfully diffused those challenges'.[33]

Stevens argues that bioethics rose to prominence because it helped legitimate research and clinical practice, formulating guidelines 'for the use of procedures and technologies that it largely accepted as inevitable'.[34] John Evans, meanwhile, claims that politicians valued bioethics because its reliance on formal philosophical principles, such as respect for autonomy, beneficence and justice, was congenial to policymaking.[35] But he also contends that these principles were divorced from the socio-economic conditions that shaped moral dilemmas, as well as the actual expectations of patients and the public. This ensured, he argues, that bioethicists failed to appreciate the depth of questions asked about issues such as genetic engineering, leading to an unfortunate 'thinning' of debates from the 1960s onwards that ultimately favoured professional interests.[36]

These conclusions are replicated in critiques of contemporary bioethics, which present it as a 'legitimation device' that insulates researchers from criticism and routinises 'the processes whereby they obtain "ethical clearance" for what they do'.[37] The political theorist Francis Fukuyama struck a now familiar chord in 2002 when he complained that 'bioethicists have become nothing more than sophisticated (and sophistic) justifiers of whatever it is the scientific community wants to do, having enough knowledge of Catholic theology or Kantian metaphysics to beat back criticisms by anyone ... who might object strenuously'.[38] In the same vein, Jonathan Imber dismissed bioethics as little more than 'the public relations division of modern medicine'.[39]

But these critiques fail to identify the mechanisms that underpin the emergence of bioethics in specific times and places, and how these lead to some issues and not others being designated as 'bioethical'. 'The mere presence of illness, death, medical technology, and professional decision making do not in and of themselves necessitate bioethics', Nikolas Rose argues, so 'why should informed consent in reproductive technology be "bioethical" and the rising rate of female infertility not?'[40] This is perhaps the most pressing question to be raised regarding bioethics and provides the greatest incentive for more historical research. Rose continues that 'alongside the urge to critique, perhaps we need to attend to what it is that this demand for bioethics manifests'.[41] Richard Ashcroft similarly claims that a more fruitful analysis would begin by wondering: 'If bioethics is the answer, *what was the question*?'[42] This would allow us to investigate precisely what interests were served and linked by external involvement with science and medicine, and to identify the various parties who benefited from the designation of certain issues as 'bioethical'.

Despite the emergence of bioethics in many countries, we are only familiar with the American story. But when looking to account for what made certain issues 'bioethical' and generated a demand for outside involvement elsewhere, we cannot fall back on those factors that have been highlighted for the United States. Pointing to the inherently controversial nature of new technologies or medical practices does not suffice. Issues such as animal and human experimentation, compulsory vaccination and reproductive medicine were seen to raise moral dilemmas long before the 1960s, yet doctors and scientists continued to regulate themselves throughout the nineteenth and for much of the twentieth century.[43] Neither can we fall back on references to countercultural and civil rights politics. Although they may partly account for the growth of bioethics in the United States during the late 1960s and 1970s, they cannot explain its emergence in locations where these political movements did not exist or lacked influence.

We therefore need to appreciate that there is no universal explanation for the emergence of bioethics, and that the influences and determining factors vary for different times and places. Instead, Cooter argues, we should always look to locate bioethics within 'the social, political and ideological context in which it is conducted'.[44] Rose also calls for more empirical studies, which 'need to be nuanced in relation to empirical studies of the actual role of different aspects of the discourses, practices, forms of expertise, and strategic engagement of bioethics in different places and practices'.[45]

The sociologist David Reubi has recently demonstrated how specific factors shape what counts as 'bioethics' in different times and places. He highlights how the recent development of bioethics in Singapore was part of a broader 'will to modernize', in which politicians sought economic growth by encouraging foreign investment in biomedical research. The presence of bioethics, Reubi argues, was designed to ensure the credibility of Singapore's biomedical research sector by reassuring incoming scientists and companies that it had good ethical standards and was a safe place to invest.[46]

Reubi also details how bioethics in Singapore is not a fixed field or discipline, but is 'an assemblage of knowledge, experts and techniques' that performs various social and political roles.[47] This insight prevents us from mistakenly identifying bioethics as a monolithic entity with a single perspective and mode of inquiry. As the authors of a recent volume state, we now recognise that 'bioethics is a plural noun and its plurality is multiple' – not simply between countries, but within them too.[48] Viewing bioethics in this way helps us acknowledge how contours of the field and the issues it designates as 'bioethical' are constantly 'fluid and changing with context ... the product of history, social organization and culture'.[49]

Bioethics in Britain

More work is needed, however, as bioethics is influential in many locations not yet covered by these empirical studies. This is certainly the case in Britain, where bioethicists are sought-after 'ethics experts' with important positions on regulatory committees and considerable public authority.[50] But our appreciation of how and why they attained this status is sketchy at best. Existing accounts, such as a chapter in the *World History of Medical Ethics*, adhere to the 'origin myth' model and claim that bioethics emerged in Britain after new technologies and radical politics fostered greater discussion of science and medicine during the 1960s and 1970s.[51] But while issues such as clinical research and *in vitro* fertilisation (IVF) were certainly discussed in this period, and while some individuals called for outside involvement in the development of regulatory guidelines, doctors and scientists continued to police themselves. This was clear in 1978, when the *British Medical Journal* diagnosed bioethics as an 'American trend', in which philosophers, lawyers, theologians and others 'acted as society's conscience in matters once left to the medical profession'.[52]

Bioethics did not gain currency in Britain until the 1980s, when increasing numbers of philosophers, lawyers and theologians became actively involved in the public discussion of medicine and biology, the teaching of professional ethics and the development of regulatory guidelines. This outside involvement was undoubtedly influential. In a 1991 article detailing the establishment of the Nuffield Council on Bioethics, the *Guardian* claimed that Britain was seeing the growth of an 'ethics industry' in which bioethicists led 'a national debate on ethical questions arising from modern developments in medicine'.[53] The architects of this 'ethics industry' had become respected public and political figures within a short space of time: chairing public inquiries and regulatory committees, occupying seats in the House of Lords, appearing regularly in the media and receiving knighthoods for 'services to bioethics'.[54]

This status and authority was captured by a 1994 *Sunday Telegraph* profile of the philosopher Mary Warnock, who became 'synonymous with British bioethics' following her spell as chair of a public inquiry into IVF and embryo research between 1982 and 1984.[55] Published to mark Warnock's retirement as mistress of Girton College, Cambridge, the *Telegraph* claimed that she had an 'extraordinary influence' over public and political life. After outlining how her inquiry recommended that embryo experiments should be permitted up to fourteen days after fertilisation, subject to regulatory approval, the *Telegraph* argued that Warnock's appointment as chair marked a pivotal moment in the 'liberalisation of medical ethics'.[56] It also predicted that fellow members of a House of Lords Select Committee on euthanasia were likely to be influenced by her own views that 'human life is not sacred'.[57]

The profile was accompanied by a striking portrait, which illustrates just how bioethics provides a decisive shift in the location and exercise of biopower. The *Telegraph* pictured Warnock in a classic philosophical profile, contemplating a human embryo as if it were Yorick's skull in *Hamlet* (see Figure 1). Lest anyone miss the Shakespearian overtones, the portrait sat above a caption that read 'To be or not to be?' This quote is from one of *Hamlet*'s most memorable soliloquies and evokes the classic existential dilemma of whether or not it is better to choose life or death. But in picturing Warnock holding an embryo the *Telegraph* was clear that, unlike previous generations of philosophers, she no longer simply contemplated what Shakespeare called 'the slings and arrows of outrageous fortune'.[58] Following the emergence of bioethics, and in a privilege hitherto reserved for doctors and scientists, she now helped determine the fate of *in vitro* embryos. For Warnock and other British

Figure 1. 'To be or not to be?' Illustration to a 1994 *Sunday Telegraph* profile of Mary Warnock. Reproduced with the permission of Edward Collet.

bioethicists, 'To be or not to be?' had become an important practical question.

This book draws on a wide range of sources to detail how and why bioethics became so influential in Britain, including the archives of government departments, public inquiries, universities and professional organisations, as well as private papers, published materials, press reports, television programmes and interviews. I use this material to chart the professional, social and political factors that underpinned the making of British bioethics: to show how certain individuals fashioned themselves into authorities on bioethics; to identify the various sites in which bioethics emerged; and to outline how it fulfilled different roles for various groups and professions.

My analysis centres largely on specific individuals, such as Mary Warnock, who represent the different disciplines and approaches that constitute and helped shape British bioethics. At first glance, this might seem indicative of the 'great man' (or woman) approach to history that has been rightly criticised for being overly simplistic and hagiographic.[59] In his work on the development of research ethics committees in the 1960s, for example, Adam Hedgecoe claims that focusing on individuals replicates a major flaw of the 'origin myths' by overlooking how social contexts play a major role in the development of ethical thought and policies.[60]

But studying individuals can shed important light on the history of bioethics by highlighting how it arose thanks to the *engagement* between sociopolitical contexts and personal or professional agendas. This approach draws on the work of scholars such as Sheila Jasanoff, who show how social structures do not simply shape ethical thought, but are themselves produced by individual and collective actions. Denying this measure of agency to specific individuals or groups, Jasanoff notes, 'operates with a much reduced, mechanistic model of human behaviour [and] overlooks the potential for altering the terms and conditions of political debate'.[61] The solution, she argues, is to adopt an 'actor-centred' outlook that investigates how states sought to utilise bioethics for particular functions while, at the same time, charting how specific figures understood, intervened in and even helped create the demand for bioethics.[62] This interplay is conveyed by this book's title. As will become clear, the 'making' I refer to is an active and ongoing process that owes as much to agency as to broader political changes.[63]

Studying individuals also brings to light different perspectives and opinions that move us beyond the misleading view that bioethics is one field or approach. It shows instead how bioethics is a pluralistic set of activities that ranges from the abstract to the practical and includes academic writing and teaching, public discussion, and developing regulatory standards for medicine and the biological sciences.[64] It also highlights the individual views and internal politics that lead to certain topics and approaches being framed as 'bioethical'. As Imber notes, 'strongly held personal opinions' often play a critical role in determining popular representations of bioethics and 'certain ways of being bioethical'.[65] Looking at the work of individuals, and their agreement or disputes with colleagues, is thus vital to explaining the public demand for bioethicists and determining why some appeared to fill the role better than others.

The following chapters detail how certain individuals were integral to a growing demand for outside scrutiny of science and medicine, to the development of regulatory guidelines and to the emergence of the various committees, groups and academic centres that make up British bioethics. Chapter 1 looks at why doctors and scientists came to regulate themselves throughout the nineteenth and for much of the twentieth century. Drawing on work that focuses on relations between professions, I show how this form of governance, which Michael Moran terms 'club regulation', stemmed not only from the professionalising tactics of doctors and scientists, but was compounded by the 'hands-off' approach of politicians and professionals in fields such as law, philosophy and theology.[66] I outline how these attitudes persisted into the 1960s, ensuring that club regulation survived a 'backlash against professional society' and criticism of medical research by the 'whistleblower' Maurice Pappworth.

Chapter 2 examines why outsiders increasingly joined debates on medical procedures such as IVF during the late 1960s and 1970s, and shows how this was led by Anglican theologians. I detail how these theologians argued that 'trans-disciplinary groups' were vital to discussing medical ethics, and outline how this formed part of efforts to stay relevant in the face of a decline in religious belief. I outline how theologians such as Ian Ramsey argued that 'trans-disciplinary groups' were needed to meet the challenges posed by secular and increasingly pluralistic societies, and examine their links with influential figures in the early history of American bioethics. I close by detailing why bioethics continued to be seen as 'an American trend' throughout the 1970s, showing that while British theologians clarified the moral aspects of certain issues, they offered no challenge to club regulation and believed that the 'final decisions remain medical ones'.[67]

Chapter 3 examines why this situation changed in the 1980s, when certain figures successfully promoted external involvement in the development of standards for medicine and the biological sciences. The chapter centres on the work of the academic lawyer Ian Kennedy, who was the most high-profile advocate of the approach he explicitly termed 'bioethics'. I detail how Kennedy's endorsement of bioethics was influential because it dovetailed with the Conservative government's neo-liberal belief that professions should be exposed to outside scrutiny to make them publicly accountable. I also show how Kennedy's arguments appealed to senior doctors, who acknowledged that bioethics was necessary to counter declining political trust in professions.

In a similar vein, chapter 4 looks at Mary Warnock's engagement with IVF and embryo research to chart how philosophers became increasingly involved with bioethics, and how the Conservative government prioritised 'non-expert' involvement in public inquiries into science and medicine during the 1980s. I show how Warnock echoed governmental calls for external oversight and, like Kennedy, promoted bioethics as beneficial to doctors and scientists. I also detail how difficulties in formulating an acceptable cut-off for embryo research led Warnock to dismiss claims that bioethics should be a vehicle for 'moral experts', and to present it as an interdisciplinary 'meeting ground'.

Chapter 5 examines the growth of bioethics in British universities during the 1980s and 1990s. I show how figures such as Kennedy claimed that 'non-medical' input in ethics teaching would benefit student doctors during the early 1980s. This stance ensured that senior doctors supported new interdisciplinary courses in medical ethics, which were predominantly aimed at student doctors and healthcare professionals. I also show how the emergence of dedicated centres for bioethics was consolidated by government cuts in university funding, which encouraged academics in the humanities and social sciences to work on 'applied' topics such as bioethics, and to promote their work's benefits to doctors, students and 'the community as a whole'.

Chapter 6 details how some senior doctors and bioethicists led calls for a politically funded national bioethics committee during the 1980s. I detail why politicians rejected these proposals on the grounds it would impede research and politicise bioethics, and show how this led to the establishment of the independent Nuffield Council on Bioethics in the early 1990s. I then detail the continued growth of bioethics under a 'New Labour' government that shared the neo-liberal enthusiasm for oversight and 'empowered consumers', and show how figures such as Ian Kennedy were increasingly central to the enactment of government policies for the National Health Service (NHS). The chapter concludes by charting how some doctors and public figures began to argue that bioethics actually damaged trust in medicine and science at the beginning of the twenty-first century, which led Kennedy to complain that external regulation was increasingly seen as 'part of the problem rather than part of the solution'.[68]

The final chapter details how recent debates on assisted dying highlight the authority and influence of British bioethicists. I discuss what this can tell us about the impact that bioethics has had on the exercise of biopower, and outline differences between Britain and the United States. I close by outlining how bioethics may look different in the future, thanks to political and economic changes that threaten it with 'retrenchment and decline'.[69]

Linking bioethics to the form of government that writers identify as either neo-liberal or 'advanced liberal' is not intended as pejorative.[70] I am not saying that bioethicists endorsed these political ideologies, but outline how their arguments for outside involvement mapped on to, and reinforced, the growing political demand for public accountability and 'empowered consumers'. Neither am I presenting an examination of any flaws and imperfections in bioethics, or offering a set of criticisms designed to make it better. My aim, to quote Hayden White, simply lies in 'discerning the time-and-place specificity of a thing', that is, bioethics, 'identifying the ways in which it relates to its context or milieu, and determining the extent to which it is both enabled and hamstrung by this relationship'.[71] This approach is more fruitful than simply linking bioethics to neo-liberalism in order to criticise it. General critiques of this kind often take a top-down approach and fail to scrutinise how neo-liberal or advanced liberal climates are made up of often divergent ideas and practices.[72] Taking a bottom-up approach, by contrast, gives a more rounded picture by allowing us to identify how particular forms of power have come into being, what relationships they have helped constitute and who benefits from this.[73]

Examining the ongoing relationship between bioethicists and their broader climate is one of several broad themes that underpin the following chapters. Each one outlines how the discussion of particular ethical issues was both influenced by, and influenced, broader concerns, highlighting the 'co-production' of social and ethical norms.[74] We see, in other words, that that while sociopolitical contexts mattered in the shaping of bioethics, the arguments of various bioethicists also helped shape their broader climate.

This mutual interplay extends to the production of guidelines that regulate specific practices and objects. Ethical guidelines categorised the legal and ontological status of entities such as *in vitro* human embryos by combining scientific theories and moral frameworks such as utilitarianism; and this categorisation subsequently reaffirmed or challenged existing notions of human development, personhood and rights.[75] Sheila Jasanoff defines this categorisation of biomedical objects and practices as 'ontological surgery', and states that 'by sorting new entities (and sometimes old ones) into ethically manageable categories, bioethics helps define the ontologies, or facts of life, that underpin legal rights and condition scientific and social behaviour'.[76] I extend that claim by showing how the bioethical categorisation of entities such as *in vitro* embryos not only involves interaction between scientific and moral norms, but is also structured by sociopolitical contexts and the preferences of specific individuals.[77] Showing how the co-production of bioethical advice was a socially contingent process challenges those authors who claim that it simply involves the application of abstract philosophical principles. And it also contributes to the recent 'empirical turn' in bioethics, in which a growing body of work shows how ethical proposals result from the interplay between theories, objects, professional motivations and social networks.[78]

Several chapters also engage with authors who claim that the main purpose of bioethics is to legitimate research on 'sensitive and controversial subject matters'.[79] They show that while bioethical guidelines may offer one form of legitimation, by setting out how research can be pursued without incurring criminal sanctions, they do not necessarily resolve public debates or controversy. As we shall see, practices such as embryo research were contentious long after guidelines had been issued, and bioethicists played a major role in continuing to generate controversy on this and other topics. This demonstrates that bioethics is not just a narrow activity that concludes with the production of regulatory guidelines, but is a constantly evolving and high-profile enterprise.

Several chapters also confront the dichotomous presentation of bioethics as either a legitimating device for biomedicine or a radical critique on behalf of patients and the public. I show instead how it functioned as both a critique and a safeguard, with bioethicists positioning themselves between politicians, scientists and the public, and moderating their arguments according to particular audiences. Prominent figures in this history certainly criticised self-regulation and argued that bioethics was needed to make science and medicine publicly accountable. But they also claimed that it would benefit scientists and doctors by maintaining public confidence and freeing them from having to make difficult moral choices. Rather than simply challenging or protecting the authority of scientists and doctors, then, British bioethicists presented them with a new means of legitimacy in a changed political climate. Their arguments ensured that many senior doctors endorsed bioethics and supported the appointment of bioethicists to professional organisations. In light of this evidence, we see that bioethics became a valued approach in Britain because it provided an *essential intermediary* between the political demand for accountability and a professional desire for legitimacy.

Each chapter also looks at the question of who counts as an expert: at what professions and groups were deemed competent to discuss and resolve certain ethical issues, and how this process was socially negotiated. This is not to say that questions about expertise are specific to the twentieth century, for the social status of the expert has long been questioned and varies significantly across professions, time and place.[80] As Harry Collins and Robert Evans note, the ancient question of 'Who guards the guardians?' has regularly been used to highlight 'unresolved tensions between expertise and democracy'.[81] But while these are recurring concerns, we need to understand why the question of 'Who guards the guardians?' has been particularly aimed at medicine and science in recent decades – with many groups using it to critique the authority of doctors, scientists and, more recently, bioethicists themselves.

Following on from this, several chapters scrutinise where, if at all, bioethics affected the location and exercise of biopower in Britain. Those accounts that portray the emergence of bioethics as a decisive change in biopower often lack empirical detail to substantiate their bold claims.[82] To redress this and fully understand the relationship between bioethics and changes in biopower, we need to ask the following questions: Did bioethics impact equally across the different locations in which biomedical knowledge was produced and deployed, such as the clinic, regulatory committees, the courtroom and the public sphere? If it did not impact equally in these settings, how can we account for the differences? And if bioethicists did change the exercise of biopower, how did they benefit from their new influence and what were the consequences for doctors and scientists?

In answering these questions, the following chapters connect the making of bioethics to major themes in recent British history, including declining trust in experts, the promotion of consumer-focused approaches to professions and the rise of the 'audit society'.[83] And by charting how bioethics both reflected and contributed to these trends, they offer an important perspective on some of the individuals, ideas and public arguments that have helped reshape 'the politics of life' in recent decades.

Notes

1 'Bioethics', *Oxford English Dictionary* (online edition, September 2011). Available online at www.oed.com/view/Entry/19200. 'Bioethics' first appeared in the 1989 edition of the *Oxford English Dictionary*.

2 Hans-Martin Sass has shown that Potter was not actually the first individual to coin the term 'bioethics'. During the 1920s and 1930s, the Protestant pastor and ethicist Fritz Jahr, from Halle an der Saale in Germany, wrote several articles on what he called '*Bio-Ethik*'. Jahr defined 'Bio-Ethik' as the assumption of new moral duties towards plants and animals based on advances in biological science, particularly experimental psychology and physiology. See Hans-Martin Sass, 'Fritz Jahr's 1927 Concept of Bioethics', *Kennedy Institute of Ethics Journal*, Vol. 17, no. 4 (2008) pp. 279–95. There is no evidence to suggest that Potter was aware of Jahr's work when he wrote on 'bioethics' in 1970. In an interview, he claimed that the word came to him one day with 'a Eureka feeling' and had 'no doubt' that he was the first to use it. See Warren Reich, 'The Word "Bioethics": Its Birth and the Legacies of Those Who Shaped It', *Kennedy Institute of Ethics Journal*, Vol. 4, no. 4 (1994) pp. 319–55 (p. 322).

3 Van Rensselaer Potter, 'Bioethics, the Science for Survival', *Perspectives in Biology and Medicine*, Vol. 14 (1970) pp. 127–53 (p. 130). See also Warren Reich, 'The Word "Bioethics": The Struggle Over Its Earliest Meanings', *Kennedy Institute of Ethics Journal*, Vol. 5, no. 1 (1995) pp. 19–34; Reich, 'The Word "Bioethics": Its Birth'.

4 Potter, 'Bioethics, the Science for Survival', p. 130.

5 Ibid, p. 131.

6 The Georgetown Institute was named after Joseph and Rose Kennedy, the parents of Shriver's wife, Eunice Kennedy Shriver, as well as President John F. Kennedy and Senators Robert and Edward Kennedy. In an interview with Warren Reich, Shriver claimed 'none of us had ever heard of Potter … I was not familiar with the word [bioethics]'. See Reich, 'The Word "Bioethics": Its Birth', p. 325. See also LeRoy Walters, 'The Birth and Youth of the Kennedy Institute of Ethics', in

Jennifer K. Walter and Eran p. Klein (eds), *The Story of Bioethics: From Seminal Works to Contemporary Explorations* (Washington, DC: Georgetown University Press, 2003) pp. 215–31.

7 Reich, 'The Word "Bioethics": The Struggle Over Its Earliest Meanings', p. 21.

8 Reich, 'The Word "Bioethics": Its Birth', pp. 330–1; Daniel Callahan, 'Bioethics as a Discipline', *Hastings Center Studies*, Vol. 1, no. 1 (1973) pp. 66–73. Callahan argued that bioethics was 'not yet a full discipline', but claimed this nevertheless ensured it was 'not yet burdened by encrusted traditions and domineering figures'. See Ibid, pp. 68–70.

9 Warren Reich, 'Introduction', in Warren Reich (ed), *Encyclopedia of Bioethics* (New York: Macmillan, 1978) pp. xv–xxii (p. xix).

10 Callahan, 'Bioethics as a Discipline', pp. 67, 71.

11 Ibid, p. 72.

12 Renée Fox and Judith Swazey, *Observing Bioethics* (Oxford: Oxford University Press, 2008) pp. 128–45; Albert Jonsen, *The Birth of Bioethics* (Oxford: Oxford University Press, 1998) pp. 99–106.

13 Senator Edward Kennedy, quoted in David Rothman, *Strangers at the Bedside: A History of How Law and Bioethics Transformed Medical Decision Making* (New York: Basic Books, 1991) p. 188.

14 Fox and Swazey, *Observing Bioethics*, pp. 133–4; Jonsen, *The Birth of Bioethics*, pp. 106–15.

15 Rothman, *Strangers at the Bedside*, p. 189.

16 Sheila Jasanoff, 'Making the Facts of Life', in Sheila Jasanoff (ed), *Reframing Rights: Bioconstitutionalism in the Genetic Age* (Cambridge, MA: MIT Press, 2011) pp. 59–85 (p. 60).

17 On the international growth of bioethics, see Fox and Swazey, *Observing Bioethics*, pp. 215–85. For specific examples, see Eduardo Diaz-Amado, 'Bioethicization and Justification of Medicine in Colombia in the Context of Healthcare Reform of 1993', PhD thesis, University of Durham, 2011; David Reubi, 'The Will to Modernize: A Genealogy of Biomedical Research Ethics in Singapore', *International Political Sociology*, Vol. 4 (2010) pp. 142–58; Farhat Moazam and Aamir M. Jafarey, 'Pakistan and Biomedical Ethics: Report from a Muslim Country', *Cambridge Quarterly of Healthcare Ethics*, Vol. 14, no. 3 (2005) pp. 249–55; Machel L. Gross and Vardit Ravitsky, 'Israel: Bioethics in a Jewish-Democratic State', *Cambridge Quarterly of Healthcare Ethics*, Vol. 12, no. 3 (2003) pp. 247–55; Darryl R. J. Macer, 'Bioethics in Japan and East Asia', *Turkish Journal of Medical Ethics*, Vol. 9 (2001) pp. 70–7.

18 Anglo-American writers have only recently utilised Foucault's work on biopower, following the posthumous translation of his lectures at the Collège de France. See Roger Cooter and Claudia Stein, 'Cracking

Biopower', *History of the Human Sciences*, Vol. 23 (2010) pp. 109–28. Foucault argued that biopower was a specifically modern form of power that first emerged in the eighteenth century and rose to prominence during the nineteenth and twentieth centuries. In contrast to earlier sovereign forms of power, which operated repressively, it was concerned with preserving and maximising individual or population health. See Michel Foucault, *The Birth of Biopolitics: Lectures at the Collège de France, 1978–1979* (Basingstoke: Palgrave Macmillan, 2008); Michel Foucault, *Society Must be Defended: Lectures at the Collège de France, 1975–1976* (Harmondsworth: Penguin, 2004); Michel Foucault, *The Will to Knowledge. The History of Sexuality: Volume One* (Harmondsworth: Penguin, 1998).

19 Paul Rabinow and Nikolas Rose, 'Introduction: Foucault Today', in Michel Foucault, Paul Rabinow and Nikolas Rose, *The Essential Foucault* (New York and London: New Press, 2003) pp. vii–xxxiii (p. xxx).

20 Paul Rabinow and Nikolas Rose, 'Biopower Today', *Biosocieties*, Vol. 1 (2006) pp. 195–217 (p. 203).

21 Brian Salter and Charlotte Salter, 'Bioethics and the Global Moral Economy: The Cultural Politics of Human Embryonic Stem Cell Science', *Science, Technology and Human Values*, Vol. 32 (2007) pp. 554–81 (p. 564). See also Brian Salter and Mavis Jones, 'Biobanks and Bioethics: The Politics of Legitimation', *Journal of European Public Policy*, Vol. 12, no. 4 (2005) pp. 710–32.

22 Brian Salter, 'Cultural Biopolitics, Bioethics and the Moral Economy: The Case of Human Embryonic Stem Cells and the European Union's Sixth Framework Programme', Working Paper I of the project 'The Global Politics of Human Embryonic Stem Cell Science' (November 2004). Emphasis added. Available online at www.york.ac.uk/res/sci/projects/res340250001salter.htm (accessed 6 February 2014).

23 See, for example, Cooter and Stein, 'Cracking Biopower'; Roger Cooter, 'After Death/After-"life": The Social History of Medicine in Post-Postmodernity', *Social History of Medicine*, Vol. 20 (2007) pp. 441–64.

24 Nikolas Rose, *The Politics of Life Itself: Biomedicine, Power and Subjectivity in the Twenty-First Century* (Princeton, NJ, and Oxford: Princeton University Press, 2007) p. 30.

25 Tomislav Bracanovic, 'Against Culturally Sensitive Bioethics', *Medical Health Care and Philosophy* (advance access 2013) doi 10.1007/s11019-013-9504-2.

26 Peter Singer, 'Introduction', in Peter Singer (ed), *Applied Ethics* (Oxford: Oxford University Press, 1986) pp. 1–9 (p. 4). See also John Harris, 'The Scope and Importance of Bioethics', in John Harris (ed), *Bioethics* (Oxford: Oxford University Press, 2001) pp. 1–25; Stephen Toulmin,

'How Medicine Saved the Life of Ethics', *Perspectives in Biology and Medicine*, Vol. 25 (1982) pp. 736–50. For book-length participant histories, see Fox and Swazey, *Observing Bioethics*; Jonsen, *The Birth of Bioethics*.

27 Robert B. Baker, 'From Meta-Ethicist to Bioethicist', *Cambridge Quarterly of Healthcare Ethics*, Vol. 11 (2002) pp. 369–79 (p. 369).

28 Rothman, *Strangers at the Bedside*, pp. 10–11.

29 On participant histories as 'origin myths', see Atwood D. Gaines and Eric T. Juengst, 'Origin Myths in Bioethics: Constructing Sources, Motives and Reason in Bioethics(s)', *Culture, Medicine and Psychiatry*, Vol. 32 (2008) pp. 303–27.

30 Robert Baker, Dorothy Porter and Roy Porter, 'Introduction', in Robert Baker, Dorothy Porter and Roy Porter (eds), *The Codification of Medical Morality: Historical and Philosophical Studies of the Formalization of Western Medical Morality in the Eighteenth and Nineteenth Centuries* (Dordrecht: Kluwer Academic Publishers, 1993) pp. 1–14 (p. 2).

31 Roger Cooter, 'The Ethical Body', in John V. Pickstone and Roger Cooter (eds), *Medicine in the Twentieth Century* (Amsterdam: Harwood Academic Press 2000) pp. 451–67 (p. 464).

32 Roger Cooter, 'Inside the Whale: Bioethics in History and Discourse', *Social History of Medicine*, Vol. 23 (2010) pp. 662–73. See also Roger Cooter, 'The Resistible Rise of Medical Ethics', *Social History of Medicine*, Vol. 8, no. 2 (1995) pp. 257–70.

33 M. L. Tina Stevens, *Bioethics in America: Origins and Cultural Politics* (Baltimore, MD, and London: Johns Hopkins University Press, 2000) p. xii.

34 Ibid, p. 158.

35 John H. Evans, 'A Sociological Account of the Growth of Principalism', *Hastings Center Report*, Vol. 30, no. 5 (2000) pp. 31–8.

36 John H. Evans, *Playing God: Human Genetic Engineering and the Rationalization of Public Bioethical Debate* (Chicago and London: University of Chicago Press, 2002).

37 Rose, *The Politics of Life Itself*, pp. 255–7; see also Salter and Jones, 'Biobanks and Bioethics'.

38 Francis Fukuyama, *Our Posthuman Future: Consequences of the Biotechnology Revolution* (London: Profile, 2002) p. 158.

39 Jonathan Imber, 'Medical Publicity Before Bioethics: Nineteenth Century Illustrations of Twentieth Century Dilemmas', in Raymond de Vries and Jonathan Subedi (eds), *Bioethics and Society: Constructing the Ethical Enterprise* (Upper Saddle River, NJ: Prentice Hall, 1998) pp. 16–38 (p. 33).

40 Rose, *The Politics of Life Itself*, p. 31. See also Giovanni Berlingeur,

'Bioethics, Health and Inequality', *Lancet*, Vol. 364 (2004) pp. 1086–91.

41 Rose, *The Politics of Life Itself*, p. 256.

42 Richard Ashcroft, 'Bioethics and Conflicts of Interest', *Studies in the History and Philosophy of the Biological and Biomedical Sciences*, Vol. 35 (2004) pp. 155–65 (p. 158). Emphasis in original.

43 See, for example, Nadja Durbach, *Bodily Matters: The Anti-Vaccination Movement in England, 1853–1907* (Durham, NC: Duke University Press, 2005); Susan Squier, *Babies in Bottles: Twentieth-Century Visions of Reproductive Technology* (New Brunswick, NJ: Rutgers University Press, 1994); Richard D. French, *Antivivisection and Medical Science in Victorian Society* (Princeton, NJ, and London: Princeton University Press, 1975).

44 Cooter, 'The Ethical Body', p. 466.

45 Rose, *The Politics of Life Itself*, p. 256. For a critical review of Rose's work, see Cooter and Stein, 'Cracking Biopower'.

46 Reubi, 'The Will to Modernize'.

47 Ibid, p. 144; see also Rabinow and Rose 'Foucault Today'.

48 Raymond de Vries, Leigh Turner, Kristina Orfali and Charles L. Bosk, 'Social Science and Bioethics: The Way Forward', in Raymond de Vries, Leigh Turner, Kristina Orfali and Charles L. Bosk (eds), *The View from Here: Bioethics and the Social Sciences* (Oxford: Blackwell Publishing, 2007) pp. 1–13 (pp. 2–3).

49 Ibid, pp. 5, 11.

50 Claire Dyer, 'Ethics Expert Calls for Legal Euthanasia', *Guardian*, 26 April 1994, p. 3.

51 Kenneth Boyd, 'The Discourses of Bioethics in the United Kingdom', in Robert B. Baker and Laurence B. McCullough (eds), *The World History of Medical Ethics* (Cambridge: Cambridge University Press, 2009) pp. 486–90.

52 Barbara J. Culliton and William K. Waterfall, 'The Flowering of American Bioethics', *British Medical Journal*, Vol. 2 (1978) pp. 1270–1 (p. 1270).

53 Nigel Williams, 'To the Heart of a Clinical Matter', *Guardian*, 17 April 1991, p. 21.

54 Anon, 'New Year Honours', *Guardian*, 31 December 2001, p. 10.

55 Sheila Jasanoff, *Designs on Nature: Science and Democracy in Europe and the United States* (Princeton, NJ, and Oxford: Princeton University Press, 2005) p. 152.

56 Anon, 'To be or not to be?' *Sunday Telegraph*, 2 January 1994.

57 Ibid.

58 William Shakespeare, *Hamlet* (Harmondsworth: Penguin, 2007) III.i, 55.

59 See, for example, Susan Reverby and David Rosner, 'Beyond the Great Doctors', in Susan Reverby and David Rosner (eds), *Healthcare in America: Essays in Social History* (Philadelphia, PA: Temple University Press, 1979) pp. 3–16.
60 Adam Hedgecoe, '"A Form of Practical Machinery": The Origins of Research Ethics Committees in the UK', *Medical History*, Vol. 53 (2009) pp. 331–50 (p. 333).
61 Jasanoff, *Designs on Nature*, p. 20.
62 Ibid, p. 172.
63 In addition to modifying his title, I have paraphrased E. P. Thompson here, who claimed his famous study of the working class was a 'study in an active process, which owes as much to agency as to conditioning'. See E. P. Thompson, *The Making of the English Working Class* (Harmondsworth: Penguin, 2013) p. 8.
64 See also De Vries et al., 'Social Science and Bioethics'.
65 Imber, 'Medical Publicity before Bioethics', p. 21.
66 Michael Moran, *The British Regulatory State: High Modernism and Hyper-Innovation* (Oxford: Oxford University Press, 2003).
67 Culliton and Waterfall, 'Flowering of American Bioethics', p. 1270.
68 Charlotte Santry, 'Healthcare Regulator Longed for Government's Embrace', *Health Service Journal*, 12 November 2008, available online at www.hsj.co.uk (accessed 6 February 2014).
69 Leigh Turner, 'Does Bioethics Exist?', *Journal of Medical Ethics*, Vol. 35 (2009) pp. 778–80 (p. 779).
70 Nikolas Rose, 'Government, Authority and Expertise in Advanced Liberalism', *Economy and Society*, Vol. 22 (1993) pp. 283–99.
71 Hayden White, 'Afterword: Manifesto Time', in Keith Jenkins, Sue Morgan and Alun Munslow (eds), *Manifestos for History* (London and New York: Routledge, 2007) pp. 220–31 (p. 224).
72 This point was recently made by Bronwyn Parry, at a University of Durham symposium. For an overview, see Abi McNiven, 'Critical Medical Humanities Symposium – Review'. Available online at http://medicalhumanities.wordpress.com (accessed 6 February 2014).
73 Joan Scott, 'History-Writing as Critique', in Jenkins et al. (eds), *Manifestos for History*, pp. 19–39.
74 See Sheila Jasanoff (ed), *States of Knowledge: The Co-Production of Science and the Social Order* (London: Routledge, 2004).
75 On the co-production of biological and ethical norms, see Jasanoff, 'Making the Facts of Life'; Giuseppe Testa, 'More than Just a Nucleus: Cloning and the Alignment of Scientific and Political Rationalities', in Jasanoff (ed), *Reframing Rights*, pp. 85–105.
76 Jasanoff, 'Making the Facts of Life', p. 77.
77 Jasanoff has undertaken a comparative study of bioethics in Britain, the

United States and Germany, but acknowledges that this is a preliminary sketch rather than a 'full-scale cross-national ethnography of bioethics'. See Jasanoff, *Designs on Nature*, p. 172.

78 See, for example, Steven P. Wainwright, Claire Williams, Mike Michael, Bobbie Farsides and Alan Cribb, 'Ethical Boundary-Work in the Stem Cell Laboratory', in De Vries et al. (eds), *The View from Here*, pp. 67–83.

79 Doris T. Zallen, 'Regulating Research: A Tale of Two Technologies', *Technology and Society*, Vol. 11 (1989) pp. 377–86 (p. 377). See also Rose, *The Politics of Life Itself*.

80 Andrew Abbott, *The System of Professions: An Essay on the Division of Expert Labor* (Chicago and London: University of Chicago Press, 1988).

81 Harry Collins and Robert Evans, *Rethinking Expertise* (Chicago and London: University of Chicago Press, 2007) p. 4. Collins and Evans attribute the question 'Who guards the guardians?' to Plato, although none of his works contain a phrase that resembles it.

82 Nikolas Rose, for instance, acknowledges that his own work lacks the 'familiar tropes of social critique' and calls for greater empirical work on who gains and loses from shifts in biopower. See Rose, *The Politics of Life Itself*, pp. 258–9.

83 Michael Power, *The Audit Society: Rituals of Verification* (Oxford: Oxford University Press, 1997).

1

Ethics 'by and for professions': the origins and endurance of club regulation

Doctors and scientists successfully argued that they should be left to determine their own conduct during the nineteenth and much of the twentieth centuries, in a form of self-governance that Michael Moran terms 'club regulation'.[1] They portrayed medical and scientific ethics as internal concerns in this period – produced 'by and for' colleagues and mainly concerned with limiting intra-professional conflicts.[2] This view of ethics functioned as what Harold Perkin calls a 'strategy of closure'.[3] It helped doctors and scientists consolidate their professional expertise by delineating boundaries, excluding unqualified groups and positioning themselves as the only people capable of providing an essential service to government and the public.

This, of course, is not a new insight and several historians have shown how members of professions set their own standards so as to exclude others.[4] In looking to explain why professions such as medicine gained control of their own practices and codes of conduct, these studies adopt a largely internalist view, focusing on the professions in question and portraying them as blocs or monopolies. When they look to external factors to explain club regulation, historians generally chart how notions of professional self-governance resonated with the *laissez-faire* ideals of nineteenth-century politicians. But this does not tell the whole story. As the sociologist Andrew Abbott argues, professions do not emerge or evolve in isolation and we need to move from 'an individualistic to a systemic view of professions'.[5] Abbott endorses a more relational model, in which the acquisition of professional authority involves mediating jurisdictional claims between different professions.

In following Abbott, we see that the history of club regulation hinges on the *interdependence* of professions. We cannot fully account for the emergence and strength of club regulation without studying the 'hands-off' approach that other professionals adopted when they considered medical and scientific practices. This is especially true of those professions and academic fields that later constituted bioethics, such as law, philosophy and theology. On the rare occasions that individuals from these fields did engage with science or medicine in the nineteenth and early twentieth centuries, they sought to consolidate the authority of doctors and scientists.

This stance persisted into the 1960s, despite a growing 'backlash against professional society'.[6] Criticism of medical research came instead from professional 'whistleblowers' such as Maurice Pappworth, who broke with club regulation when he publicly rebuked doctors for experimenting on patients without consent and, crucially, demanded that outsiders should play a role in formulating and administering a new statutory code for medical research. These factors have led some to claim that Pappworth is a significant figure in 'the birth of British bioethics'.[7] But while his work attracted public attention, it ultimately had little impact on the continuing support for club regulation among doctors, politicians and other professions. Despite Pappworth's best efforts, outside involvement was dismissed as 'quite impracticable' and doctors were left, as before, to determine their own conduct and ethical standards.

Enshrining club regulation: medical and scientific ethics as professional concerns

The emergence of club regulation in medicine and other professions resulted from social and economic changes during the nineteenth century. Before this, doctors and other professionals operated under a system of 'lay patronage', in which their actions were determined by a relatively small band of aristocratic and wealthy clients.[8] Ivan Waddington argues that lay patronage fostered a model 'not of colleague control, but of client control', in which the patient's superior social status allowed them to dictate their own needs 'and the manner in which those needs are to be met'.[9] As Roy Porter states, this ensured that 'for authority and status, reward and advancement, doctors looked not to collective professional paths to glory, but to the personal favour of grandees'.[10]

Lay patronage also meant that medical practitioners showed greater loyalty to their clients than to their colleagues, and that the 'solidarity of the occupational group was relatively underdeveloped'.[11] As part of a thriving 'medical marketplace', orthodox practitioners such as physicians, surgeons and apothecaries competed for patients with each other and with a variety of alternative healthcare providers, such as homeopathists, mesmerists and bone-setters.[12] In a period marked by high consumer choice and 'low professionalisation', when distinctions between 'regular and irregular' practitioners were unclear and a new division between general practitioners and hospital consultants threatened the old tripartite structure of medicine, disputes between physicians were commonplace.[13]

These disputes and rivalries, which hinged on arguments over competition and the new division of labour, led some physicians to write professional codes of ethics during the late eighteenth and early nineteenth centuries.[14] Their guidelines notably differed from previous recommendations for medical practitioners, which were indistinguishable from the general 'advice to gentlemen' published in conduct manuals.[15] While these early modern codes focused on individual manners and conduct, especially in client–patron relations, newer guidelines such as Thomas Percival's 1803 *Medical Ethics* dwelt far more on smoothing relations between practitioners in order to forestall professional conflict. Percival's code is notable for introducing the term 'medical ethics', but it is perhaps more significant in another respect.[16] In order to restrict the power of lay hospital governors, who physicians believed were interfering in running the Manchester Infirmary, Percival's *Medical Ethics* stressed the collective autonomy of medical practitioners and the need for 'collaborative self-regulation'.[17] To Percival, 'medical ethics' denoted a set of professional, not public, concerns.

Percival's view of medical ethics was adopted by a later generation of reformers who sought from the 1820s to portray medicine as a discrete and socially valuable profession. These reformers dwelt less on notions of gentlemanly virtue and more on their possession of specialist knowledge and authority.[18] They promoted their 'scientific' training in anatomy, chemistry and pathology, and argued that they alone possessed the expertise to care for the changing 'social body' created by industrialisation and urbanisation.[19] Calling for government restrictions on alternative practitioners, whose services were popular among the urban population, they argued that their reward for combating diseases such as cholera should be freedom to practise without outside interference. Physicians exploited the social capital they gained through public health measures by arguing that the state should restrict care of the population 'to those with recognised qualifications, talents and abilities', and these arguments were later helped by advances in anaesthetics and germ theory.[20] Codes and associations that bore the term 'medical ethics' were integral to this reforming campaign, helping to strengthen professional unity, consolidating expertise *vis-à-vis* the public and politicians and excluding unorthodox practitioners. This meant that when doctors established regional associations such as the Manchester Medico-Ethical Society, they functioned as 'a trade union in disguise'.[21]

Arguments for professional self-control resonated with Victorian politicians who espoused *laissez-faire* ideals of liberal self-governance.[22] Political support for medical reform was also strengthened by the fact that orthodox doctors, like many other professionals, became increasingly central to the machinery of a growing Victorian state from the mid nineteenth century onwards.[23] In an era when professional expertise was 'inextricably linked to the formal political apparatus of rule', doctors worked as Poor Law officers, factory medical inspectors or prison doctors, and were later central to the administration of the 1853 Compulsory Vaccination Act and the 1864 Contagious Diseases Act.[24] The demand for medical expertise, in turn, led politicians to recruit doctors into the expanding civil service. The first Chief Medical Officer (CMO) was appointed in 1855 and was soon supported by a team of medically qualified civil servants. In addition to providing expert advice, these civil servants furthered professional interests by ensuring that the state directed funds to medical programmes without compromising the independence of doctors.[25]

With doctors increasingly central to government policy, and politicians committed to notions of self-governance, it was no surprise that the 1858 Medical Act recognised medicine as a unitary and autonomous profession. The Medical Act distinguished orthodox from alternative practitioners by requiring the creation of a register of qualified doctors (though it did not forbid alternative practitioners from practising). It also granted doctors a significant degree of 'self-governing authority' by leaving them in charge of the new General Medical Council (GMC) that controlled registration, education and discipline.[26] Politicians then withdrew from the issue of medical regulation and only intervened on the rare occasions when doctors requested it themselves.[27]

Although politicians granted the GMC formal disciplinary powers, it did not issue a binding set of ethical guidelines. Registered doctors were given no written guidance on professional conduct until 1883, when the GMC began to issue a series of 'warning notices'.[28] These arose from disciplinary rulings and specified conduct that the GMC considered unacceptable enough to warrant the removal of a doctor from the medical register. By the turn of the twentieth century, the warning notices encompassed improper or fraudulent acquisition of qualifications, advertising or canvassing, sexual misconduct such as committing adultery with patients, publishing indecent work, abortion, drunkenness and improperly disclosing confidential patient information.[29] The twenty-four members of the GMC who were eligible to reach these decisions were all medically qualified, reaffirming that medical ethics was seen as a solely professional matter.[30]

The reluctance to issue binding ethical guidelines was mirrored by the British Medical Association (BMA), which represented the interests of doctors after its formation in 1836.[31] During the 1850s the BMA appointed two ethics committees and instructed them to produce similar codes to the American Medical Association, which had produced an 1847 set of guidelines based on Percival's *Medical Ethics*.[32] Neither group actually met or produced a code, but Jukes de Styrap, a member of the second BMA committee and chair of the Shropshire medico-ethical association, updated Percival's guidelines to produce his own *Code of Medical Ethics* in 1878. Like Percival, de Styrap aimed 'to promote harmony and prevent disputes within the profession'.[33] His main ethical precept, the so-called 'Golden Rule', drew on the biblical injunction to 'do unto others as you would have them do unto you' – although this applied to a doctor's relations with their colleagues far more than it did to their patients.[34] De Styrap viewed the Golden Rule as vital to establishing a 'generous *esprit de corps*' and stressed it should be every doctor's aim to 'raise our profession, not only by our scientific labours, and the careful and accurate study of disease and its remedies, but by our feeling of brotherhood and mutual support – so that the public may respect us as a body at unity within itself'.[35]

As this quote indicates, the stress on professional relationships did not mean that writers on medical ethics ignored the interests of patients and the public. For de Styrap, the professional and the public interest were firmly linked, and patients were best served by a unified medical profession that avoided 'public rancour', refrained from activities such as advertising and was clearly distinguished from 'tradesmen and quacks'.[36] By highlighting their 'scientific' training and professional authority, de Styrap also stressed that patients were best served by leaving decisions to doctors, since they alone possessed the expertise to evaluate the benefits or drawbacks of specific procedures.

This view of medical ethics persisted well into the twentieth century. The only writers on the subject were doctors such as Robert Saundby, who continued to argue that medical professionals were the best judges of a patient's interests.[37] It was also evident in new committees and survived reforms that admitted laypeople to the GMC. In 1902 the BMA underwent reform that resulted in the creation of a Central Ethical Committee (CEC), following tensions between doctors and mutual aid societies, who provided healthcare in return for members' contributions and were eventually organised under government control by the 1911 National Health Insurance Act.[38] The CEC was established to issue guidance to local BMA branches, and to draw up reports or sets of rules for difficult issues. Its creation notably was 'the first time doctors in Britain had a national body to examine questions of conduct without resorting to extremes of hearings before the GMC'.[39]

But while it was new a body, the CEC embodied the traditional view that medical ethics was produced 'by and for' doctors.[40] Its meetings focused on advertising, contract disputes and confidentiality, and its members were drawn from the senior ranks of the medical profession. This ethos also persisted in the GMC, despite the appointment of the former politician Sir Edward Young as its first layman following a public outcry at the treatment of F. W. Axham, who was removed from the medical register for working with the osteopath Sir Herbert Barker. Young and successive lay members were generally the only non-doctors on the GMC and exerted little, if any, influence over its decision-making.[41]

Yet we should not presume that relations between doctors and patients were completely paternalistic. They were certainly not as one-sided as is implied by some historical accounts, particularly those written by bioethicists. We need to see these participant histories as rhetorical efforts to differentiate bioethics from 'old' and problematic styles of medical ethics and, having done so, to 'open up a space for intervention and reform of unsatisfactory relationships'.[42] During the late nineteenth and early twentieth century, in a partially regulated 'medical marketplace', patients were certainly free to select doctors for private practice and mutual aid schemes. Jukes de Styrap reminded readers of this in his third edition of *Medical Ethics*, where he claimed that 'the right of a patient to change or to discard his doctor is unquestionable'.[43] This principle also applied in National Insurance schemes from 1911 onwards, which incorporated provisions for 'free choice of doctor' at the BMA's insistence.[44]

But this measure of autonomy and the emphasis on professional expertise did not guarantee public trust. The connections between orthodox doctors and the Poor Law authorities ensured that many working-class people viewed them with suspicion following the 1858 Medical Act, and continued to rely on alternative therapies when they fell ill. Alternative practitioners played on this by claiming that 'orthodox medicine was a tyrannical system of state-sanctioned interference with the lives and health of an oppressed people'.[45] These suspicions found expression in the anti-vaccination movement, in which supporters of alternative medicine joined with large sections of the working and middle classes to argue that the 1853 Compulsory Vaccination Act infringed on an individual's right to govern their own homes and families.[46]

Doctors also faced resistance thanks to their association with the 1864 Contagious Diseases Act, which permitted compulsory examination of any suspected prostitute, and their detention in 'lock' hospitals should they be infected.[47] Feminist and socialist reformers argued that these Acts represented state-sanctioned infringement upon the bodies and rights of working-class women. Many of these campaigners also opposed the increase in vivisection from the 1870s, linking the plights of defenceless animals and women, and portraying medical researchers as indifferent to the suffering they caused to those less fortunate.[48] While these movements differed in some respects, they all resisted the growing authority of doctors and scientists, and criticised the fact that politicians increasingly gave them licence to 'dictate morality and personal behaviour'.[49]

But this ultimately had little impact on state support for medical or scientific expertise. Indeed, the political response to these popular movements effectively consolidated club regulation. When the government convened a Royal Commission on Vaccination in 1889, for example, they filled it 'with eminent medical practitioners who almost unanimously supported vaccination'.[50] In 1896 the Commission proposed the introduction of a conscientious objection clause that significantly weakened the anti-vaccination movement, since individuals could now simply 'opt out' by obtaining an exemption certificate.[51]

The government's 1875 Royal Commission on Vivisection also increased professional authority in the biological sciences, albeit less directly. This Commission was more balanced between scientists, representatives from the Royal Society for the Protection of Animals (RSPCA) and individuals 'uncommitted to either side'.[52] Its composition reflected how researchers in fields such as physiology lacked a 'meaningful professional identity' in the 1870s, with less political influence than doctors or campaign groups such as the RSPCA.[53] But their response to the Commission's recommendations, which underpinned the 1876 Cruelty to Animals Act, galvanised biological scientists into acting as a more coherent and influential body from the 1880s onwards. The Cruelty to Animals Act angered many biologists by ruling that Home Office officials should decide whether or not to issue licences permitting animal experiments.[54] Figures such as Richard Owen, who had previously opposed vivisection for teaching purposes, now condemned politicians for undermining 'the expertise and thus authority of his profession'.[55] Like doctors before them, biologists formed groups to endorse self-regulation, such as the Association for the Advancement of Medicine by Research (AAMR).

Members of the AAMR, which united physiologists, botanists and zoologists, argued that they were 'better judges than an average person in matters of research and its moral aspects, "because they possessed the additional knowledge indispensible to form a correct judgement"'.[56] Their efforts were certainly influential. In 1883 the government decided the AAMR should review all licence applications before they were passed to the Home Office, which led to a significant increase in licence approvals.[57] Professional control over animal experiments increased further after 1913, when a second Royal Commission, now weighted in favour of scientists, recommended that a new advisory body should consider licence applications. Members of this Home Office advisory committee were selected by the Home Secretary from a list of names submitted by solely professional bodies such as the Royal Society and the Royal Colleges of Physicians and Surgeons. From 1913 to the late 1970s, the advisory committee always consisted of ten scientists and one lawyer.[58] Perhaps unsurprisingly, it approved the vast majority of licence applications and played a major role in encouraging the growth of biological disciplines such as pathology, pharmacology and bacteriology.[59]

Biologists also benefited from increasing control over how the government distributed funding for research. In 1913, following concerns over infant mortality raised by a Royal Commission on Tuberculosis, the government formed a Medical Research Committee that was administered under the 1911 National Insurance Act. Although the committee was expected to focus on tuberculosis, it soon became dominated by Cambridge physiologists who helped 'establish a presence within government for the elites of British science and education'.[60] In its early years, and following its reconstitution as the Medical Research Council (MRC) in 1919, these influential scientists were able to distribute money without political interference, using it to free biological sciences from clinical concerns and encourage research into basic problems.[61]

The interwar period also saw a decline in organised public opposition to medical or scientific authority. Conscientious objection clauses effectively killed off the anti-vaccination movement, while the anti-vivisection cause was dealt a blow after scientists argued that new drugs such as Salvarsan proved the value of animal experiments. At the same time, although conventional treatments were expensive and often ineffective, orthodox medicine gradually won public acceptance.[62] With increasing state investment and declining public opposition, some doctors and biological scientists promoted their expertise with greater confidence during the 1920s and 1930s. They not only objected to involving laypeople in professional debates, but now asserted a 'far more comprehensive authority [in] determining the shape of things to come'.[63] A new generation of 'public' biologists such as Julian Huxley, Conrad Waddington and J. B. S. Haldane used popular outlets such as newspapers, magazines, radio and science-fiction stories to assert that human progress could only be ensured by giving them a greater say in social and even moral affairs.[64]

Calls for greater professional influence over social and moral issues permeated the eugenics movement, in which scientists and doctors sought to counter evolutionary 'degeneration' by controlling the reproduction of supposedly inferior groups.[65] They were also evident in works such as Conrad Waddington's *Science and Ethics*, which originally appeared as an essay in *Nature* and argued that biologists with knowledge of evolution and the human mind could make a decisive contribution 'to the study of ethics'.[66] Waddington claimed that biologists were in a better position to study ethics than philosophers or theologians, since they possessed the expertise to reposition notions such as 'good' as 'facts of the kind with which science deals'.[67] He outlined how scientists could define ethical principles as 'actual psychological principles derived from experience', and could also demonstrate how 'the real good cannot be other than that which has been effective, namely that which is exemplified in the course of evolution'.[68]

Not everyone welcomed these incursions into social and moral affairs. Some scientists maintained that they simply studied natural phenomena and argued it was not their place to assert their work's relevance to 'questions of personal or corporate morality'.[69] This view proved attractive to many 'because it protected the freedom of scientists to pursue their work without fear of external controls'.[70] Criticism also came from a small but high-profile group of elite critics, such as F. R. Leavis and Hilaire Belloc, who extolled traditional ways of life and equated science with moral and political decline. Lamenting the waning influence of 'humanist' scholars, they argued that scientists had narrow expertise and were ill-equipped to discuss matters outside their specialism. For them, 'questions concerning both the ends of scientific applications and the desirability of progress, were to the humanist's mind not for the scientist *qua* scientist to answer'.[71] Yet this criticism again hinged on the belief that professionals should stay within their bounds of expertise. Critics such as Leavis and Belloc accepted the judgements of scientists '*in their own sphere*', and did not believe that outsiders should determine scientific or medical conduct.[72]

Perhaps the only advocate of external involvement with medicine or science in this period was the playwright George Bernard Shaw, who remained a committed anti-vivisectionist and supporter of alternative medicine until his death in 1950. In his 1909 play *The Doctor's Dilemma* and a series of later essays, Shaw argued that doctors were motivated by profit and 'professional trade interest' rather than a concern for patients and the public.[73] This, he concluded, led to a 'dogmatic' exclusion of alternative therapies and ensured that 'what is called scientific progressive medicine is thus seen to be largely dictated by the hygiene of the pocket'.[74] Shaw was not criticising professional authority *per se* here. He advocated professional expertise providing it was harnessed for the greater social good, as evidenced by his support for the eugenics movement and membership of the socialist Fabian Society, which 'embraced a scientistic form of politics'.[75] This desire for socially useful expertise led Shaw to propose reforms that he believed would foster a more 'disinterested' and trustworthy medical profession. These included establishing a 'state medical service' and, notably, reconstituting the GMC so that it included 'a majority of laymen'.[76] Shaw argued that this latter measure was vital since 'all trade union experience shows that the doors of a trade or profession must not be guarded, either for entrance or exit, by the members inside'.[77] In contrast to de Styrap and others, who believed they were mutually enforcing, Shaw concluded that the 'protection of the laity' and 'the progress of science' were incompatible with club regulation.[78]

Despite Shaw's profile, doctors and politicians overwhelmingly rejected any form of outside involvement. Following the Second World War, for example, the BMA and the *British Medical Journal* often portrayed Nazi medical crimes as a direct result of outside interference.[79] When Clement Attlee's Labour government sought to implement its 1946 National Health Service Act, doctors agreed to reform on the condition that 'there should be as little scrutiny as possible of their privileged clinical position or research practices'.[80] Politicians were wary of challenging a profession that had a high standing in the eyes of the public, and gave doctors a significant degree of autonomy when they established the NHS in 1948.[81] This agreement ensured that while the state allocated resources for the care of the population in the NHS, doctors retained control over their own practices and how resources were allocated.[82]

As before, this control encompassed clinical treatment and medical research. Doctors presumed that citizens would support biomedical research and contribute to medical progress by willingly offering their bodies in exchange for the 'protection against deprivation, ignorance and disease' they received from the welfare state.[83] There was little discussion of whether patient consent was needed for research, or whether doctors required any outside supervision. With the creation of the NHS boosting public trust, the doctor was widely perceived not only 'as an expert but also a gentleman whose inherent integrity and good character prevent him or her from any wrongdoing'.[84]

So while Shaw's vision of a 'state medical service' was realised in his lifetime, he did not get his wish for greater outside involvement in setting standards for doctors. Club regulation also persisted in science, despite the arguments of Marxist scientists such as J. D. Bernal. In his 1939 book *The Social Function of Science*, Bernal had claimed that *laissez-faire* attitudes were not conducive to scientific progress and argued that the solution lay in central planning of science, 'as is already occurring in the Soviet Union'.[85] Although Bernal received support from left-wing scientists such as Joseph Needham, many others argued that scientific progress could only be guaranteed by giving researchers the freedom to make their own decisions and regulate their own conduct.[86] Figures such as the émigré chemist Michael Polanyi, who co-founded a Society for Freedom in Science in 1940, maintained that science could only thrive in a liberal society and free from outside interference.[87] These arguments were strengthened following the Second World War, when it became clear that Soviet efforts to control genetics involved the arrest and execution of scientists opposed to Trofim Lysenko, who fraudulently claimed to have perfected a way of increasing crop yields and transmitting acquired characteristics to later generations.[88] Supporters of scientific freedom argued that the collapse of Soviet genetics and agriculture proved just how harmful external interference was for science.

Support for club regulation was strengthened further during the 1950s, thanks to advances in biological and medical research such as the development of effective anti-tuberculosis drugs, open-heart surgery, kidney transplantation and the discovery of DNA's helical structure. These successful projects involved no external planning and were all 'developed through the single-minded efforts of a few dedicated individual scientists and doctors'.[89] At a time when professions were highly regarded, this research further increased public confidence in science and medicine.[90] Celebratory media coverage portrayed doctors and scientists as pioneering figures who were central to a 'new Elizabethan era' of progress and discovery.[91] When 'science and expertise were synonymous', both in public and in government, the future of club regulation seemed more assured than ever.[92]

Compounding club regulation: other professions and 'doctor knows best'

Historians have thus far explained the growth of club regulation by detailing how the professional desire for autonomy mapped on to the *laissez-faire* attitude of politicians, and examining how the expertise of doctors and scientists became central to government policy from the mid nineteenth century onwards. But we cannot fully account for club regulation without also examining attitudes in other professions. As we shall see, medical and scientific ethics were also seen as professional concerns thanks to the overwhelmingly 'hands-off' approach in the fields that later constituted bioethics, such as law, philosophy and theology.

This stance partly reflected the broad support for technical expertise during the nineteenth and twentieth centuries; but it also reflected factors that were specific to each of these fields. This was certainly the case in law, where *laissez-faire* attitudes to medicine were most evident. From the eighteenth century onwards it was extremely rare for the courts to adjudicate in medical malpractice cases. This was largely because the legal system adapted itself to the rules of the market and, in doing so, 'became unwilling to interfere with the freedom of trade'.[93] This stance was compounded in the nineteenth century when lawyers, like doctors, reorganised themselves to 'control competition in the new markets opened up by industrialism'.[94] They exploited statute and common laws to establish monopolies and rebuilt their governing institutions, such as the Inns of Court and the Law Society, to organise training and discipline with a high degree of autonomy from the state. One bastion of club regulation was hardly likely to interfere with the affairs of another, especially after the 1858 Medical Act formally entrusted doctors with the power to regulate themselves.

In the rare instances when the courts did consider medical practices, they sought not to challenge but to strengthen medical authority. Abortion was the only operative procedure governed by law during the nineteenth century, with the 1861 Offences Against the Person Act specifying that any attempt to induce miscarriage was punishable by life imprisonment. But this was less about regulating doctors, who were free to perform an abortion if they believed it would save a woman's life, and more about prohibiting the activities of 'backstreet' abortionists who offered competing systems of healthcare.[95] While doctors believed that the law should 'interfere as little as possible with clinical practice', they nevertheless supported legislation to 'retain medical control of abortion and to exclude the "racketeer who has brought such discredit upon our profession"'.[96]

The decisions from two 1950s medical negligence cases demonstrate how lawyers and judges continued to believe that 'the medical profession should be held in special regard and interfered with by the law as little as possible'.[97] The first, *Hatcher* v. *Black*, arose in 1954 after a patient claimed that they were not informed about possible nerve damage in thyroid surgery. Ruling in favour of the doctors, the judge, Alfred Denning, argued that 'we should be doing a disservice to the community at large if we were to impose liability on hospitals and doctors for everything that goes wrong'. Denning warned that giving courts the power to decide what constituted negligent behaviour would lead to 'defensive medicine', where doctors thought 'more of their own safety than of the good of their patients'. [98] This, he predicted, would stifle innovation, cost lives and ultimately harm public confidence in the NHS.

The second case, *Bolam* v. *Friern Hospital Management Committee*, arose in 1957 when a patient sued doctors for injuries that arose after they failed to restrain him during electroconvulsive therapy.[99] Here, as in *Denning*, the courts ruled in favour of the doctors rather than the patient. Their decision hinged not on the possibility of 'defensive medicine' but on the argument that the patient's treatment conformed to standard medical practices. This ruling became known as the '*Bolam* test' and was applied to all subsequent medical negligence claims.[100] As the lawyer Margaret Brazier notes, by deciding that medical conduct should be judged according to professional norms, and not the expectations of patients or the public, the *Bolam* test affirmed that 'the underlying trend in the English courts was that "doctor knows best"'.[101]

While philosophers took a similarly 'hands-off' stance, they did so for different reasons. During the eighteenth and nineteenth centuries, work on ethics had formed a major component of philosophy. British philosophers such as David Hume, Jeremy Bentham and John Stuart Mill claimed that acts should be guided by notions of sympathy, natural or individual rights and the utilitarian faith in increasing the happiness of the greatest number of people; and some of these ideas, especially Hume's work on sympathy, influenced codes of medical ethics *circa* 1800.[102] During the early twentieth century, however, philosophers abandoned work on ethics and refused to state how things ought to be. This shift involved a rejection of the previous belief that notions such as 'good' or 'right' could be objectively determined, which John Dewey had encapsulated when he defined ethics as 'the science that deals with conduct, in so far as this is concerned as right or wrong, good or bad'.[103]

In his 1903 book *Principia Ethica*, the Cambridge philosopher G. E. Moore argued that ethics was not a science since 'good' and 'right' were indefinable categories that could not be empirically verified.[104] Moore coined the term 'naturalistic fallacy' to describe the seemingly mistaken belief that a certain action could be objectively shown to be 'good' in the same way that, say, blood could be shown to flow around the body.[105] He argued that while we may recognise that something is intrinsically good, just as we recognise something is yellow, we cannot then *prove* it really was 'good' in order to specify what kinds of actions we should perform.

Moore's argument underpinned the redefinition of philosophy as a more objective field that was free of any political, nationalistic or religious bias.[106] Following *Principia Ethica*, philosophers adopted an approach that Bertrand Russell called 'modern analytical empiricism', which centred solely on clarifying the properties of logical or moral propositions. Russell argued that this method distanced philosophy from the doctrinaire and incommensurable notions of 'good' that had been disastrously employed during the First World War, and gave it the objective 'quality of science … by which I mean the habit of basing our beliefs upon observations and inferences as impersonal, and as much divested of local and temperamental bias, as is possible for human beings'.[107] Rather than challenge science, then, prominent philosophers such as Russell and Ludwig Wittgenstein sought to emulate it. They viewed philosophy as a 'disinterested search for truth', and believed that anyone who made a normative statement was committing 'a kind of treachery'.[108]

This position was reaffirmed by the young Alfred J. Ayer, whose 1936 book *Language, Truth and Logic* was widely credited with having 'a huge influence on people's notion of what ethics is all about'.[109] Ayer took Moore and Russell's stance to its logical conclusion when he endorsed a highly subjectivist view of ethics, claiming that moral statements were 'simply expressions of emotion that can be neither true nor false'.[110] Since philosophers only studied verifiable and 'genuine propositions', he argued, 'a strictly philosophical treatise on ethics should therefore make no ethical pronouncements'.[111] What was more, Ayer also believed that since ethical statements were unverifiable expressions 'with no objective validity', and since there was 'no relevant empirical test' to resolve competing claims, it was misleading for a philosopher or anyone else to 'set themselves up as arbiters of right and wrong'.[112]

As he increasingly became a 'public intellectual' and appeared on television and the radio from the 1950s, Ayer found himself in the ironic position where 'the authority of his public role rested on his professional identity as a philosopher, but his declared philosophical position was that philosophy could have little to say on issues that were of public interest'.[113] Ayer made this clear in his 1965 *Philosophical Essays*, when he stated that 'to analyse moral judgements is not itself to moralise' and warned that members of the public would be disappointed if 'they mistakenly look to the philosopher as a champion of virtue'.[114] Over fifty years after *Principia Ethica* had been published, Ayer ensured that this austere view of ethics remained paradigmatic. As Mary Warnock outlined in 1960, it 'seemed as if there were no other virtue in a moral philosopher except that he should avoid the naturalistic fallacy'.[115] Moral philosophy had become defined, she argued, by 'the refusal of philosophers in England to commit themselves to moral opinions'.[116]

This gave scientists and doctors freedom to discuss ethics in their own fields and more generally. On the rare occasions that philosophers responded to the ethical work of scientists or doctors, they simply affirmed why they avoided normative issues. For instance, when Conrad Waddington told Ludwig Wittgenstein that he was writing *Science and Ethics*, the horrified philosopher replied that it was 'a terrible business – just terrible! You can at best stammer when you talk of it.'[117] C. E. M. Joad was the only philosopher who responded to Waddington's essay, in the journal *Nature*, yet this was only to criticise him for presuming that notions such as 'good' could be objectively measured.[118] And when the CIBA Foundation convened a 1963 symposium on 'Man and His Future', which examined whether biological research might 'reshape traditional grounds for ethical beliefs', there were no philosophers in attendance.[119]

Religious figures, on the other hand, were more prepared to discuss science and medicine. While no philosophers or lawyers attended the 'Man and His Future' symposium, the predominantly scientific audience was joined by the Revd H. C. Trowell, curate of Stratford-Sub-Castle, who discussed food allocation and family planning in the developing world.[120] Theologians and the clergy were also second only to scientists and doctors in responding to *Science and Ethics*. In line with the complexity that had long characterised relations between religion and science, attitudes here were less uniform than in law or philosophy. Some religious figures opposed what they saw as Waddington's attempts to portray science as a secular religion. The Dean of St Paul's Cathedral, for example, claimed that *Science and Ethics* was a 'disastrous error' and asserted that morality came 'from a Source deeper and more intimate than the course of evolution'.[121] Others, meanwhile, claimed that science and religion could not conflict because 'they were quite separate provinces'.[122] This position was endorsed by philosophers such as Ayer, who claimed in *Language, Truth and Logic* that 'there was no logical ground for antagonism between religion and natural science', because 'since religious utterances are not genuine propositions at all, they cannot stand in logical opposition to the propositions of science'.[123]

But a significant proportion also sought to assimilate religious and scientific worldviews. This had been a longstanding tactic within the Church of England, especially in efforts to reconcile religion and evolutionary theories, and the tendency increased after the 1920s when modernising figures such as William Temple, later Archbishop of Canterbury, argued that theologians needed to engage with contemporary issues to ensure they were not 'isolated from the mainstream of public life'.[124] This belief led to greater discussions of how Christian faith related to political, economic and scientific concerns, and was evident when Ernest Barnes, the Bishop of Birmingham and a former mathematician, wrote to *Nature* expressing his 'fundamental agreement' with *Science and Ethics*. There was no reason, Barnes argued, why evolutionary and ethical progress could not both be seen as evidence of God's 'progressive revelation of Himself'.[125]

From the late 1930s onwards many clergy and Christian intellectuals believed this 'synthesis' could be achieved by working with doctors, scientists and others to discuss common concerns, and endorsed collaboration in small interdisciplinary groups.[126] In 1938, for example, the ecumenist J. H. Oldham co-founded the 'Moot' group with Anglican clergymen such as Alec Vidler and Daniel T.

Jenkins, Christian intellectuals such as the poet T. S. Eliot, the sociologist Karl Mannheim and the educationalist Walter Moberly.[127] The Moot group discussed a wide range of issues, including relations between science and religion, and sought to ensure that Christian values were at the forefront of postwar social reconstruction. Despite its illustrious background, however, the Moot's emphasis on elite leadership was unfashionable in the egalitarian welfare state and it disbanded after Mannheim died in 1947.[128]

While William Temple was not a member of Moot, his enthusiasm for interdisciplinary groups was evident shortly before his death in 1944, when he established the Churches' Council of Healing (CCH) 'to bring together the churches and the medical profession'.[129] Temple saw collaboration here as vital since the physical, mental and spiritual aspects of healing were 'so interdependent that successful treatment of disease in one was not possible without consideration of the others'.[130] He also argued that doctors stood to benefit from cooperating with theologians 'in the study and performance of their respective functions in the work of healing', as they would receive valuable help in assisting those patients who believed that 'religious ministrations will conduce to health and peace of mind'.[131]

The BMA initially questioned the 'propriety of the association of doctors with clergy as unqualified persons', and sought assurances that the CCH had no desire to 'overlap the realm of physical or psychiatric medicine' and was not advocating 'unscientific' methods such as faith healing.[132] But after meeting a deputation headed by the Bishop of Croydon, they claimed that there was 'no ethical reason to prevent medical practitioners co-operating with the clergy' and supported appointing BMA representatives as *ex officio* CCH members.[133] The BMA council also broadened this proposal and endorsed 'fuller co-operation', in which 'medicine and the Church working together should encourage a dynamic philosophy of health which would enable every citizen to find a way of life based on moral principle and a sound knowledge of the factors which promote health and well-being'.[134]

The BMA's belief that collaboration with theologians was 'necessary and desirable' might appear surprising, as club regulation was particularly strong in the late 1940s. But doctors were happy to collaborate because they believed that religious figures ultimately strengthened their professional authority. This partly stemmed from a hope that they would reconcile doctors to the changing landscape of the new welfare state. In a letter to the *British Medical Journal*, which followed a report on the CCH, one doctor outlined how they and their colleagues feared being 'grossly overworked' in the new NHS because 'patients will be entitled to medical advice without a fee and will consult their doctor far more readily'.[135] These worries were also expressed by medical practitioners in early meetings of the Frontier Medical Group (FMG), which was co-founded by Christian doctors and some Moot clergymen, such as Daniel Jenkins and Alec Vidler.[136] The meetings prompted Jenkins to write a 1949 book on *The Doctor's Profession*, in which he claimed that the 'establishment of a National Health Service' forced doctors 'to reckon with even greater interest in [their] activities on the part of the community', but left them too overworked to fully consider ethical issues.[137] While Jenkins acknowledged that it was unusual for a book on medicine to 'be written not by a doctor but by a theologian', he argued that doctors were now simply 'too busy to write books of this kind'.[138]

Jenkins outlined how Christian doctors increasingly sought advice from theologians because the NHS placed new demands on 'an already overcrowded life'.[139] These concerns ensured that *The Doctor's Profession* was one of several books written for denominational audiences in the late 1940s, which aimed to show that it was possible 'to be a doctor and a good Christian' in the welfare state.[140] This trend increased during the 1950s, as Christian doctors 'demanded to know what their options were' in the face of growing public demand for contraceptives, an increase in artificial insemination and the questions raised by new artificial respirators about whether withdrawing treatment from 'hopeless' cases conflicted with the 'Christian's reverence for life'.[141]

At the same time, doctors also welcomed the input of theologians because they positioned themselves as ancillaries to the medical profession. They saw their job as to clarify religious views on particular issues, not to criticise doctors or influence decision-making. William Temple, for one, argued that theologians should elucidate general principles 'according to which precise policy might be formulated', and held that it was not for them to 'argue how principles should be put into practice'.[142] This stance was also clear in *The Doctor's Profession*, in which Jenkins provided no direct advice and stressed that it was 'clearly not the function of a book of this kind to pass judgement'.[143] Like philosophers and lawyers, Temple, Jenkins and other theologians ultimately believed that medical decisions were for doctors alone to make.

Criticising club regulation and 'the birth of bioethics'?

At the start of the 1960s no-one argued that scientists or doctors required any external supervision. But this was to change over the course of the decade, which witnessed the beginnings of what Harold Perkin calls a 'backlash against professional society'.[144] Scientists and doctors were no longer seen as 'the god-like functionaries, beyond questioning much less criticism, they had once been', and public debates increasingly centred on the drawbacks as much as the benefits of research.[145] While distrust of medical or scientific authority was nothing new, of course, it had previously come from specific campaign groups or elite critics such as George Bernard Shaw. But several linked factors ensured that it was far more widespread in the 1960s and arose from a broader social base than before. These included horror at the neonatal disabilities caused by the morning sickness drug Thalidomide, which came to light in 1962 and burst 'the bubble of postwar optimism' surrounding medical research.[146] At the same time, in their reports on Thalidomide and other issues, the media adopted a more critical 'watchdog' stance in which they focused on social and ethical issues instead of deferring to professional experts.[147]

Criticism also reflected the emergence of a 'new politics' in the 1960s and 1970s, in which concerns over class identity and economic security were replaced by an emphasis on human rights and individual autonomy.[148] Change was often driven by the activities of the many 'new social movements' that incorporated civil rights and libertarian ideologies to campaign on behalf of marginal groups.[149] These movements increasingly criticised professions as obstacles to empowerment, as unaccountable and self-serving power blocs. Some of the more radical ones drew inspiration from leftist academics such as the Austrian philosopher Ivan Illich, who claimed that medical control over definitions of health and illness fostered a 'debilitating' client mentality among patients and was itself a major threat to health.[150] This was certainly the case with the National Association for Mental Health, which rebranded itself as MIND under the leadership of the American lawyer Larry Gostin and began to expose professional misconduct and campaign for a 'rights-based' approach to mental illness.[151]

But the 'backlash against professional society' did not emanate solely from new social movements, a critical media or a disaffected public. Indeed, one of the earliest and strongest critiques of medicine, which contributed to public unease and influenced campaign groups, came from the medical 'whistleblower' Maurice Pappworth. Born Maurice Papperovitch in 1910, before his family changed their name in the 1930s, Pappworth claimed that anti-semitism prevented him from obtaining consultant positions in London after he graduated from the University of Liverpool in 1932.[152] Rather than take a 'peripheral' hospital post, he decided to earn a living by tutoring junior doctors looking to pass the diploma that controlled entry to the Royal College of Physicians (RCP). It was here that Pappworth learned of questionable research practices, after his students told him that they were often expected to undertake experiments on NHS patients without their full knowledge or consent.[153] While informed consent had been prioritised as 'absolutely essential' by the Nuremberg Code that was drawn up during the Nazi medical trials, it was routinely ignored by researchers in Britain, the United States and elsewhere, who believed the guidelines were designed to prosecute 'barbarians' and did not apply to them.[154]

Pappworth was certainly not the first doctor to have misgivings about the lack of consent in medical research, but he was the first to go public. In line with club regulation, doctors had previously kept their views 'in house' and refused to openly criticise their colleagues. This was clear when Pappworth wrote letters to journals that published work which he found to be ethically dubious, but the editors refused to publish them.[155] Frustration with these rejections led Pappworth to break with protocol in 1962, when he published a short piece in the popular *Twentieth Century* magazine. His article drew on a sample of published studies to claim that researchers often exposed patients to risky experiments, including liver biopsies and withholding of insulin, without their 'full consent, after honest and detailed explanation of what was to be meted out to them'.[156]

After listing fourteen questionable experiments, Pappworth detailed how animal experiments 'were rigorously controlled and supervised' whereas 'doctors can indulge in human vivisection without let or hindrance'.[157] He argued that it was no longer sufficient to claim that 'only the clinician in charge could say what was right and proper and what safeguards were needed'.[158] But while he called for 'proper safeguards' to be introduced, Pappworth did not detail what changes he felt were needed, other than recommending that 'the investigator who is also the practising physician in control of the patient cannot be the person best qualified to judge objectively the risk involved in any experiment'.[159]

The *Twentieth Century* article was published in the same year that the public learned the full scale of the Thalidomide tragedy, and both played a significant role in generating disquiet over medical research. Pappworth's call for a 'battle to defend the rights of all patients against the whims and ambitions of some doctors' prompted the teacher Helen Hodgson to establish the Patients Association in January 1963.[160] The Patients Association was one of the earliest and most high-profile 'new social movements' concerned with healthcare, and regularly challenged medical paternalism in letters to newspapers and professional journals.[161] Like other new social movements, the Patients Association emphasised individual autonomy and claimed that patients had a fundamental right to choose whether or not they were subjected to research.[162] It also, notably, demanded greater public involvement in the development of regulatory guidelines for clinical research. In a 1963 letter to the *British Medical Journal*, Hodgson warned that patients 'would not be willing for much longer to submit blindly their health and their lives to any arbitrary code of ethics in which they have no say'.[163]

Pappworth's work also caused unrest among doctors, who believed that he should have confined his critique to the medical community.[164] Many tried to dissuade him from making further public claims, warning that he would seriously undermine people's faith in medicine. One senior doctor summarised these views in a letter to Pappworth several years later, when he claimed that 'in common with many people, I disliked your tactics as much as I approved of your message'.[165] But despite the attempts of other doctors, Pappworth went ahead and published a longer book, entitled *Human Guinea Pigs*, in 1967.

Human Guinea Pigs was similar to Pappworth's earlier article in many respects, providing a long list of British and American experiments that had been undertaken without valid consent, carried no therapeutic benefit and were often dangerous. But it also differed thanks to a long final chapter that set out proposed legal changes and, notably, endorsed outside involvement in the development and enforcement of new guidelines. At the outset of the book Pappworth explained why he had contravened one of the main tenets of club regulation and encouraged 'discussion outside professional circles'.[166] Drawing on his own experiences, he argued that 'little heed has been paid by the experimenters themselves to the occasional voices raised in protest against these practices, and there has been, on the part of editors of professional journals, some censorship of the expression of protest – presumably for fear of offending some of their readers'.[167] The only way to adequately 'stir the consciences of doctors', Pappworth concluded, was to 'enlighten the public about what is going on in such experiments'.[168]

But Pappworth also publicised his work because he believed, like Hodgson, that 'the medical profession must no longer be allowed to ignore the problems or assert, as they often do, that this is a matter to be solved by doctors themselves'. He instead claimed that ethical issues in clinical research could only be solved 'by frank discussion among informed people, lay as well as medical'.[169] While Pappworth advocated 'frank discussion', he also called for new and legally binding guidelines for clinical research. 'After careful thought over many years', he wrote, 'I have reluctantly come to the conclusion that the voluntary system of safeguarding patients' rights has failed and new legislative procedures are absolutely necessary.'[170]

Pappworth saw outside involvement as vital here. He argued that in order to fully protect patients, 'who are at present exposed to dangers and indignity', it was essential that 'our laws do not place the entire authority to decide what is permissible and what is not in the hands of a professional class'.[171] He recommended that Parliament should formulate an Act that established 'consultation committees', which would review all research applications and 'judge objectively ... whether or not any proposed experiment is legally and ethically justifiable'.[172] Pappworth proposed that every regional hospital board should include a 'consultation committee' that was answerable to the GMC and Parliament. Although he did not specify how many members they should have, he stressed that one 'must be a clinician who is not involved in research, and there should be at least one lay member, preferably but not necessarily a lawyer'.[173]

By endorsing lay involvement in deciding whether research was 'ethically justifiable', Pappworth was clear in his belief that medical ethics should no longer be a matter for doctors alone. This has led some to claim that his work was a critical moment in 'the birth of British bioethics'.[174] But this is far from the case. Pappworth's arguments had little, if any, impact on the continuing support for club regulation among doctors, politicians and other professions. While all these groups agreed that aspects of clinical research were problematic, they maintained that responsibility for implementing reforms should continue to rest with the medical profession.

Writing in the *Times Literary Supplement*, for example, the renowned geriatrician Lord Basil Amulree stated that Pappworth had been 'right to draw attention to this disquieting trend in medicine' and acknowledged that it was 'surely undesirable to carry out any experiment on patients without their consent'. But Amulree disagreed with Pappworth in his firm insistence that 'it is the members of the profession itself … who can do most to ensure that this undesirable and unethical form of experimentation ceases to be practised'.[175] Involving outsiders in developing guidelines, he argued, would simply ensure that they were 'difficult to draft and equally difficult to enforce'.

The *Lancet*, too, claimed that the best way to protect patients was by ensuring that 'the difficult and important decisions that research doctors have to make must be kept under constant review by other doctors'. Implementing Pappworth's recommendations, it continued, would 'only lead to another ineffectual code of vague ethics'.[176] And in a review for *World Medicine*, the doctor and epidemiologist Charles Fletcher, who was a longstanding critic of Pappworth, pointedly dismissed his calls for lay involvement as a 'quite impracticable' measure that 'could not seriously have been proposed by anyone engaged in medical research'.[177]

Parliament also continued to endorse *laissez-faire* attitudes to regulation. The vast majority of politicians echoed Amulree, Fletcher and other doctors by rejecting outside involvement in clinical research. Members of Harold Wilson's Labour government, which had promised to turn scientific innovation into economic and material prosperity when it won the 1964 election, were reluctant to interfere with professional expertise and believed the best solution was for 'the medical profession to put its house in order'.[178] This was made clear during a Commons debate that followed the publication of *Human Guinea Pigs* in May 1967. The government's Minister for Health, Kenneth Robinson, rejected the Labour MP Joyce Butler's call for a public inquiry and claimed that hospital authorities and the MRC already provided researchers with 'comprehensive guidance'.[179] The government reiterated its position the following year, when the Ministry of Health rebuffed the Patients Association's demands for a public inquiry and claimed that ethical questions were 'for the profession to consider'.[180] In 1969 the Conservative MP Quintin Hogg, who had previously endorsed *laissez-faire* approaches as Minister for Science, told Pappworth that external regulation was highly unlikely as 'I do not myself think that Parliament is in the position in which positive legislation can be imposed without detriment to the freedom of the medical profession'.[181]

Pappworth's recommendations also found little support from other professions, who maintained their 'hands-off' stance into the 1960s despite the 'backlash against professional society'. Reviewing *Human Guinea Pigs* for the BBC's *Listener* magazine, the philosopher Bernard Williams said nothing about Pappworth's call for lay involvement and statutory regulation, and dwelt instead on whether the 'Golden Rule' was an appropriate ethical safeguard for research: that is, whether it was sufficient to argue that doctors should not submit patients to a procedure they would not be willing to undertake on themselves or their families.[182] In line with the *Bolam* test, lawyers also maintained that doctors should be left to determine their own conduct and standards of care. In a long letter to Pappworth, the lawyer Cecil Clothier dismissed his demands for 'full informed consent', since 'nobody in a hospital *ever* consents in the sense you suggest'.[183] Clothier also claimed that legal guidelines would be overly restrictive, as notions of acceptable risks and safeguards differed between individual patients and specific research projects. He outlined how one patient might demand full information while another might not care, and stated that prioritising informed consent was inappropriate when a doctor was faced with an unconscious patient whose only chance of survival 'could include trying a newly-devised drug if nothing else had done any good'. These complications, Clothier argued, ensured that 'individual assessment' remained the best form of governance for doctors.[184]

Under no pressure to implement change, either from politicians or other professions, doctors largely ignored Pappworth's recommendations. This was clear in 1967 when an RCP committee, comprised solely of doctors, produced a short report that proposed the formation of research ethics committees (RECs) to review applications for projects 'where the subject, be he a patient or a normal person, cannot expect clinical benefit'.[185] The apparent similarity between this proposal and *Human Guinea Pigs* led Pappworth to claim that he had influenced the RCP.[186] But as one doctor informed him years later, the RCP's decision '*antedated* your book' and was prompted by changing grant policies in the United States, where the Public Health Service (PHS) stated that it would only fund research if an applicant's institution had conducted a prior ethical review.[187] The RCP report clearly stated that RECs should be established at hospitals where researchers 'were in receipt' of or were likely to seek PHS money. They also predicted that RECs would assess proposals from British funding bodies once established, since 'it is unlikely they will feel they can sensibly confine their attentions solely to cases where research is sponsored by a foreign country'.[188]

The RCP committee's attitude to outside involvement highlights the extent to which doctors ignored Pappworth. In marked contrast to *Human Guinea Pigs*, it proposed that RECs should be composed of 'a group of doctors including those experienced in clinical investigation'.[189] When 'difficult ethical problems arise', it claimed, 'even the most experienced workers would often welcome the opinion and advice of their peers'. The RCP committee dismissed any outside involvement or formal regulation when it argued that it was

> of great importance that clinical investigation should be free to proceed without unnecessary interference and delay. Imposition of rigid or central bureaucratic controls would be likely to deter doctors from undertaking investigations, and if this were to happen, the rate of growth in medical knowledge would inevitably diminish with resultant delay in advances in medical care.[190]

The responses to a 1971 survey show that the vast majority of hospitals followed the RCP's proposals when they established RECs. Only one-fifth of those set up after 1967 included a lay member, who was generally the hospital or group secretary, and none included more than one.[191] If this were not testament enough to the continued strength of club regulation, it was officially endorsed by a government inquiry into the structure and function of the GMC, which had been established in 1972 following professional unrest at the decision to 'strike off' any doctor who did not pay a new annual retainer fee.[192] When the committee's report was published in 1975, it unanimously agreed that staffing the GMC predominantly with doctors safeguarded the public, since 'it is the essence of professional skill that it deals with matters unfamiliar to the layman'.[193] Despite Pappworth's efforts, and to his continued frustration, responsibility for deciding ethical issues continued to rest 'firmly on the shoulders of the medical profession'.[194]

Conclusion

This evidence undermines claims that Maurice Pappworth made a 'significant contribution to the development of medical research ethics' and that '*Human Guinea Pigs* is a major milestone on the journey towards the modern system of research ethics committee review'.[195] While Pappworth's work alerted the public to the ethical issues associated with clinical experiments, and contributed to a broader critique of professional expertise, it had little impact on the governance of medical research or treatment. Several writers have sought to explain Pappworth's lack of influence by claiming that his confrontational manner 'alienated most of his audience' and that he 'was not an authoritative figure in medical circles'.[196] These are certainly valid points. Journal editors and correspondents were often irritated by the strident tone of Pappworth's correspondence, while Cecil Clothier argued that he might have a more sympathetic audience if he moderated his 'candour'.[197] It is also clear that senior doctors often used Pappworth's lack of professional status to dismiss his work, with Charles Fletcher, for one, claiming that it was the product of an embittered outsider and would not have arisen from 'anyone engaged in medical research'.[198]

But while his manner and status did not help, I believe that Pappworth was mainly ignored because his calls for outside involvement conflicted with the longstanding and continued support for club regulation among doctors, politicians and other professions.[199] This makes it hard to portray him as a significant figure in 'the birth of British bioethics'. Sections of *Human Guinea Pigs* certainly resemble later work in bioethics, not least its calls for patient empowerment and lay involvement, but portraying Pappworth as significant to the development of bioethics involves reading history backwards and reduces bioethics to little more than a public critique of medicine and science, which is far from the case. Bioethics is a multi-sited and interdisciplinary set of activities, and we cannot attribute its emergence to a single event or figure.

What is more, the fact remains that bioethics only became a recognised term and approach in Britain once politicians, doctors, lawyers, philosophers and others came to believe that external involvement with medicine and science benefited patients, professions and the public. This was clearly not the case in the 1960s, when they all agreed that doctors should retain 'jurisdictional control' over their own practices.[200] Instead of mistakenly trying to see Pappworth as influential to bioethics, then, we need to concentrate on the broad changes that led politicians, lawyers, philosophers and theologians to adopt a more 'hands-on' approach in decades to come.

Notes

1 Moran, *British Regulatory State*, pp. 38–66.

2 On ethics produced 'by and for professions', see Andrew A. G. Morrice, '"Honour and Interests": Medical Ethics and the British Medical Association', in Andreas-Holger Maehle and Johanna Geyer-Kordesch (eds), *Historical and Philosophical Perspectives on Medical Ethics* (Aldershot: Ashgate, 2002) pp. 11–37 (p. 12).

3 See Harold Perkin, *The Rise of Professional Society: England Since 1800* (London and New York: Routledge, 1990) p. 4.

4 For a recent example, see Michael Brown, *Performing Medicine: Medical Culture and Identity in Provincial England, c. 1760–1850* (Manchester: Manchester University Press, 2011). For an overview of historical work on professionalisation and expertise, see John C. Burnham, 'How the Concept of a Profession Evolved in the Work of Historians of Medicine', *Bulletin of the History of Medicine*, Vol. 70 (1996) pp. 1–24.

5 Abbott, *The System of Professions*, p. 2.

6 Perkin, *The Rise of Professional Society*, p. 472.

7 See, for example, Rachel McAdams, 'Human Guinea Pigs: Maurice Pappworth and the Birth of British Bioethics', MSc thesis, University of Manchester, 2005; Boyd, 'Discourses of Bioethics in the United Kingdom', p. 487.

8 Ivan Waddington, 'The Development of Medical Ethics – A Sociological Analysis', *Medical History*, Vol. 19 (1975) pp. 36–51 (p. 37). See also Nicholas Jewson, 'The Disappearance of the Sick Man from Medical Cosmology', *Sociology*, Vol. 10 (1976) pp. 225–44; Ivan Waddington, 'Medical Knowledge and the Patronage System in Eighteenth Century England', *Sociology*, Vol. 8 (1974) pp. 369–85. See also Perkin, *The Rise of Professional Society*, pp. 117–18.

9 Waddington, 'Development of Medical Ethics', p. 37. For a detailed analysis of the social status of eighteenth-century medical practitioners, see Brown, *Performing Medicine.*
10 Roy Porter, 'Laymen, Doctors and Medical Knowledge in the Eighteenth Century: The Evidence of the *Gentleman's Magazine*', in Roy Porter (ed), *Patients and Practitioners: Lay Perceptions of Pre-Industrial Society* (Cambridge: Cambridge University Press, 1985) pp. 283–315 (p. 287).
11 Waddington, 'Development of Medical Ethics', p. 37.
12 On the 'medical marketplace', see Mark Jenner and Patrick Wallis (eds), *Medicine and the Marketplace in England and its Colonies, c.1450–c.1850* (Basingstoke: Palgrave Macmillan, 2007); Roy Porter, 'Before the Fringe: "Quackery" and the Eighteenth Century Medical Market', in Roger Cooter (ed), *Studies in the History of Alternative Medicine* (Basingstoke: Macmillan, 1988) pp. 1–27.
13 Porter, 'Before the Fringe', p. 3. See also Baker et al., 'Introduction', p. 5. On challenges to the tripartite hierarchy posed by a new division between general practitioners and hospital consultants, see Waddington, 'Development of Medical Ethics', pp. 41–2. Michael Brown has recently claimed that much literature on the 'medical marketplace' presents a misleading portrayal of medicine in the seventeenth and eighteenth centuries by focusing largely on competition and conflict. He argues that many historians have ignored how orthodox medical practitioners also acted cooperatively in this period and displayed 'collective identities or shared values'. See Brown, *Performing Medicine*, p. 5.
14 Baker et al., 'Introduction', p. 7.
15 Mary E. Fissell, 'Innocent and Honourable Bribes: Medical Manners in Eighteenth Century Britain', in Baker et al. (eds) *Codification of Medical Morality*, pp. 19–47.
16 Although several historians have noted how Percival's guidelines proved valuable for nineteenth-century medical reformers, John Pickstone details how they were very much a product of eighteenth-century Manchester. They were written to settle a dispute in the local Infirmary involving an increase in honorary staff and perceived lay interference, and drew on the prior tradition of gentlemanly advice and moralising. See John Pickstone, 'Thomas Percival and the Production of *Medical Ethics*', in Baker et al. (eds), *Codification of Medical Morality*, pp. 161–78. On Percival's later influence, see Andreas-Holger Maehle, 'Medical Ethics and the Law', in Mark Jackson (ed), *The Oxford Handbook of the History of Medicine* (Oxford: Oxford University Press, 2011) pp. 543–60.

17 Baker et al., 'Introduction', p. 9.
18 Brown, *Performing Medicine*, pp. 138–40, 150–83.
19 Perkin, *The Rise of Professional Society*, p. 117. See also Brown, *Performing Medicine*, pp. 150–83; Christopher Lawrence, *Medicine in the Making of Modern Britain, 1700–1920* (London: Routledge, 1994) pp. 27–51; John Pickstone, 'The Professionalisation of Medicine in England and Europe: The State, the Market and Industrial Society', *Journal of the Japanese Medical Society*, Vol. 25, no. 4 (1979) pp. 550–21.
20 Brown, *Performing Medicine*, p. 195; Pickstone 'The Professionalisation of Medicine', p. 526.
21 Pickstone, 'The Professionalisation of Medicine', p. 529.
22 Brown, *Performing Medicine*, p. 175; Moran, *The British Regulatory State*, pp. 40–1.
23 Perkin, *The Rise of Professional Society*, p. 117; Lawrence, *Medicine in the Making of Modern Britain*, pp. 55–7.
24 Rose, 'Government, Authority and Expertise in Advanced Liberalism', p. 285.
25 Sally Sheard, 'Quacks and Clerks: Historical and Contemporary Perspectives on the Structure and Function of the British Medical Civil Service', *Social Policy and Administration*, Vol. 44 (2010) pp. 193–207.
26 Brown, *Performing Medicine*, p. 226.
27 Moran, *The British Regulatory State*, p. 49.
28 Russell G. Smith, 'The Development of Ethical Guidance for Medical Practitioners by the General Medical Council', *Medical History*, Vol. 37 (1993) pp. 56–67 (p. 58).
29 The initial punishment for misconduct was permanent removal from the medical register, but temporary suspensions later become more common. See Maehle, 'Medical Ethics and the Law', p. 547; Smith, 'The Development of Ethical Guidance', pp. 58–61.
30 Smith, 'The Development of Ethical Guidance', p. 57.
31 The BMA was originally known as the Provincial Medical and Surgical Association, before changing its name in 1856. See Lawrence, *Medicine in the Making of Modern Britain*, pp. 77–8.
32 Morrice, 'Honour and Interests', p. 16.
33 Ibid, p. 16.
34 Jukes de Styrap, *A Code of Medical Ethics: With General and Special Rules for the Guidance of the Faculty and the Public in the Complex Relations of Professional Life* (London: Lewis, 1890) p. 14.
35 Ibid, pp. 16–17. Emphasis in original
36 For more detail, see Morrice, 'Medical Ethics and the British Medical Association', pp. 24–5.

37 Robert Saundby, *Medical Ethics, A Guide to Professional Conduct* (London: Charles Griffin, 1907).
38 On 'friendly societies' and the 1911 Act, see Lawrence, *Medicine in the Making of Modern Britain*, pp. 78–81.
39 Morrice, 'Medical Ethics and the British Medical Association', p. 19.
40 Ibid, pp. 19–24.
41 Smith, 'The Development of Ethical Guidance', pp. 25–6.
42 David Reubi, 'The Human Capacity to Reflect and Decide: Bioethics and the Reconfiguration of the Research Subject in the British Biomedical Sciences', *Social Studies of Science*, Vol. 42 (2012) pp. 348–68 (p. 356). See also Laurence B. McCullough, 'Was Bioethics Founded on Historical and Conceptual Mistakes about Medical Paternalism?' *Bioethics*, Vol. 25 (2011) pp. 66–74. For a history that McCullough claims overstates the paternalistic nature of doctor–patient relations in this period, see Jay Katz, *The Silent World of Doctor and Patient* (New York: The Free Press, 1984).
43 De Styrap, *Medical Ethics*, p. 46. Emphasis added.
44 Morrice, 'Medical Ethics and the British Medical Association', p. 30. See also Alex Mold, 'Patient Groups and the Construction of the Patient-Consumer in Britain: An Historical Overview', *Journal of Social Policy*, Vol. 39 (2010) pp. 505–21 (pp. 506–9).
45 Durbach, *Bodily Matters*, p. 28.
46 Ibid, pp. 69–91.
47 Ibid, pp. 7–9.
48 See, for example, Hilda Kean, '"The Smooth Cool Men of Science": The Feminist and Socialist Response to Vivisection', *History Workshop Journal*, Vol. 40 (1995) pp. 16–38. On links between the anti-vivisection and anti-contagion movements, see Mary Ann Elston, 'Women and Vivisection in Edwardian England', in Nikolaas Rupke (ed), *Vivisection in Historical Perspective* (London and New York: Routledge, 1990) pp. 259–95; Coral Lansbury, *The Old Brown Dog: Women, Workers and Vivisection in Edwardian England* (Madison, WI, and London: University of Wisconsin Press, 1985).
49 French, *Antivivisection and Medical Science*, p. 229; Elton, 'Women and Anti-Vivisection', p. 274.
50 Durbach, *Bodily Matters*, p. 177.
51 The first conscientious objection clause was added to the 1898 Vaccination Act, and exempted any parent who satisfied two petty justices or a police magistrate that they honestly believed vaccination would damage their child's health. This, however, proved difficult to implement thanks to uncertainty over what constituted an 'honest conscientious objection', and a revised 1907 Vaccination Act allowed

any parent to get exemption without being questioned. See Ibid, pp. 171–97.

52 French, *Antivivisection and Medical Science*, p. 93.

53 Ibid, pp. 151–2. The Commission's more balanced composition also stemmed from the fact that Queen Victoria, a patron of the RSPCA, was 'strongly against' vivisection and had personally instructed Prime Minister Disraeli to convene an inquiry into the procedure. See Ibid, pp. 123–4; Brian Harrison, 'Animals and the State in Nineteenth Century England', *English Historical Review*, Vol. 88 (1973) pp. 786–820 (p. 791).

54 French, *Antivivisection and Medical Science*, pp. 177–90.

55 See Nikolaas Rupke, 'Pro-Vivisection in England in the 1880s: Arguments and Motives', in Rupke (ed), *Vivisection in Historical Perspective*, pp. 188–209 (p. 202).

56 Ibid, p. 204. Emphasis added.

57 French, *Antivivisection and Medical Science*, p. 208.

58 Details of the composition of the Home Office Advisory Committee are given in Sydney Littlewood (chair), *Report of the Departmental Committee on Experiments on Animals* (London: HMSO, 1965) p. 156. See also Richard Ryder, *Victims of Science* (London: National Anti-Vivisection Society, 2nd edn, 1983) p. 157.

59 French, *Antivivisection and Medical Science*, p. 215.

60 David Edgerton and John Pickstone, 'Science, Technology and Medicine in the United Kingdom, 1750–2000' (2009) pp. 1–35 (p. 19). Available online at: https://workspace.imperial.ac.uk/humanities/Public/files/Edgerton%20Files/edgerton_science_technology_medicine.pdf (accessed 7 February 2014).

61 John Gummett, *Scientists in Whitehall* (Manchester: Manchester University Press, 1980). For more on the MRC, see Joan Austoker and Linda Bryder (eds), *Historical Perspectives on the Role of the MRC* (Oxford: Oxford University Press, 1989). For specific examples of how the MRC used government money to encourage particular fields and approaches, see Michael Bresalier, 'Uses of a Pandemic: Forging the Identities of Influenza and Virus Research in Interwar Britain', *Social History of Medicine*, Vol. 25 (2012) pp. 400–24; Duncan Wilson, *Tissue Culture in Science and Society: The Public Life of a Biological Technique in Twentieth Century Britain* (Basingstoke: Palgrave Macmillan, 2011) pp. 8–27.

62 On interwar attitudes to orthodox medicine, see Martin Pugh, *We Danced all Night: A Social History of Britain Between the Wars* (London: Vintage, 2009) pp. 37–42.

63 Anna K. Mayer, 'A Combative Sense of Duty: Englishness and the Scientists', in Christopher Lawrence and Anna K. Mayer (eds),

Regenerating England: Science, Medicine and Culture in Inter-War Britain (Amsterdam: Rodopi, 2000) pp. 67–107 (p. 93).

64 See, for example, Julian Huxley, 'Science and Health', *Listener*, 8 November 1933, pp. 706–8; J. B. S. Haldane, *Daedalus, Or Science and the Future* (London: Kegan Paul, Trench, Trubner, 1924). For more background on these popularising activities, see Wilson, *Tissue Culture in Science and Society*; Peter J. Bowler, *Science for All: The Popularization of Science in Early Twentieth Century Britain* (Chicago: University of Chicago Press, 2009).

65 Daniel Kevles, *In the Name of Eugenics: Genetics and the Uses of Human Heredity* (Cambridge, MA: Harvard University Press, 2004).

66 Conrad H. Waddington, *Science and Ethics* (London: Unwin Brothers, 1942) p. 18.

67 Ibid, pp. 18–19.

68 Ibid, p. 18.

69 John Hedley Brooke, *Science and Religion: Some Historical Perspectives* (Cambridge: Cambridge University Press, 1991) p. 336.

70 Ibid.

71 Mayer, 'Combative Sense of Duty', p. 83.

72 Ibid, p. 89. Emphasis in original.

73 George Bernard Shaw, *Doctor's Delusions, Crude Criminology and Sham Education* (London: Constable, 1932) p. 9.

74 Ibid, p. 10.

75 Edgerton and Pickstone, 'Science, Technology and Medicine in the United Kingdom', p. 21.

76 Shaw, *Doctor's Delusions*, p. 46.

77 Ibid, p. 53.

78 Ibid.

79 Paul Weindling, 'Human Guinea Pigs and the Ethics of Experimentation: The BMJ's Correspondent at the Nuremberg Medical Trial', *British Medical Journal*, Vol. 313 (1996) p. 1467.

80 Ibid.

81 Perkin, *The Rise of Professional Society*, p. 346.

82 See Rudolf Klein, *The New Politics of the NHS: From Creation to Reinvention* (Oxford and New York: Radcliffe Publishing, 2010), especially 'The Politics of Creation', pp. 1–21.

83 Reubi, 'The Human Capacity to Reflect and Decide', p. 353. See also Peter Miller and Nikolas Rose, *Governing the Present: Administering, Economic, Social and Personal Life* (Oxford: Polity Press, 2008); Phillipe Fontaine, 'Blood, Politics and Social Science: Richard Titmuss and the Institute of Economic Affairs, 1957–1973', *Isis*, Vol. 93 (2002) pp. 401–34.

84 Reubi, 'The Human Capacity to Reflect and Decide', pp. 353–4; Jenny Hazelgrove, 'The Old Faith and the New Science: The Nuremberg Code and Human Experimentation Ethics in Britain, 1946–1973', *Social History of Medicine*, Vol. 15 (2002) pp. 109–35 (pp. 120–3).
85 See Andrew Brown, *J. D. Bernal: The Sage of Science* (Oxford: Oxford University Press, 2005) pp. 158–64.
86 Brown, *J. D. Bernal*, pp. 162–3; Gary Werskey, 'The Marxist Critique of Science: A History in Three Movements?', *Science as Culture*, Vol. 16 (2007) pp. 397–461.
87 On Polanyi and the Society for Freedom in Science, see Werskey, 'The Marxist Critique of Science', pp. 412–13.
88 Brown, *J. D. Bernal*, pp. 300–6; David Joravsky, *The Lysenko Affair* (Chicago: University of Chicago Press, 1986).
89 Brown, *J. D. Bernal*, p. 163.
90 On welfare states and professional expertise, see Moran, *The British Regulatory State*, pp. 60–1; Rose, 'Government, Authority and Expertise in Advanced Liberalism', pp. 292–4. On the welfare state and public support for professions, see Tony Judt, *Ill Fares the Land* (London: Allen Lane, 2010); Perkin, *The Rise of Professional Society*.
91 See, for example, Robert Bud, 'Penicillin and the New Elizabethans', *British Journal for the History of Science*, Vol. 31 (1998) pp. 305–33.
92 Elizabeth Fisher, *Risk Regulation and Administrative Constitutionalism* (Oxford: Hart Press, 2007) p. 66.
93 Baker et al., 'Introduction', p. 4.
94 Moran, *The British Regulatory State*, p. 84.
95 Barbara Brookes, *Abortion in England, 1900–1967* (London: Croom Helm, 1988) p. 54.
96 Gayle Davis and Roger Davidson, '"A Fifth Freedom" or "Hideous Atheistic Expediency"? The Medical Community and Abortion Law Reform in Scotland, *c.*1960–1970', *Medical History*, Vol. 50 (2006) pp. 29–48 (pp. 39–40).
97 Michael Davies, *Textbook on Medical Law* (London: Blackstone Press, 2nd edn, 1998) p. 94. See also Margaret Brazier, *Medicine, Patients and the Law* (Harmondsworth: Penguin, 3rd edn, 2003).
98 Alfred Denning, quoted in Davies, *Textbook on Medical Law*, p. 95.
99 *Bolam v Friern Hospital Management Committee* [1957] 2 All Er 118. See also Kim Price, 'The Art of Medicine: Towards a History of Medical Negligence', *Lancet*, Vol. 375 (2010) pp. 192–3.
100 See Brazier, *Medicine, Patients and the Law*, pp. 102–3.
101 Ibid, pp. 102, 146-7.
102 Alasdair MacIntyre, *A Short History of Ethics* (London: Routledge, 2002). For an example of how moral philosophers influenced medical ethics, see Laurence B. McCullough, 'John Gregory's Medical Ethics

and Humean Sympathy', in Baker et al. (eds), *Codification of Medical Morality*, pp. 145–61.

103 John Dewey, *Ethics* (New York: Henry Holt, 1909) p. 1.

104 MacIntyre, *Short History of Ethics*, pp. 242–5; Mary Warnock, *Ethics since 1900* (Oxford: Oxford University Press, 1960) pp. 18–26.

105 I have taken this analogy from Ben Rogers, *A. J. Ayer: A Life* (London: Chatto and Windus, 1999) p. 48.

106 On how philosophers promoted objectivity and the search for a 'neutral language' in this period, see Lorraine Daston and Peter Galison, *Objectivity* (New York: Zone Books, 2007) pp. 289–307.

107 Bertrand Russell, *History of Western Philosophy* (London: Routledge, 4th edn, 2008) pp. 743–4. See also Bertrand Russell, *Sceptical Essays* (London: Routledge, 2002); Rogers, *A. J. Ayer*, pp. 80–9.

108 Russell, *History of Western Philosophy*, p. 743. Although Russell wrote books on science, marriage and politics, he believed they were not philosophy because they simply offered his own subjective views on a particular subject. They were, he argued, written 'as a human being who suffered from the state of the world and wished to find some way of improving it … [speaking] in plain terms to others who had similar feelings'. Philosophers such as Charles Pidgen have recently claimed, however, that these books do represent important contributions to moral philosophy and are 'no worse … than the moral and political writing of, say, Sartre, Nietzsche or Voltaire'. See Charles R. Pidgen, 'Bertrand Russell: Moral Philosopher or Unphilosophical Moralist?', in Nicholas Griffin (ed), *The Cambridge Companion to Bertrand Russell* (Cambridge: Cambridge University Press, 2003) pp. 475–505 (p. 476).

109 Alastair Campbell, telephone interview with the author (May 2009).

110 A. J. Ayer, *Language, Truth and Logic* (Harmondsworth: Penguin, 2001) p. 99.

111 Ibid.

112 Rogers, *A. J. Ayer*, p. 145; see also Ayer, *Language, Truth and Logic*, pp. 103–4.

113 Stefan Collini, *Absent Minds: A History of Intellectuals in Britain* (Oxford: Oxford University Press, 2007) p. 398.

114 A. J. Ayer, *Philosophical Essays* (London: Macmillan, 1965) pp. 246–7.

115 Warnock, *Ethics since 1900*, p. 31.

116 Ibid, p. 203.

117 Ludwig Wittgenstein, quoted in Waddington, *Science and Ethics*, p. 7.

118 C. E. M. Joad, in Ibid, pp. 26–9.

119 Francis Crick, quoted in Gordon Wolstenholme (ed), *Man and His Future* (London: J & A Churchill, 1963) p. 364. The CIBA Foundation was established in 1949 and funded by the CIBA chemical company.

Its aim was to promote interdisciplinary collaboration in medical and chemical research. The 1963 meeting on 'Man and His Future' was the first of a series of symposia dedicated to ethical issues in science and medicine. For more background, see Peter Woodford, *The CIBA Foundation: An Analytical History, 1949–1974* (Amsterdam: Elsevier, 1974).

120 H. C. Trowell, in Wolstenholme (ed), *Man and His Future*, pp. 63–6.

121 The Very Revd W. R. Matthews, Dean of St Paul's, in Waddington, *Science and Ethics*, pp. 22–4.

122 Ibid, p. 37.

123 Ayer, *Language, Truth and Logic*, p. 110.

124 Keith Clements (ed), *The Moot Papers: Faith, Freedom and Society* (London: T&T Press, 2010) p. 8. See also Adrian Hastings, 'Temple, William (1881–1944)', *Oxford Dictionary of National Biography*, http://www.oxforddnb.com/view/article/36454 (accessed 7 February 2014). On religious efforts to accommodate evolutionary theories, see Brooke, *Science and Religion*, pp. 275–320.

125 The Right Revd E. W. Barnes, Bishop of Birmingham, in Waddington, *Science and Ethics*, pp. 20–2 (p. 22).

126 Clements, *The Moot Papers*, pp. 6–8.

127 Matthew Grimley, 'Moot' (*act.* 1938–1947), *Oxford Dictionary of National Biography*, www.oxforddnb.com/view/theme/67745 (accessed 7 February 2014).

128 The politically diverse nature of its participants also ensured that the Moot failed to reach agreement on any concrete proposals, and it wound up shortly after Mannheim's death in 1947. See Grimley, 'Moot'.

129 Anon, 'Introduction', *Bulletin of the Churches' Council of Healing*, no. 1 (1967) pp. 3–4 (p. 3). Held in the Ian Ramsey Papers at Durham Cathedral Archives (uncatalogued at time of writing). Henceforth Ramsey archives.

130 Anon, 'British Medical Association: Proceedings of Council', *British Medical Journal*, Vol. 1 (1947) pp. 103–14 (p. 112).

131 Ibid, p. 112.

132 Ibid, p. 105.

133 Ibid, p. 112.

134 Ibid.

135 George T. Allerton, 'Working Hours in the NHS', *British Medical Journal*, Vol. 1 (1947) p. 112.

136 The FMG held regular meetings in London and was part of the broader Christian Frontier Movement, which aimed to promote 'effective thought and action in relation to the life of Christians in secular society'. See Daniel T. Jenkins, *The Doctor's Profession*

(London: SCM Press, 1949) p. 7. See also Alec R. Vidler, *Scenes from a Clerical Life* (London: Collins, 1977) pp. 133–5.

137 Jenkins, *The Doctor's Profession*, p. 9.

138 Ibid, p. 7.

139 Ibid, p. 123.

140 Ibid, p. 61. See also Father Alphonsus Bonnar, *The Catholic Doctor* (London: Burns & Oates, 1948).

141 Robert M. Veatch, *Disrupted Dialogue: Medical Ethics and the Collapse of Physician-Humanist Communication, 1770–1980* (Oxford: Oxford University Press, 2005) p. 177. See also Jenkins, *The Doctor's Profession*, p. 58.

142 Hastings, 'Temple, William'.

143 Jenkins, *The Doctor's Profession*, p. 51.

144 Perkin, *The Rise of Professional Society*, pp. 472–506. See also Mold, 'Patient Groups and the Construction of the Patient-Consumer', p. 510.

145 Perkin, *The Rise of Professional Society*, p. 477.

146 Cooter, 'The Ethical Body', p. 458. For more on Thalidomide, see Stefan Timmermanns and Valerie Leiter, 'The Redemption of Thalidomide: Standardizing the Risk of Birth Defects', *Social Studies of Science*, Vol. 30 (2000) pp. 41–71.

147 On changing media attitudes in the 1960s, see Ayesha Nathoo, *Hearts Exposed: Transplants and the Media in 1960s Britain* (Basingstoke: Palgrave Macmillan, 2009) pp. 86–102; Jane Gregory and Steve Miller, *Science in Public: Communication, Culture and Credibility* (Cambridge, MA: Perseus Publishing, 1998) pp. 44–5.

148 See, for examples, Dominic Sandbrooke, *White Heat: A History of Britain in the Swinging Sixties* (London: Little, Brown, 2006) pp. 491–515; Arthur Marwick, *The Sixties: Cultural Revolution in Britain, France, Italy and the United States c.1958–1974* (Oxford: Oxford University Press, 1998).

149 Mold, 'Patient Groups and the Construction of the Patient-Consumer', p. 511. See also Nick Crossley, *Making Sense of Social Movements* (Buckingham: Open University Press, 2002).

150 See, for example, Ivan Illich (ed), *Disabling Professions* (London: Marion Boyars, 1977); Ivan Illich, *Limits to Medicine, Medical Nemesis: The Expropriation of Health* (Harmondsworth: Penguin, 1975).

151 Nick Crossley, *Contesting Psychiatry: Social Movements in Mental Health* (Oxford: Routledge, 2006).

152 Stephen Lock, 'Pappworth [formerly Papperovitch], Maurice Henry (1910–1994)', *Oxford Dictionary of National Biography*, www.oxforddnb.com/view/article/55242 (accessed 7 February 2014).

Rachel McAdams notes that while some London hospitals did have a reputation for anti-semitism in this period, there was also a shortage of clinical positions in the 1940s thanks to a surplus of qualified doctors returning from the war and the creation of the NHS. See McAdams, 'Human Guinea Pigs'.

153 Hazelgrove, 'The Old Faith and the New Science', p. 118.

154 On medical disregard for the Nuremberg Code, see Ibid, pp. 111–18; Noortje Jacobs, 'Which Principles, Doctor? The Early Crystallization of Clinical Research Ethics in the Netherlands, 1947–1955, MA thesis, University of Utrecht, 2012; Mary Dixon-Woods and Richard E. Ashcroft, 'Regulation and the Social Licence for Medical Research', *Medical Health Care and Philosophy*, Vol. 11 (2008) pp. 381–91; Jay Katz, 'The Consent Principle of the Nuremberg Code: Its Significance Then and Now', in George J. Annas (ed), *The Nazi Doctors and the Nuremberg Code* (Oxford: Oxford University Press, 1992) pp. 227–40. For a detailed study of why informed consent was prioritised in the Code, see Paul Weindling, 'The Origins of Informed Consent: The International Scientific Commission on Medical War Crimes, and the Nuremberg Code', *Bulletin of the History of Medicine*, Vol. 75 (2001) pp. 37–71.

155 Lock, 'Pappworth, Maurice Henry'.

156 Maurice H. Pappworth, 'Human Guinea Pigs: A Warning', *Twentieth Century*, Vol. 171 (1962–63) pp. 66–75 (p. 67).

157 Ibid, p. 72.

158 Ibid, p. 71.

159 Ibid, p. 74.

160 Ibid, p. 75.

161 Mold, 'Patient Groups and the Construction of the Patient-Consumer', p. 510; Hazelgrove, 'The Old Faith and the New Science', p. 120.

162 Mold, 'Patient Groups and the Construction of the Patient-Consumer', p. 511. See also Crossley, *Making Sense of Social Movements*.

163 Helen Hodgson, 'Medical Ethics and Controlled Trials', *British Medical Journal*, Vol. 1 (1963) pp. 1339–40 (p. 1340). See also Hazelgrove, 'The Old Faith and the New Science', p. 128.

164 Hazelgrove, 'The Old Faith and the New Science', p. 120.

165 Sir Francis Avery Jones to Maurice H. Pappworth, 30 November 1973. Wellcome archives, Maurice Henry Pappworth papers: PP/MHP/C/5.

166 Maurice H. Pappworth, *Human Guinea Pigs: Experiments on Man* (London: Routledge and Kegan Paul, 1967) p. x

167 Ibid, p. 3.

168 Ibid, p. 5.

169 Ibid, p. ix.

170 Ibid, p. 200.
171 Ibid, p. 204.
172 Ibid, p. 208.
173 Ibid.
174 McAdams, 'Human Guinea Pigs'; Boyd, 'The Discourses of Medical Ethics in the United Kingdom', p. 487.
175 Lord Basil Amulree, 'Doctors and Patients', *The Times Literary Supplement*, 8 June 1967, p. 506.
176 Anon, 'Responsibilities of Research', *Lancet*, Vol. 289 (1967) p. 1144.
177 Charles Fletcher, cited in Anon, 'Controversy: Human Guinea Pigs', *World Medicine*, 6 June 1967, pp. 86–7 (p. 87).
178 Dixon-Woods and Ashcroft, 'Regulation and the Social Licence for Research', p. 382. On Wilson's Labour government, see Sandbrooke, *White Heat*.
179 Kenneth Robinson, MP, quoted in *Parliamentary Debates: House of Commons*, Vol. 747 (31 May 1967) col. 35.
180 Cited in Hedgecoe, 'A Form of Practical Machinery', p. 342.
181 The Rt Hon. Quintin Hogg QC to Maurice H. Pappworth, 22 January 1969. Wellcome archives PP/MHP/C/5. On Hogg's endorsement of self-regulation in science, see Gummett, *Scientists in Whitehall*, p. 41.
182 Bernard Williams, 'What Doctors Were Supposed to Know', *Listener*, 15 June 1967, p. 795. Williams was dubious about applying the 'Golden Rule' to clinical research, arguing that while a doctor may reasonably presume that their family or friends would consent to a procedure, 'the unhappy patient, beneath his nervous consent', may not.
183 Cecil Clothier to Maurice Pappworth, 29 March 1978. Wellcome archives PP/MHP/C/5. Emphasis in original.
184 Ibid.
185 Sir Max Rosenheim (chair), *Report of the Committee on the Supervision of the Ethics of Clinical Investigation in Institutions* (London: Royal College of Physicians, 1967) p. 1. Wellcome archives PP/MHP/C/5.
186 See, for example, Maurice Pappworth, '"Human Guinea Pigs" – a History', *British Medical Journal*, Vol. 301 (1990) pp. 1456–60. See also Hazelgrove, 'The Old Faith and the New Science', p. 130.
187 Desmond Laurence to Maurice H. Pappworth, 10 May 1990. Emphasis in original. Wellcome archives PP/MHP/C/5. Several historians have also noted how the RCP was motivated by changes in US funding. See Hedgecoe 'A Form of Practical Machinery'; Hazelgrove, 'The Old Faith and the New Science'. On the origins of the PHS policy, see Laura Stark, *Behind Closed Doors: IRBs and the Making of*

Ethical Research (Chicago and London: University of Chicago Press, 2012).

188 Rosenheim, *Report of the Committee on the Supervision of the Ethics of Clinical Investigation in Institutions*, p. 2.

189 Ibid, p. 1.

190 Ibid.

191 As Hedgecoe notes, of the 126 hospitals that formed ethics committees, only 25 contained a lay member. Of the 30 committees formed in non-teaching hospitals, 6 had a lay member. See Hedgecoe, 'A Form of Practical Machinery', p. 344.

192 For more background, see Moran, *British Regulatory State*, pp. 82–3.

193 Anon, 'Merrison Committee: Report of Inquiry', *British Medical Journal*, Vol. 2 (1975) pp. 183–8 (p. 188).

194 Hedgecoe, 'A Form of Practical Machinery', p. 338.

195 Allan Gaw, 'Searching for Pappworth', *South African Medical Journal*, Vol. 101, no. 9 (2011) p. 608. See also Pappworth, '"Human Guinea Pigs" – A History'.

196 Lock, 'Pappworth, Maurice Henry'; Hazelgrove, 'The Old Faith and the New Science', p. 118.

197 Clothier to Pappworth, 29 March 1978. Wellcome archives PP/MHP/C/5.

198 Anon, 'Controversy: Human Guinea Pigs', p. 87. See also McAdams, 'Human Guinea Pigs'.

199 In this sense Pappworth was at odds with the American whistleblower Henry Beecher, who shared his interest in exposing unethical research but is widely acknowledged as being far more influential. We can partly explain their differing fortunes by examining their contrasting attitudes to club regulation. While Beecher's position as research professor at Harvard gave him more authority than Pappworth, and while he was also a more conciliatory figure, he importantly claimed that responsibility for improving ethical standards should rest with doctors alone. For more on Beecher, see Rothman, *Strangers at the Bedside*, pp. 70–85. On his support for self-regulation, see Nathoo, *Hearts Exposed*, p. 159; Fox and Swazey, *Observing Bioethics*, p. 45.

200 Abbott, *System of Professions*, p. 5.

2

Ian Ramsey, theology and 'trans-disciplinary' medical ethics

During the 1960s and 1970s Anglican theologians increasingly endorsed 'trans-disciplinary' discussion of new procedures such as IVF in societies and journals dedicated to medical ethics.[1] Although theological engagement with medical ethics was by no means new, it increased from the 1960s thanks to a decline in religious belief. Figures such as Ian Ramsey, an Oxford theologian and later Bishop of Durham, endorsed greater engagement with social and moral issues to maintain the Church's relevance in the face of increasing secularisation. Ramsey and other theologians did not claim that interdisciplinary debates were necessary because procedures such as IVF raised unprecedented moral dilemmas. They instead believed that IVF touched on longstanding moral questions such as 'respect for life', but argued that collaboration was needed because these questions had become hard to resolve in secular societies that lacked 'a common morality'.[2]

Crucially, these theologians emulated their predecessors by positioning themselves as ancillaries to doctors. They did not criticise procedures such as IVF and did not seek to involve themselves in medical decision-making. They also believed that the new 'trans-disciplinary' societies and journals should be considered as medical bodies and should work to 'safeguard the doctor's role'.[3] This stance ensured that while discussion of medical ethics increasingly involved professions other than doctors, it was still undertaken primarily for their benefit. Interdisciplinary debates in Britain consequently differed from those that were termed 'bioethics' in the United States, where outsiders publicly questioned the expertise of doctors and scientists, and took an active role in determining professional conduct.

'Brave new medical world': IVF, ethics and 'the biological revolution'[4]

By the late 1960s clinical research was not the only procedure that aroused public concern or prompted calls for outside involvement with medicine and science. Research on DNA and the induction of genetic mutations, the creation of cross-species hybrid cells in tissue culture, advances in organ transplantation and work on IVF all contributed to media reports on a so-called 'biological revolution', in which researchers had acquired 'vast control of our physical environment' and were able to manipulate life on an unprecedented scale.[5] But in line with the 'backlash against professional society', this coverage was largely ambivalent and questioned the social and moral implications of research. As the playwright Dennis Potter claimed in *The Times*, it had become 'the taste of the times to look around the laboratory, then to look ahead and shudder'.[6]

Popular writers such as Gordon Rattray-Taylor warned of a 'biological time-bomb', whose dangers equalled those of nuclear weapons and threatened 'nothing less than the break-up of civilization as we know it'.[7] Broadsheet and tabloid newspapers also linked biological research to fears over nuclear weapons, claiming that biologists were 'taking over where the physicists left off' and questioning whether they could be trusted to 'handle the properties of life, death and destruction … as casually as if they were sunflower seeds'.[8] Television coverage was similarly foreboding. A BBC documentary screened as part of the *Towards Tomorrow* series, for example, presented cell fusion, genetic engineering and IVF as an 'Assault on Life' which raised 'grave legal, social, religious, philosophical and spiritual questions'. The documentary's narrator claimed that the public was right to distrust scientific claims that research posed no dangers, 'because Rutherford also said splitting the atom would serve no practical purpose'.[9]

These suspicions formed the central premise of the BBC drama *Doomwatch*, which was the brainchild of the writer and former scientist Kit Pedler. First screened in 1970, *Doomwatch* centred on the work of ex-scientists in a fictitious Department for the Observation and Measurement of Science, who protected society from human–animal hybrids, artificial viruses and genetically modified rats. Its largely negative portrayal of scientists offered a telling contrast to the 'new Elizabethans' who were celebrated in popular coverage during the 1950s. The scientists in *Doomwatch* consistently ignored or refused to consider the social implications of their research, often with disastrous consequences for the public and themselves. In the episode 'Tomorrow the Rat', for example, a scientist released a laboratory strain of intelligent and flesh-eating rats that proceeded to attack the public and eventually devoured their creator. Surveying these pessimistic attitudes for *The Times* in 1971, the Labour politician Shirley Williams claimed that programmes such as *Doomwatch* embodied 'a growing suspicion about scientists and their discoveries, and a widespread opinion that science and technology need to be brought under greater control'. It was clear, she argued, that 'for the scientists, the party is over'.[10]

IVF often featured in popular coverage throughout the 1960s, following its application in animals and false reports of human successes. But it became synonymous with the 'biological revolution' in February 1969 after the Cambridge physiologists Robert Edwards and Barry Bavister, and the Oldham obstetrician Robert Steptoe, announced the successful formation of seven pro-nuclear zygotes among thirty-four mature human oocytes fertilised *in vitro*.[11] An editorial in the edition of *Nature* that carried their paper attempted to forestall negative reports, claiming that Edwards, Bavister and Steptoe were 'not perverted men in white coats doing nasty experiments on human beings, but reasonable scientists carrying out perfectly justifiable research'.[12] As Jon Turney notes, 'the first responses in the press suggested that *Nature*'s argument might carry the day'.[13] Some reports were positive, claiming that IVF would shed crucial light on human development and might 'offer new hope for the childless'.[14]

Others, however, struck a more ambivalent tone. The *Guardian* aligned IVF with concerns over the 'biological time-bomb' in a cartoon that portrayed a scientist cultivating a baby in a test-tube, before it emerged, grew into a monster and imprisoned him.[15] Similar concerns appeared in the *Daily Mail*, which printed a cartoon that showed a 'Doctor Frankenstein' horrified to find that he had accidentally cloned the Prime Minister, Harold Wilson. *The Times*, meanwhile, highlighted the eugenic implications of IVF when it warned that politicians in totalitarian states might use it to 'concentrate on breeding a race of intellectual giants'.[16]

Although IVF did not feature in *Doomwatch*, Kit Pedler also claimed that it could allow despotic generals to 'mass produce' troops 'without the advent of a mother at all'.[17] Turning his attention away from clinical experiments, Maurice Pappworth similarly noted that IVF had 'eugenic' possibilities and could be used to 'produce a race of supermen free from physical and mental taints'.[18] Pappworth hinted at the possibility of external control over IVF when he questioned whether it was any longer 'acceptable' to let scientists claim that 'this is all pure science and pure research and if others put their findings to undesirable uses that is not their fault'.[19]

Others were more explicit and used reports on IVF to call for external control over biological research. In a piece for the *New Statesman*, the medical writer Donald Gould warned that researchers such as Robert Edwards tended to be 'single-minded enthusiasts, blind to the implications of their work'. Gould argued that *laissez-faire* forms of self-regulation were 'no longer enough ... and is it time that society took a hand in deciding what is meet and what is not'.[20] In a series of columns for the *New Scientist* and a book titled *What is Science For?*, the journalist Bernard Dixon also outlined how discussion of 'potential social problems as malevolent exploitation of "test tube babies"' had reinforced 'public suspicions of the scientist as a sinister and irresponsible figure'.[21] Dixon claimed that public unease reflected a growing belief that 'experts *do not* always know best', which he endorsed by arguing that 'science can all too easily be out of touch with the needs, goals and aspirations of the society that nourishes it'.[22] Public confidence could no longer be assured by 'contentions that science is best left to scientists', he warned, 'and we need to try to create channels through which the pressure of the citizen can influence decisions more directly'.[23]

Dixon reassured scientists that he was not proposing 'communal *control* over science', since this would probably foster 'short-sightedness and a failure to understand the importance of speculative research as against the application of immediately useful techniques'. He instead endorsed 'more public *influence* – if only as a healthy corrective to the present autonomy and internal politicking of the scientific community'.[24] Dixon also sought to reassure scientists by dwelling on the possible benefits of 'greater democracy in decision-making about science and technology'.[25] He predicted that while 'some research projects would probably be killed, and rightly so', another consequence 'of a wider public debate might be to demand more science, not less'.[26] Dixon argued that exposing science to 'wider democratic influence' would increase public confidence by ensuring that the predominantly 'negative' stance adopted by anti-vivisectionists and other groups would be replaced by a scenario in which campaigners 'work more positively for funds to go into particular areas ... such as better kidney machines, or money to build a new hospital'.[27]

Like Gould, Dixon did not specify how 'channels' might be created to increase public influence over science. But he acknowledged a clear debt to members of the recently established British Society for Social Responsibility in Science (BSSRS), whose work did outline how the public might influence scientific decisions. The BSSRS was established after a small group of radical scientists organised a 1968 conference to oppose the British government's support for research into chemical and biological weapons.[28] Participants at the meeting decided to form 'a continuing and more radical group' which became the BSSRS.[29] These founding members quickly showed their 'libertarian socialist tendencies' by admitting non-scientists such as the American philosopher and historian of science Jerry Ravetz.[30] At the same time, to heighten their profile within science and the wider public, they also admitted more senior and elite figures, including forty-four Fellows of the Royal Society.

The presence of these elite figures secured press coverage and allowed the BSSRS to hold its inaugural meeting at the Royal Society in April 1969. But it also ensured that this 'informal coalition of old Left, liberal, and more radical scientists' had differing views on what the BSSRS should achieve.[31] Letters inviting elite scientists, such as the cell biologist Dame Honor Fell, to join showed that senior members viewed the BSSRS as a means of countering 'the loss of esteem for science in the community at large'. Citing declining university admissions for science and fears over the 'biological time-bomb', they claimed that the 'future of science is threatened by the hostility now felt by young people', and argued that this could only be overcome by encouraging scientists to 'become more aware of the social and cultural role of science and play a more responsible role in society'.[32] This aim was made clear when the crystallographer Maurice Wilkins, the first president of the BSSRS, argued that it should help combat the widespread 'breakdown in confidence' among the public. In a long letter to *The Times* and a paper at a 1969 conference on 'The Social Impact of Modern Biology', Wilkins argued that BSSRS members should rebuild confidence in science by publicly discussing the benefits as well as the social and ethical aspects of their research.[33]

The younger and more radical members of the BSSRS, however, pursued a different and more politically engaged agenda. In addition to supporting 'self-education for scientists concerning the control and abuses of science', and opposing the use of CS gas in Northern Ireland, they also explored the possibility of a 'socialist science' in which laypeople would play a major role in developing scientific policies and guidelines.[34] In their 1969 book *Science and Society*, which Bernard Dixon cited extensively, Steven and Hilary Rose claimed that public suspicion of 'the men in white coats' could only be overcome by ensuring that 'decision-making processes [were] opened at all levels'.[35] Arguing that public involvement had become as important as 'the fostering of professional ethics among scientists', they claimed that lay representatives should be allowed to vote on the allocation of 'resources between disciplines and fields in the basic sciences', and should also be appointed to managerial positions 'at every research institute and university'.[36]

Their opposition to 'the present oligarchies' in science led some of these radical BSSRS members to undertake a series of interventions at the 1970 meeting of the British Association for the Advancement of Science, which was held in Durham.[37] In addition to holding a teach-in entitled 'science is not neutral', they handed out a leaflet that proclaimed that 'science is in crisis' and commissioned a theatre group to perform a 'nerve gas charade' outside Durham Cathedral, where the chemist Alexander Todd had just given his presidential address.[38] While many attendees criticised these tactics as 'unexpected and embarrassing', they were notably praised as 'significant' in a sermon by Ian Ramsey, the Bishop of Durham.[39]

This was a telling endorsement, for by 1970 Ian Ramsey was the most influential and high profile of the Anglican theologians who were increasingly engaged with the ethics of medicine and science. Ramsey and other religious figures believed that the answer to public unease over procedures such as IVF was to encourage 'professional cooperation' in the discussion of ethical issues. As we shall see, their calls for interdisciplinary debates reflected, and linked, social and religious concerns, demonstrating again how ethical debates are 'part of larger processes and larger histories, which shape and mutually influence each other'.[40]

'The way ahead for Christian thinking': Ian Ramsey, Anglican theology and 'trans-disciplinary' medical ethics[41]

Ian Ramsey was born in Kersal, Lancashire, in 1915. He won a scholarship to Farnworth grammar school at the age of ten and another to Christ's College, Cambridge, at the age of eighteen.[42] During his time as an undergraduate he developed an interest in science, mathematics and philosophy, but decided to be ordained after a long spell in hospital with tuberculosis. After a curacy near Oxford, Ramsey returned to Christ's College in 1943, the same year that Charles Raven, the college master and regius professor of divinity, sought to reconcile scientific and theological world-views in his book *Science, Religion and the Future*. Ramsey was appointed lecturer in the philosophy of religion the following year, and although Raven's influence ensured that he was interested in the relations between science, medicine and religion, much of his early work engaged with linguistic philosophy and sought to refute A. J. Ayer's claim that religious assertions were 'not genuine propositions at all'.[43]

Ramsey moved to Oxford to take the Nolloth chair of philosophy in 1951, and it was here that he came to believe that 'his own job in life was to build bridges between Christian theology and modern problems'.[44] His work increasingly looked less at the properties of religious language and more at contemporary issues, 'particularly the frontiers between religion, medicine and law'.[45] This was evidenced by his membership of the CCH and of the Warneford and Park hospitals management committee, which he joined in 1954 and chaired between 1960 and 1966. Aware of his interest in practical issues, in 1956 the theologian Gordon Dunstan invited Ramsey to work with doctors, civil servants, demographers and others in developing a Church of England report on the relation of contraception to marriage and population control.[46]

The committee's report on *The Family in Contemporary Society* was formally adopted by Anglican bishops at the 1958 Lambeth Conference, and its success encouraged Ramsey to play a major role in helping the Church of England's Board for Social Responsibility establish interdisciplinary groups on the morality of issues such as suicide, sterilisation, the artificial prolongation of life and abortion. Many of the proposals in their reports might be best described as 'cautiously liberal' and argued, for instance, that suicide should be decriminalised, that abortion might be justified to save a mother's life and that discontinuing treatment for unconscious patients with no hope of recovery need not conflict with Christian reverence for life.[47] These conclusions chimed with and helped shape the increasingly liberal climate in Britain, with the 1959 report on suicide acknowledged as influencing the government's decision to decriminalise suicide in the 1960 Suicide Act.[48]

Ramsey's involvement with these groups strengthened his conviction that theologians had a duty to confront what he called 'the uncertainties and anxieties of our own day'.[49] They also reaffirmed his belief that theology was not 'a subject apart', and that discussing ethical issues required a 'dialogue with other disciplines, and making possible their cross-fertilization ... on teasing and stubborn problems of contemporary thought and behaviour'.[50] This, of course, was not a new conviction. Analysis of contemporary issues had been integral to Anglican moral theology throughout the twentieth century, as the work of William Temple and groups such as the FMG demonstrates. Ramsey acknowledged his debt to Temple, to whom he was often compared, and publicly praised 'the outstanding character of his Christian social concern'.[51] But he argued that theologians had an increasing duty to engage with practical issues, and especially 'medical moral problems', in the 1960s.[52]

Several linked factors underpinned Ramsey's call for greater engagement with medical ethics. One major factor was what Perkin and others identify as a profound 'decline in religious faith' during the 1960s.[53] While theologians had voiced concerns at the apparent secularisation of Britain throughout the early twentieth century, Callum Brown has shown that the 1940s and 1950s actually 'witnessed the greatest church growth that Britain had experienced since the mid nineteenth century'.[54] Church attendance, Sunday school enrolment and confirmations increased significantly, and huge crowds flocked to see the American evangelist Billy Graham tour football stadia in 1954 and 1956. In the 1960s, however, 'the institutional structures of cultural traditionalism started to crumble', thanks to the *Lady Chatterley* trial and the ending of moral censorship, the legalisation of abortion and homosexuality, the facilitation of easier divorce, the emergence of the women's liberation movement, the loss of domestic ideologies in youth culture and growing rebellion against traditional sources of authority.[55] Attendance at Protestant churches, Sunday schools and religious rites of passage fell away dramatically, and a young generation were less concerned with the ethics surrounding faith, God and the afterlife, and more with issues that many religious figures had traditionally ignored, such as the environment, gender and racial equality, nuclear weapons, political activism and, crucially, science and medicine.

Some Anglican figures, such as John Robinson, the Bishop of Woolwich, responded to the decline in religious faith by endorsing an 'agnostic theology' that rejected the idea of God as a supernatural being living 'up there', questioned the veracity of the virgin birth and the Holy Trinity, and encouraged their replacement with belief in a non-anthropomorphic life force.[56] But instead of making the Church of England appear more relevant, this led many to 'question the integrity of an endowed priesthood that did not believe in God'.[57] As a writer from *TIME* magazine claimed in a 1966 letter to Ramsey, the decline of traditional Christianity, both outside and within the Anglican Church, ensured that atheism had 'become a basic premise of a new generation' and 'the distinguishing feature of contemporary civilization is the almost universal loss of the sense of God'.[58]

In addition to proclaiming the 'death of God', religious figures responded to secularisation in a variety of ways. Traditionalists such as the vicar and poet R. S. Thomas blamed science and medicine for the decline in religious belief. Writing for the *Times Literary Supplement* in 1966, Thomas criticised the 'scientific age' as a 'mechanised and impersonal age', where 'under the hard gloss of affluence there can be detected the murmuring of the hard heart and the heavy spirit'.[59] But liberal theologians such as Ramsey argued that the only way to counter their weakening authority was by 'expressing continuing concern with problems that are of significance to everyone – believer and unbeliever unlike'.[60] It was only 'by wrestling with such problems in a co-operative venture of scholarship with other academic disciplines', he claimed, 'that theology may find a new prospect and new relevance'.[61]

Ramsey acknowledged that the Church of England reports he was involved with were accused of undermining religious authority and encouraging the 'permissive society', by adopting a liberal stance on ethical issues instead of 'holding fast to absolute rules and even stiffer authoritarianism'.[62] By refusing to argue that suicide 'should have been said to be always wrong, abortion to be always forbidden, and artificial means never withdrawn from those who are being kept alive by them', he noted, 'those concerned with these reports have thus been blamed for the erosion in moral standards'.[63] But while Ramsey believed that engaging with practical issues increasingly provided theology with a 'secular task', he stressed that just 'because it might be called a secular theology does not mean it is an irreligious theology'. He argued that by working on practical issues with other disciplines, 'there can arise from the co-operative venture a common vision' that highlighted people's obligations to one another and encouraged 'a shared sense of the infinite wonder and grandeur of that which we are all endeavouring better to understand'.[64]

Ramsey also believed that theology's 'new task' gave it perhaps *the* vital role in the discussion of practical issues. He viewed theologians as the 'common link' who facilitated debates between 'experts in different disciplines and from different occupations'. This was especially the case for discussions of medical and biological research, which Ramsey considered to be the major source of 'frontier problems' in the 1960s and 1970s.[65] Throughout the 1960s this belief led Ramsey to extend his work with the Church of England reports and play a 'prominent part' in efforts to promote collaboration between doctors, scientists, the clergy and others in a range of settings.[66]

During 1962 and 1963, for instance, he was on a working party set up by the Archbishop of Canterbury, Michael Ramsey, which recommended the establishment of an Institute of Religion and Medicine (IRM) 'devoted to the study and advancement of mutual interest to clergy and doctors'.[67] The IRM was formally established in 1964 and sought to encourage cooperation between those professions involved in the 'promotion of health and the healing of the sick', which a brochure argued was 'not only the particular concern of doctors but is something which also concerns members of associated therapeutic professions and religious leaders'.[68] In regional and national meetings and its dedicated journal *Contact*, the IRM encouraged collaboration between doctors and clergymen on subjects such as 'decisions about life and death, the care of the dying, the role of religion in mental health, abortion, medical education, casework and counselling, ethical decision-making, counselling the bereaved, groups and teams in medicine and ministry'.[69]

Ramsey was also involved with the London Medical Group (LMG), which was one of the most significant examples of 'doctor clergy co-operation' in the 1960s.[70] The LMG originated in 1963, when the Student Christian Movement (SCM), the arm of the ecumenical movement concerned with higher education in Britain, commissioned the doctor and chaplain Andrew Mepham to survey the needs of medical school students. Mepham found that while theology students received lectures and seminars from visiting doctors, with some attending hospital courses on subjects such as mental health care, teaching hospitals made no systematic effort to allow medical students to receive lectures from theologians or other non-doctors on 'subjects such as the care of the dying patient'.[71] His report concluded that medical education undermined 'the care of the patient as a man' by focusing predominantly on scientific training and pathology to the neglect of social and ethical issues.[72]

The SCM responded to Mepham's report by commissioning Edward Shotter, a young university chaplain, to look into medical education in the twelve London teaching hospitals. With a budget of only forty pounds, later in 1963 Shotter organised four informal lectures during which medical and nursing students could discuss social and ethical issues with specialists from various professions.[73] Encouraged by attendance at these early lectures, he established a student council to select topics for a longer programme in 1964. This 'representative council' chose a wide range of topics in 1964 and 1965, including the management of pain and terminal illness, suicide, drug addiction, birth control, reforming the laws for homosexuality, patient confidentiality and marriage guidance.[74]

A 'consultative council' comprising senior doctors from the London teaching hospitals then liaised with Shotter to select appropriate speakers, which included doctors and prominent religious figures, as well as representatives from pressure groups and charities such as the Samaritans.[75] The LMG talks were free and open to the public, and attendances averaged one hundred people by the mid 1970s.[76] The LMG's success, and the student demand for lectures on social and ethical issues, also led to the establishment of medical groups elsewhere in Britain. In 1967 the ecumenical chaplain Kenneth Boyd and the religious philosopher Alastair Campbell established the Edinburgh Medical Group (EMG), while other groups were established in Newcastle, Sheffield, Glasgow, Birmingham, Bristol, Liverpool and Manchester during the late 1960s and early 1970s.[77]

In addition to giving a paper at a 1967 symposium on 'Decisions about Life and Death', Ian Ramsey was involved with the LMG in several ways. As Gordon Dunstan acknowledged, his call for greater engagement with practical issues and collaboration across disciplines influenced Shotter, who 'began to fulfil what Ramsey knew ought to be done' when he established the LMG.[78] At the same time, Shotter emulated Ramsey's claim that theologians should act as a 'common link' between professions by acting as 'a "catalyst" who facilitated and helped coordinate dialogue'.[79]

Ramsey also had a more direct role as a member of the LMG's governing body, which was established in 1966 after financial constraints and the need to appear 'non-partisan' led Shotter to end links with the SCM and establish the LMG as an independent charity.[80] Gordon Dunstan advised Shotter to recruit senior doctors and theologians 'for counsel and repute', and in addition to Ramsey and Dunstan himself, the governing body included renowned clinicians such as Lord Basil Amulree and the surgical endocrinologist Ronald Welbourne.[81] The presence of respected doctors helped dispel suspicions that the LMG was a 'pincer movement on the profession ... by its cadets and senators'.[82] Senior theologians such as Ramsey, meanwhile, helped secure money from religious bodies and discussed the possibility of establishing formal ties between the LMG and the IRM, although the governing body eventually rejected the proposal.[83]

Following his 1966 appointment as Bishop of Durham, Ramsey often had to apologise for 'being a very unprofitable servant' and missing LMG meetings due to commitments in his new diocese.[84] Yet despite his increased workload and responsibilities, he continued to encourage interdisciplinary debates on ethical issues. During 1967 and 1968, for instance, he organised and chaired a series of CIBA symposia on 'Personality and Science', where participants discussed the ethics of altering an individual's personality through surgery or new psychotropic drugs. Ramsey again displayed his preference for 'trans-disciplinary' work by assembling speakers from medicine, science, law, philosophy and theology. During the group's first meeting in May 1967, he admitted that their diverse backgrounds departed 'from the normal CIBA Foundation pattern' which relied heavily on medical and scientific expertise.[85]

In justifying this shift, Ramsey notably highlighted another factor that led Anglican figures to support interdisciplinary work on medical ethics during the 1960s and 1970s. His introductory talk outlined how understanding moral issues in secular and increasingly pluralist societies 'required the creative meeting of all the relevant disciplines which are needed for an adequate appraisal of such problems'.[86] Ramsey argued that the ethical issues that medicine and science raised were not novel, but now appeared 'more complex' in the absence of a common morality and obviously 'correct' answers. This ensured, he claimed, that it was 'a mark of immaturity for any discipline to think it has ready-made, copy-book answers – whether that discipline is psychology, or economics or sociology or philosophy or theology'.[87]

This stance was endorsed in a Church of England document that explained why Ramsey and Dunstan convened interdisciplinary groups to investigate ethical issues such as abortion, euthanasia and sterilisation. It acknowledged that 'inevitably, on some issues agreement remains difficult or impossible. An obvious example is abortion.'[88] But it continued that 'once the plunge is taken, agreement *is* possible over a very wide area'.[89] Agreement was only possible, however, if specialists from different fields worked together to 'ensure that the widest cross-section of informed opinion is brought to bear on moral questions'. This was essential, the document concluded, because '*modern pluralist societies depend for their life on such co-operation*'.[90]

By the early 1970s, Ramsey was considered an authority on medical ethics thanks to his work with various interdisciplinary groups and House of Lords speeches on subjects such as euthanasia.[91] This was clear in 1970, when the BMA invited him to give the opening address at its 1972 Annual Clinical Meeting, which was to be held in Nicosia, Cyprus. As a letter informed him, the opening address was a prestigious lecture given by 'an authority on some general subject affecting medicine', and BMA members viewed Ramsey as an expert on the 'moral problems facing the medical profession at the present time'.[92]

Ramsey's speech reiterated many of the claims he made throughout the 1960s and illustrated why theologians increasingly engaged with medical ethics. He began by arguing that issues such as the artificial prolongation of life and IVF did not cause public concern because they raised unprecedented ethical dilemmas. 'For instance', he stated, 'medical treatment to save and prolong life; the need and duty to ease pain; the conception and birth of a child – these are all situations which, overall and in outline, are the same as they have always been.'[93] Ramsey instead claimed that prolonging life and IVF touched on longstanding questions such as 'respect for life', but stressed that questions of how best to treat ventilated patients or *in vitro* embryos were now problematic because 'traditional moral absolutes are being questioned ... and society lacks a common morality'.[94] 'In other words', Ramsey continued, 'the very contexts in which recent developments in medicine occur are those which make the solution of the associated moral problems all the more difficult.'[95]

To Ramsey, the 'multiplicity of explanations' in pluralist societies meant that 'the old rules for dealing with these situations are too large-scale to do justice to the new detail. It is as though we tried to catch sprats in the net of a trawler.'[96] In his BMA speech and a 1970 article on Christian ethics, he maintained that 'agreement on certain moral problems of the kind that perplex society' could no longer be found by simply evoking 'absolute rules or principles'.[97] In a passage that hinted at why he had praised the actions of some BSSRS members in Durham, Ramsey argued that reliance on fixed principles evoked a 'reactionary authoritarianism' at odds with the contemporary 'morality of rebellion and revolution'.[98] He maintained that the only appropriate response to 'our new civilization' was to fashion 'creative moral decisions of a novel kind'. Ramsey closed his BMA speech by claiming that these 'creative decisions' required the formation of 'trans-disciplinary groups' in which members of various disciplines and professions did not apply principles 'in a rule-of-thumb fashion', but engaged in a 'deeper grappling with the empirical facts'.[99]

Ramsey's call for engagement with 'empirical facts' touched on a broader religious debate about situationist ethics. From the 1930s onwards, but with increasing frequency in the 1960s, theologians such as Paul Lehmann had endorsed a 'contextual' approach to morality and claimed that:

> Christians are to be obedient to the command of God. But the command of God is not given in formal, general ethics; it is not given in traditional rules of conduct. It is given by the living God in the concrete situation. It is a particular command addressed to a particular person in a particular sphere of activity.[100]

This belief was notably also shared by some of the theologians who 'presided over the birth of bioethics' in the United States.[101] In their writing on medicine and science, these figures echoed Ramsey's claims that awareness of empirical details was vital for reaching 'creative' moral decisions.

While other American theologians criticised the 'wastelands of relativism' and emphasised the value of binding principles, they nevertheless agreed that interdisciplinary collaboration was vital to understanding ethical dilemmas.[102] But these theologians, crucially, all held differing views to their British counterparts on the role of outsiders and their relation to medical expertise. This ensured that the 'trans-disciplinary' view of medical ethics that Ramsey promoted was significantly different from the emerging American field known as 'bioethics' – with profound consequences for the discussion and regulation of new procedures such as IVF and genetic engineering.

The scope and limits of outside involvement: differences between Britain and the United States in the 1960–70s

Joseph Fletcher, professor of Christian ethics at the Episcopal Theological School in Cambridge, Massachusetts, was one of first American theologians to look at scientific and medical ethics. His 1955 book *Morals and Medicine* discussed the moral aspects of contraception, euthanasia, artificial insemination, sterilisation and 'a patient's right to know the truth'. Although it is sometimes cited as a pioneering work in 'the new field' of bioethics, *Morals and Medicine* is best characterised more as 'a book that ends the past than one that opens the future'.[103] It was certainly more reminiscent of earlier books such as Jenkins's *The Doctor's Profession*, which Fletcher cited approvingly, than later work that called for outside involvement. This was clear in the preface, where Fletcher stated that:

> At no time have I ever meant to take the position, or to give any comfort to those who would like to assume, that the clergy or other moralists can or should settle problems *for* the doctors ... Physicians have an *expertise* and competence without which all non-medical discussion of the rights and wrongs of medicine will be in danger of becoming only wool-gathering.[104]

But Fletcher did pre-empt a central tenet of bioethics when he emphasised that doctors must respect their patient's 'human rights (certain conditions being satisfied) to use contraceptives, to seek insemination anonymously from a donor, to be sterilized, and to receive a merciful death from a medically competent euthanasiast [*sic*]'.[105] In his discussion of each issue, he maintained that individual choice and responsibility were 'the heart of ethics and the *sine qua non* of a man's moral status'.[106] 'Without freedom to choose and know the truth', he argued, 'patients are only puppets. And there is no moral quality in a Punch and Judy show.'[107]

Fletcher believed that for doctors to fully respect their patients' rights in specific circumstances, 'the ethics of medical care have to change, to grow, and engage constantly in self-correction'.[108] This led him to promote a 'reflective and rational morality' that was 'subject to the rule of change', and to dismiss adherence to binding principles as evidence of a 'primitive personality' that accepted moral opinions 'without much reflection'.[109] While Fletcher did not expand on these claims in *Morals and Medicine*, they formed the central premise of his popular 1966 book *Situation Ethics*. Here, Fletcher endorsed an 'empirical, fact-minded, data conscious and inquiring' approach to morality over obedience to rules and principles.[110] 'Every man must decide for himself according to his own estimates of conditions and consequences', he argued, 'and no one can decide for him or impugn the decision to which he comes.'[111]

Fletcher believed the only norm or principle that was 'binding and unexceptionable, always right and good' in situation ethics was to promote actions that maximised '"love" – the *agape* of the summary commandment to love God and thy neighbour'.[112] He asserted that 'even the most revered principles may be cast aside if they conflict in any concrete case with love', and argued that this doctrine should be translated into its secular counterpart of utilitarian beneficence.[113] Although *Situation Ethics* was not concerned primarily with medicine or science, Fletcher applied his argument to medical ethics when he claimed that a woman had a right to demand an abortion after she had been raped and if continuing the pregnancy risked harming her mental health. 'The most loving thing possible (the right thing) in this case', he claimed, was to acceded to the woman's 'responsible decision to terminate the pregnancy' – even if her doctor believed that 'killing was wrong' and never performed abortions, or only did so if continuing a pregnancy endangered a patient's life.[114]

Fletcher's libertarian outlook in *Situation Ethics* drew on the civil rights campaigns that he had become increasingly involved with during the 1960s, when he often stood in picket lines, marched in protests and was occasionally 'the victim of brutal beatings'.[115] Ian Ramsey, who also publicly supported 'the morality of revolution and rebellion', albeit less directly than Fletcher, praised the book for its anti-authoritarian message and for highlighting the 'significance of the particular context or situation in which a moral decision is made'.[116] Ramsey convinced the Christian SCM Press to publish *Situation Ethics* in Britain, and wrote to Fletcher claiming that it was the 'best statement of a view which I grant appeals to me very much'.[117]

But while Ramsey and Fletcher both endorsed a similar approach, there were also telling differences between Britain and the United States. Catholic theologians in the United States, for example, played a greater role in interdisciplinary debates and adopted a different ethical stance to their British colleagues. Young American priests such as Warren Reich and Albert Jonsen began to work on medical and scientific ethics thanks largely to 'upheavals in the 1960s that took place in Roman Catholicism'.[118] They were enthused by the civil rights emphasis on individual autonomy and became disillusioned with the Church's conservative position on issues such as euthanasia and birth control, which upheld the absolute sanctity of life in all circumstances.[119] Their work on biology and medicine expressed these growing religious doubts, re-examining traditional Catholic doctrines, showing a willingness to engage with different groups and endorsing a similarly contextual approach to Fletcher and Ramsey.[120] In Britain, by contrast, Catholic theologians were largely unaffected by civil rights campaigns and did not question the Vatican's position on medical ethics. They also enjoyed a more secure position than their Anglican counterparts, with church attendances growing thanks to immigration from Ireland and Catholic schools representing 'a seductive alternative for parents distrustful of ordinary state education'.[121] These factors meant that Catholic theologians in Britain reaffirmed traditional principles when they discussed issues such as abortion or euthanasia, irrespective of context, and felt little need to engage with non-denominational audiences throughout the 1960s and 1970s.[122]

But the most significant difference undoubtedly lay in the fact that American theologians believed that they and other 'outsiders' should play an active role in regulating medical and scientific practices. This was most evident in the work of Paul Ramsey, a Methodist professor of religion at Princeton University, who Ian Ramsey also recommended to British publishers as 'someone good and certainly worth approaching'.[123] Paul Ramsey took a different approach to ethics than Fletcher and his British namesake. While he shared Fletcher's belief that the goal of Christian ethics was to promote *agape*, Ramsey argued that it could only be achieved through obedience to binding rules and principles. By the mid 1960s, thanks to his work on rules in Christian ethics and the morality of war, Ramsey was well known as the principal opponent of 'situation ethics' and for arguing that actions should be guided by 'standards of conduct which invariably can and must be applied to all people in every situation'.[124]

Ramsey was provoked to write on science and medicine towards the end of the 1960s by Fletcher's work, which he considered to be overly secular and uncritical of the 'liberal progressivism which sustains any science-based ethics today'.[125] In his 1970 books *Fabricated Man* and *The Patient as Person*, he claimed that Fletcher and others ignored the binding norms that should underpin medical or scientific practices and simply dwelt on the 'wholly unique situations' where they believed these norms ought to be discarded.[126] Ramsey instead stressed that doctors and scientists should adhere to 'exceptionless rules', irrespective of circumstances or consequences.[127] He believed that the overriding rule for medicine and science was to 'respect the uniqueness and dignity of the human individual', which led him to claim that medical research could only ever be justified with the full consent of the patient or experimental subject.[128] He acknowledged that this meant prohibiting research on children, mentally ill adults and others 'who cannot give a mature and informed consent', but refuted utilitarian appeals to the common good and argued that 'we may have to accept the fact that some limits exist to our search for knowledge'.[129]

Despite his dismissal of 'situation ethics', Paul Ramsey shared Joseph Fletcher and Ian Ramsey's belief that discussion of ethical issues should be a 'joint venture' between doctors, theologians and other professions.[130] He also believed that the need for collaboration resulted from 'the moral pluralism of our society', which had given 'all of mankind reason to ask how much longer we can go on assuming that what can be done has to be done or should be, without uncovering the ethical principles we mean to abide by'.[131] In his preface to *The Patient as Person*, Ramsey argued that all sides benefited from interdisciplinary dialogue 'about the urgent moral issues arising in medical practice'.[132] He claimed that doctors and scientists could educate theologians 'about the technical problems' associated with particular issues, while theologians and philosophers could help 'explain some of the things that need to be asked of the science and of the ethics'.[133]

But while this appeared similar to the collaboration that British theologians endorsed, Paul Ramsey believed that 'outsiders' should do more than simply discuss ethical issues. In *The Patient as Person* and later work, he argued that they should also play an active role in determining the conduct of doctors and scientists. Ramsey was by no means the first or the most high-profile American figure to endorse outside involvement. During 1968, for instance, the senator and former vice-president Walter Mondale responded to public discussion of organ transplants and genetic research by calling for a national Commission on Health and Society, which would act as a forum where laypeople and representatives of several professions could debate 'the fundamental ethical and legal questions' raised by biomedical research.[134]

Mondale argued that external oversight was necessary because the public were consumers with a stake in federally funded research, and were therefore entitled to know its potential risks and implications. His proposals received support from theologians who appeared as witnesses at congressional hearings, such as Kenneth Vaux and Jerald Brauer, who called for a 'fresh look' at scientific regulation and supported a commission that was 'very broad in makeup'.[135] Despite this support, Mondale's proposal failed after prominent doctors claimed that outside influence would impede research, and that the public continued to be best served by leaving decisions to 'conscionable people in the profession who are struggling to advance medicine'.[136] The esteemed surgeon Owen Wangansteen encapsulated this view when he told Congress that: 'If you are thinking of theologians, lawyers, philosophers and others to give some direction ... I cannot see how they could help'. To Wangansteen, like other doctors, 'the fellow who holds the apple can peel it best'.[137]

Although Mondale's proposals were defeated, calls for external scrutiny and involvement nevertheless grew during the early 1970s thanks to scholars in the burgeoning field of 'bioethics'. Foremost among these was Paul Ramsey, who argued in *The Patient as Person* that 'the problems of medical ethics that are especially urgent in the present day are by no means technical problems on which the expert (in this case, the physician) can have an opinion'.[138] 'The doctor makes decisions as an expert but also as a man among men', he continued, and 'I hold that medical ethics is consonant with the ethics of a wider human community.' Here and throughout the 1970s, Ramsey asserted that:

> My view is that we, the people, are the final authority within constitutional limits in determining how in future we mean to be healed – when the means is human experimentation. The technical expertise of the medical research community cannot be the sole or chief arbiter in this matter, which is a question of morality and public policy.[139]

The Yale psychiatrist Jay Katz also argued that fundamental questions needed to be asked about 'the nature of authority assigned to physicians'.[140] In several articles and his 1972 book *Experimentation with Human Beings*, which ran to over a thousand pages, Katz claimed that doctors possessed no unique expertise that justified making them sole arbiters of medical ethics and asked: 'Who is to keep guard over the guardians themselves?'[141] Like Walter Mondale and Paul Ramsey, his solution was to endorse 'more active participation of non-scientists in research decisions'.[142]

Calls for outside involvement were strengthened considerably in 1972, when newspapers reported that researchers investigating the 'natural history' of syphilis had intentionally withheld treatment from 400 African Americans in Tuskegee, Alabama, since 1932.[143] These revelations 'appeared at a time of heightened concern and anger about racial discrimination and of heightened sensitivity to abuse of the poor and powerless', and seriously undermined support for self-regulation.[144] Newspapers, civil rights groups and an official inquiry, whose nine members included Jay Katz, called for federal regulation of medical research and argued that external oversight was vital to safeguard the interests of patients and experimental subjects.

In 1974 President Nixon responded to the Tuskegee study by establishing the National Commission for Protection of Human Subjects in Biomedical and Behavioral Research.[145] Politicians stipulated that philosophers, theologians, lawyers and others should play a major role in shaping policies for research, and the majority of the Commission's members were non-doctors and scientists. Their growing policy influence was evident when philosophers on the Commission, including Tom Beauchamp and the British-born Stephen Toulmin, played a major role in drafting guidelines for medical research. The guidelines, known as the *Belmont Report*, stated that research on human subjects should adhere to the principles of respect for persons, beneficence and justice, and became public law governing the activities of federally funded scientists in 1978.[146]

Heightened concern for the subjects of biomedical research was also evident in the American discussion of IVF. This led Paul Ramsey to condemn the procedure in a 1971 conference on 'Fabricated Children', which Edwards and Steptoe likened to 'a denunciation of our work as if from a nineteenth century pulpit'.[147] In this conference and a 1972 article for the *Journal of the American Medical Association*, Ramsey stated that since the risks of IVF were unknown and embryos could not provide consent, 'it constitutes unethical medical experimentation on possible future human beings, and therefore is subject to absolute moral prohibition'.[148] He claimed that these factors justified a permanent ban on IVF and argued that it 'should not be allowed by medical or public policy in the United States – not now or ever'.[149]

At the same time, well-funded and influential pro-life groups targeted research on human embryos and foetuses as part of their campaign against the 1973 *Roe* v. *Wade* case, in which the Supreme Court ruled that a woman's constitutional right to privacy included the right to have an abortion in the early stages of pregnancy.[150] They argued that research on foetuses or embryos fostered acceptance of abortion among doctors and the general public, and called for the prosecution of anyone who performed this work. In the face of this sustained criticism, and as part of its 1974 National Research Act, Congress imposed a moratorium on federal funding for research that involved human foetuses and embryos.[151]

The situation in Britain, however, differed markedly. In contrast to the United States, the theologians who discussed medical ethics here saw no real problem with IVF. This was clear in December 1971 when Ian Ramsey received a letter from several bishops expressing concern at news that Edwards and Steptoe were planning to 'implant in a human uterus an embryo fertilised in the laboratory'. The bishops claimed that this was 'unethical medical experimentation, contrary to the known laws of God and man', and vowed to pass a resolution urging Edwards and Steptoe to 'devote their scientific skills towards life-saving and disease curing projects'.[152] But Ramsey, tellingly, did not support their position and replied that his 'immediate and off-the-cuff reaction is not to be too fussed about it'.[153]

Gordon Dunstan, too, considered IVF to be unproblematic. In his 1974 book *The Artifice of Ethics*, he argued that the overriding ethical priority in IVF involved ensuring that sperm and egg were brought together responsibly *in vitro*, but claimed that this should also underpin reproduction through artificial insemination by donor (AID) as well as the actions of couples looking to conceive naturally. Dunstan also saw no problem with experiments on *in vitro* embryos, which was by far the most contentious aspect of IVF in the United States. He argued that embryo experiments were vital 'for research into recesses otherwise inaccessible ... to study embryonic growth, for instance, with a view to detecting the origin of disorders and to find, perhaps, the means to correct or prevent them'.[154]

Ramsey and Dunstan's attitude to IVF encapsulated their broader views on medical expertise. While Anglican theologians endorsed 'trans-disciplinary' debates on ethical issues in Britain, they did not criticise scientists or doctors and stressed that they were not seeking to involve themselves in professional decision-making. At the CIBA symposium on 'Personality and Science', for instance, Ramsey argued that they simply wanted to 'encourage a wider public to face important moral problems arising from contemporary developments in medicine and also to facilitate responsible debate on these topics'.[155] The role of 'mixed advisory groups', in this view, was not to challenge or replace medical authority, but was to 'safeguard the doctor's role as the advocate of his patient's interests'.[156]

Ramsey reiterated this position in his 1972 address to the BMA. He stressed that trans-disciplinary groups were not designed to 'compromise the physician's or the surgeon's responsibility' and maintained that 'the decision in the end must be taken by the person who is to carry out the action'. The involvement of theologians and other non-doctors, he concluded, 'does not in any way compromise the surgeon's or the physician's responsibility for making decisions', and it was the job of trans-disciplinary groups simply to enable 'that decision to be better informed, and therefore more responsible'.[157]

While he supported the involvement of other professions in discussing ethics, then, Ramsey still believed that it should be aimed at and primarily undertaken for the benefit of scientists and doctors. This was again clear when Shotter wrote to him in 1971 and claimed that 'a logical progression in the development' of the student medical groups was to establish 'some sort of organization concerned with medical ethics, which junior doctors could join'.[158] Ramsey supported the formation of this 'Society for the Study of Medical Ethics' (SSME), but was clear that 'it has to be seen as a medical society from the start'.[159]

Although Ramsey died from a heart attack in October 1972, his outlook continued to influence the work of theologians who engaged with medical ethics. When the SSME published a *Journal of Medical Ethics* from 1975, members decided it should focus primarily on clinical issues and initially wanted a senior doctor as editor.[160] After struggling to find a suitable candidate, they approached the religious philosopher Alastair Campbell, who was secretary of the EMG and had recently published a book on *Moral Dilemmas in Medicine*. Unlike Jay Katz or Paul Ramsey, Campbell's book did not question medical authority and sought to improve professional conduct by providing 'a rational framework for understanding the complexities of moral judgement'.[161] The SSME considered him an ideal editor because he was seen as sympathetic to doctors and 'able to communicate in a way that was intelligible to medics'.[162] While the *Journal of Medical Ethics* contained articles by doctors, theologians, lawyers and philosophers, Campbell used an early editorial to outline that its aim was to help doctors 'make more informed decisions'. Like Ian Ramsey, he stressed that 'the final decisions remain medical ones and the responsibility remains with that profession'.[163]

Perhaps unsurprisingly, doctors and scientists endorsed this stance. As one early letter to the *Journal of Medical Ethics* argued, 'if the study of moral issues does not lead to a practical outcome which helps the individual doctor – what is the point of all the discussion?'[164] The report of a British Association study group on science and ethics, which had been commissioned following events at the 1970 Durham conference, also claimed that theologians, philosophers and others had an important role to play in discussing ethical issues, but should stop short of influencing professional conduct. 'Whilst guidance to scientists in ethical matters is obviously desirable', the report argued, 'there is no substitute for sensitivity and awareness on the part of individual scientists. They are the ones who make the decisions, they are the ones who must be able to grasp the issues involved.'[165]

Robert Edwards, meanwhile, used talks, publications and symposia on IVF to reject external involvement with decision-making. In a 1974 article for the *Quarterly Review of Biology*, he outlined how some claimed that the seemingly 'formidable' ethical and legal implications of IVF meant that 'external controls should be imposed on scientists'.[166] Yet despite the fact that the MRC had refused to fund research on IVF in 1971, forcing him and Steptoe to seek private funds, Edwards claimed that 'responsibility for applying new research methods' should continue to rest with the medical profession.[167] He argued that allowing theologians, philosophers and others to make 'committee decisions' would delay the clinical application of IVF, since different professions had diverse views on issues such as how to treat embryos, and 'the chance of a united moral and ethical stance on such questions seems remote'.[168] Edwards warned that any delay would harm the 'right of couples to have their own children', which he claimed was the overriding ethical priority in IVF and outweighed the 'irrelevant' misgivings of critics such as Paul Ramsey.[169] 'Patients have the right to benefit from research', Edwards concluded, 'and there is no reason to believe that ethical advice from outsiders about their condition is sounder than their own judgement of it.'[170]

Edwards had the chance to air these views as a member of another British Association working party on the ethics of 'breakthroughs' such as IVF and genetic screening, which was chaired by the biologist Walter Bodmer and also included Gordon Dunstan, the Labour politicians Shirley Williams and David Owen, the science journalist John Maddox and the biologist Anne McLaren.[171] This group toed a now familiar line when they considered the role of outsiders. Their 1974 report agreed that 'lawyers, theologians and Members of Parliament need to be closely involved with scientists in discussions of the implications of scientific research', but again maintained that decision-making should continue to rest solely 'with the experimenter, his profession, and the local ethical committee which has to approve any line of research'.[172] The report also echoed Edwards's publications when it claimed there was 'no objection' to the development of IVF, provided that it was used by married couples and 'that only the husband's sperm will be used for fertilization of the ova removed from the wife's ovaries'.[173] Members of a CIBA symposium on reproductive medicine, which included Edwards and Steptoe, reached similar conclusions when they agreed that IVF posed fewer ethical and legal problems than AID because it did not involve a third party and did not undermine marriage by raising questions over paternity.[174]

These factors ensured that there was no sustained critique of IVF in Britain by the mid 1970s. In contrast to American figures such as Paul Ramsey, high-profile theologians such as Ian Ramsey and Gordon Dunstan voiced no concerns and claimed that decisions should be left to doctors or scientists. Their 'hands-off' stance allowed scientists such as Robert Edwards to dominate ethical debates, and in line with his presentation of IVF as a cure for infertility 'that does not pose any moral problems', ambivalent newspaper articles gave way to positive reports that claimed that it offered hope to thousands of childless couples.[175] The absence of any major criticism also stemmed from the fact that pro-life groups such as the Society for the Protection of Unborn Children (SPUC) lacked the profile and influence of their American counterparts throughout the 1970s, and did not target biomedical research as part of their campaigns against abortion.[176] Indeed, politicians who opposed the 1967 Abortion Act, such as the Conservative Norman St John Stevas, publicly supported IVF provided it was used by married couples, where a woman's ovum was fertilised by her husband's sperm *in vitro* and it was 'impossible for her to have a child in any other way'.[177]

In their discussion of IVF, politicians also continued to argue that scientists and doctors should be free to determine the course of research. Despite ambivalent newspaper reports in the late 1960s, they showed no enthusiasm for a public inquiry and were content to leave decisions to professional bodies such as the MRC and the BMA.[178] Later in the 1970s, Shirley Williams and David Owen agreed with the rest of Walter Bodmer's working party that 'the experimenter and his profession' should continue to determine the course of research. Williams, in particular, had long believed that external control of science was 'terribly dangerous' after travelling to the Soviet Union and seeing how political efforts to promote Lysenko had devastated agriculture and genetics.[179] When she responded to fears over the 'biological time-bomb' by calling for new regulatory machinery in *The Times*, Williams argued that this should consist of a committee that was not composed of politicians and other outsiders, but of 'scientific advisers, representing the various sciences and responsible to the Cabinet Office'.[180]

But the political attitude to outside involvement appeared to change later in the decade, when the Labour government established a National Consumer Council in 1975 and began to argue that the views of different stakeholders, not just professionals, should be heard in the formulation of policy for public services.[181] The Prime Minister James Callaghan promoted widened participation in a 1976 speech on the future of education, when he called for a 'great debate on the subject, conducted in every region of the country, to which representatives of industry, trade unions, parents and local authorities should be invited'.[182] Crucially, the government was also keen to hear from different interest groups in its assessment of new scientific procedures. This was made clear when Fred Mulley, the Secretary of State for Education and Science, appointed a working party on the regulation of genetic manipulation techniques in August 1975.[183] This group, which consisted solely of doctors and scientists and was chaired by the bacteriologist Sir Robert Williams, recommended the establishment of an advisory body to provide ministers with guidance on potential hazards and containment risks. At their initial meetings, they argued that this should be a non-statutory body composed of eminent scientists with expertise in genetic manipulation techniques.[184] But while ministers agreed that the proposed 'Genetic Manipulation Advisory Group' (GMAG) should not have the statutory power to forbid experiments, they believed that it should include non-scientific members to assuage public fears about the risks of genetically modified micro-organisms. Following this advice, the final report of Williams's working party proposed that: 'The membership of GMAG should include not only scientists with knowledge both of the techniques in question and of relevant safety and precaution measures but also individuals able to take account of the interests of employees and the general public.'[185]

Shirley Williams became Secretary of State for Education and Science following a Cabinet reshuffle in September 1976, and one of her first jobs was to select the members of GMAG and determine its remit 'following consultation with other Ministers concerned'.[186] GMAG's terms of reference were to advise scientists involved in genetic manipulation, undertake continuing assessment of risks and precautions, maintain contacts with the relevant government departments, keep records of containment facilities, make available advice on matters concerning genetic manipulation, including staff training, and issue annual reports.[187] In what was credited as 'a novel organizational response to fears about genetic manipulation', these tasks were to be performed not only by practising scientists and trade union representatives, many of whom were scientists themselves, but also by non-scientists whose job was to 'represent the public interest'.[188]

Shirley Williams selected four individuals to represent the public interest on GMAG: Marie Jahoda, a sociologist from the University of Sussex; Jean Lawrie, from the Women's Doctors Federation; John Maddox, the science journalist and director of the charitable Nuffield Foundation; and Jerome Ravetz, a former member of the BSSRS and author of a 1973 book on *Scientific Knowledge and Its Social Problems*.[189] The inclusion of lay members with no ties to science and industry was designed to safeguard public trust in GMAG, by showing it was not simply a vehicle for professional interests.[190] It also seemed to represent a break from the *laissez-faire* belief that scientists or doctors were the best guardians of the professional and the public interest, which had long underpinned political support for club regulation in Britain.

But this was not quite the case. GMAG was certainly not comparable to American bodies such as the National Commission for Protection of Human Subjects and the later President's Commission for the Study of Ethical Problems in Medicine, which was established by President Jimmy Carter in 1978.[191] This was clear when civil servants stressed that 'working scientists must be the backbone' of GMAG.[192] Secondly, and in a more telling contrast, the lay representatives were the least influential members of GMAG and had little say in its decision-making.[193] Most of the issues GMAG considered were highly technical, and debates were consequently led and decided by the scientific members. With the exception of John Maddox, who was competent enough in risk analysis to play some part in debates on containment, 'the public interest representatives had but limited success in mainly peripheral matters'.[194]

It was also never made clear just how the lay members were supposed to represent the 'public interest', other than by voicing possible misgivings and refereeing clashes between scientists and trade union members.[195] What was more, while the meetings of American groups were recorded and open to the public, GMAG meetings were conducted behind closed doors and bound by the Official Secrets Act.[196] This meant that the lay members were unable to publicly voice any concerns they may have held, or were censured if they did. Jerome Ravetz, for instance, was removed from GMAG after he expressed his frustrations in a conference talk and described the inclusion of lay members as nothing more than 'a cosmetic exercise'.[197]

Ravetz, a former colleague of Stephen Toulmin, was well aware of the emergence of bioethics in the United States and was frustrated at the ancillary role that British outsiders played in the oversight of science and medicine.[198] In committees such as GMAG, as elsewhere, they were viewed as useful in the discussion of ethical issues but marginal to decision-making and policy formation. Yet few people voiced any dissatisfaction at this situation, and critics such as Ravetz were firmly in the minority by the late 1970s. This was not lost on Walter Bodmer's working party, which noted that Britain lacked 'calls for research to be controlled from outside the scientific field'.[199] While calls for outside participation did emerge from the BSSRS, these declined markedly after its radical members left in 1972 because they viewed it as an 'insufficiently socialist' organisation.[200] Elite members also left the BSSRS during this period to form a more moderate Council for Science and Society, which focused more on improving public confidence than on demanding outside participation.

In contrast to American figures such as Katz and Paul Ramsey, the most influential 'outsiders' who discussed medical research in Britain, such as Ian Ramsey and Gordon Dunstan, also offered no challenge to club regulation. Perhaps unsurprisingly, then, criticism of the lack of outside involvement in Britain tended to come from elsewhere. American bioethicists such as Edmund Pellegrino claimed that the CIBA symposia needed 'wider representation', while the radical philosopher Ivan Illich told Alastair Campbell that he regarded groups such as the LMG as little more than 'medical masturbation'.[201] Casting a glance across the Atlantic in a 1978 report on IVF, and detailing how theologians such as Paul Ramsey influenced federal policies, the *British Medical Journal* rightly noted that bioethics was 'an American trend'.[202] But this was to change dramatically in the 1980s, as growing numbers of British outsiders called for external involvement with decision-making, and the policies of a new Conservative government led doctors to acknowledge that 'the era which required paternalism is past'.[203]

Conclusion

During the late 1960s and 1970s, increasing numbers of non-doctors and scientists began to discuss the ethics of new procedures such as genetic research and IVF. In Britain, as in the United States, these 'trans-disciplinary' debates were encouraged by theologians such as Ian Ramsey and Gordon Dunstan. We should not presume, as some do, that religious figures engaged with issues such as IVF simply because they raised unprecedented moral dilemmas. Ramsey and Dunstan, among others, instead claimed that IVF touched on longstanding moral questions, but argued that new approaches were needed because these questions had become hard to answer in pluralist and increasingly secular societies. Informed debates and 'more responsible decisions' could only be reached, they argued, by seeking the views of different professions. This was also the case in the United States, where Tristram Engelhardt Jr claims that 'the wide plurality of beliefs' led theologians such as Paul Ramsey to discuss medical research, and to encourage lawyers, philosophers and others to do likewise.[204] Viewing the decline in religious belief as a significant influence on these 'trans-disciplinary' debates, rather than focusing on the inherently controversial nature of new procedures, helps link the activities of these theologians to their broader social context. This is evident in the 'situational' approach that Ian Ramsey and Joseph Fletcher endorsed, which they viewed both as a response to and a cause of the increasingly 'spontaneous' climate in the 1960s and 1970s.[205]

But while links certainly existed between Britain and the United States, we must also be aware of the significant differences. The growing involvement of theologians and other outsiders in 'trans-disciplinary' debates does not mean that both countries witnessed what Richard Ashcroft identifies as a shift 'from medical ethics to bioethics'. Bioethics, Ashcroft argues, differs from medical ethics because theologians, lawyers and philosophers play a more active role in decision-making, and the issues under consideration 'move from being internal concerns of the professions to matters of public, political debate'.[206] Theologians clearly led a 'shift from medical ethics to bioethics' in the United States during the 1970s, portraying issues such as IVF as public concerns and successfully demanding a role in decision-making. But this was not the case in Britain. The theologians who discussed medical ethics there believed that ethical issues were largely professional concerns, positioned themselves as ancillaries to doctors, and did not look to shape professional conduct. In direct contrast to their American counterparts, they argued that 'responsibility for making decisions' should continue to rest with doctors and scientists.[207]

Notes

1 Ian Ramsey, 'Moral Problems Facing the Medical Profession at the Present Time', Inaugural Address to the Annual BMA Clinical Meeting, Nicosia, Cyprus (April 1972). Ramsey archives.

2 Ibid. See also Ian Ramsey, 'A New Prospect in Theological Studies', *Theology*, Vol. 67 (1964) pp. 527–35.

3 Ian T. Ramsey and Ruth Porter (eds), *Personality and Science: An Interdisciplinary Discussion* (Edinburgh and London: Churchill Livingstone, 1971) p. 143.

4 Anon, 'Brave New Medical World', *Progress in Medical Ethics*, no. 1 (1968) pp. 1–2. *Progress in Medical Ethics* was a periodical published by the London Medical Group between 1968 and 1972.

5 Robert Sinsheimer, quoted in Wilson, *Tissue Culture in Science and Society*, p. 75. See also Jon Turney, *Frankenstein's Footsteps* (New Haven, CT: Yale University Press, 1998).

6 Dennis Potter, 'Biological Revolution: But What Else?', *The Times*, 27 April 1968, p. 18

7 Gordon Rattray-Taylor, *The Biological Time-Bomb* (London: Thames and Hudson, 1968) p. 21.

8 Gerald McKnight, 'Breakthrough for Boffins Trying to Answer the Puzzle "Can Life Be Created?"' *Tit-Bits*, 24 June 1967, p. 24.

9 *Towards Tomorrow, Biology: Assault on Life* (BBC 1, screened 7 December 1967). For more on this documentary, see Wilson, *Tissue Culture in Science and Society*, pp. 78–80.

10 Shirley Williams MP, 'The Responsibility of Science', *The Times Saturday Review*, 27 February 1971, p. 15.

11 Robert G. Edwards, Barry D. Bavister and Patrick C. Steptoe, 'Early Stages of Fertilization *in vitro* of Human Oocytres Matured *in vitro*', *Nature*, Vol. 221 (1969) pp. 632–5. For background to research on IVF, including its origins in animal physiology, see Turney, *Frankenstein's Footsteps*, pp. 161–5.

12 Anon, 'What Comes After Fertilization?' *Nature*, Vol. 221 (1969) p. 613. See also Turney, *Frankenstein's Footsteps*, p. 167.
13 Turney, *Frankenstein's Footsteps*, p. 167.
14 Ibid.
15 Anthony Tucker, 'Conception in the Lab', *Guardian*, 15 February 1969, p. 9.
16 Ibid. Earlier reports also highlighted the eugenic implications of IVF. See Anthony Tucker, 'A New Breed of Men?', *Guardian*, 22 February 1966, p. 6.
17 Kit Pedler, cited in Turney, *Frankenstein's Footsteps*, p. 172.
18 Maurice Pappworth, 'Experimental Work on Early Human Embryos' (undated manuscript). Wellcome archives, Maurice Pappworth papers: PP/MHP/C/5.
19 Ibid.
20 Donald Gould, 'Guinea Pig Code', *New Statesman*, 12 November 1971, p. 26.
21 Bernard Dixon, *What is Science For?* (London: Collins, 1973) p. 14.
22 Ibid, pp. 53, 213. Emphasis in original. See also Bernard Dixon, 'Science and the Silent Citizen', *New Scientist*, 27 August 1970, pp. 410–12.
23 Dixon, *What is Science For?*, p. 214.
24 Ibid. Emphasis in original.
25 Ibid, p. 223.
26 Ibid, p. 217.
27 Ibid.
28 Ibid, p. 191. See also Werskey, 'The Marxist Critique of Capitalist Science', pp. 431–2.
29 Steven Rose and Hilary Rose, 'The Radicalization of Science', in Steven and Hillary Rose (eds), *The Radicalization of Science* (London: MacMillan Press, 1976) pp. 1–32 (p. 19).
30 Ibid, p. 20.
31 Werskey, 'The Marxist Critique of Capitalist Science', p. 431.
32 R. Smith to Honor B. Fell, 19 February 1969. Wellcome archives, Strangeways Research Laboratory Papers: SA/SRL/G.46.
33 Maurice Wilkins, 'Science, Technology and Human Values', in Watson Fuller (ed), *The Social Impact of Modern Biology* (London: Routledge and Kegan Paul, 1971), pp. 5–10 (p. 5). See also, Maurice Wilkins, 'Making Science More Socially Responsible', *The Times*, 26 November 1970, p. 26.
34 Rose and Rose, 'Radicalization of Science', pp. 19–22.
35 Hilary and Steven Rose, *Science and Society* (London: Pelican, 1969) p. 268.
36 Ibid, pp. 269, 271.

37 Ibid, p. 269.
38 Dixon, 'Science and the Silent Citizen', pp. 193–4.
39 Ibid.
40 Michael Whong-Barr, 'Clinical Ethics Teaching in Britain: A History of the London Medical Group', *New Review of Bioethics*, Vol. 1, no. 1 (2003) pp. 73–84 (p. 81).
41 Ian Ramsey, 'The Way Ahead for Christian Thinking', *Point Magazine*, No. 3 (1968) pp. 57–62.
42 David L. Edwards, 'Ramsey, Ian Thomas (1915–1972)', *Oxford Dictionary of National Biography*, http://www.oxforddnb.com/view/article/31583 (accessed 7 February 2014).
43 Ayer, *Language, Truth and Logic*, p. 103.
44 David L. Edwards, *Ian Ramsey, Bishop of Durham: A Memoir* (London: Oxford University Press, 1973) p. 3.
45 Edwards, 'Ramsey, Ian Thomas'.
46 Gordon Dunstan, 'The Authority of a Moral Claim: Ian Ramsey and the Practice of Medicine', *Journal of Medical Ethics*, Vol. 13 (1987) pp. 189–94 (p. 189).
47 Church Assembly Board for Social Responsibility, *Abortion: An Ethical Discussion* (London: Church Information Office, 1965); Church Assembly Board for Social Responsibility, *Decisions About Life and Death: A Problem in Modern Medicine* (London: Church Information Office, 1965); Church Assembly Board for Social Responsibility, *Sterilization: An Ethical Inquiry* (London: Church Information Office, 1962); Church Assembly Board for Social Responsibility, *Ought Suicide to be a Crime?* (London: Church Information Office, 1959). For background to these committees' recommendations, see Ian Ramsey, 'Christian Ethics in the 1960s and 1970s', *The Church Quarterly*, Vol. 2 (1970) pp. 221–7. Not all the reports were liberal, however, and the report on sterilisation has been described as 'more hesitant about its use in limiting the population than liberal opinion was tending to be'. See Edwards, *Ian Ramsey*, p. 59.
48 Edwards, *Ian Ramsey*, p. 59.
49 Ramsey, 'A New Prospect in Theological Studies', p. 530.
50 Ibid.
51 Ian Ramsey, cited in Edwards, *Ian Ramsey*, p. 4.
52 Ramsey, 'Christian Ethics in the 1960s and 1970s', p. 227.
53 Perkin, *The Rise of Professional Society*, pp. 477–8.
54 Callum G. Brown, *The Death of Christian Britain: Understanding Secularisation, 1800–2000* (London and New York: Routledge, 2001) p. 170.
55 Ibid, p. 176.
56 John A. T. Robinson, *Honest To God* (London: SCM Press, 1963).

For more on the background and reception to Robinson's book, see Sandbrook, *White Heat*, pp. 434–6.

57 Perkin, *The Rise of Professional Society*, p. 477.

58 Horace Judson to Ian T. Ramsey, 7 March 1966. Ramsey archives.

59 R. S. Thomas, 'A Frame for Poetry', *Times Literary Supplement*, 3 March 1966, p. 169.

60 Ramsey, 'A New Prospect in Theological Studies', p. 531.

61 Ibid, p. 532.

62 Ramsey 'Christian Ethics in the 1960s and 1970s', pp. 223–4.

63 Ibid, p. 224.

64 Ramsey, 'A New Prospect in Theological Studies', p. 533.

65 Ibid, p. 532.

66 G. E. Harding, CCH, to Ian T. Ramsey, 15 August 1963. Ramsey archives.

67 Anon, *Report of Working Party on Clergy-Doctor Co-operation* (London, 1963) p. 4. Ramsey archives.

68 *Institute of Religion of Medicine* (London, 1970) p. 2. See also Annis Gillie and Leslie D. Weatherhead, 'The Institute of Religion and Medicine', *Journal of the Royal College of General Practitioners*, Vol. 15 (1968) p. 226.

69 *Institute of Religion and Medicine*, p. 3.

70 G. E. Harding, CCH, to Ian T. Ramsey, 15 August 1963. Ramsey archives. For more on the LMG, see Whong-Barr, 'Clinical Ethics Teaching in Britain'.

71 Anon, *Report of the Working Party on Doctor-Clergy Co-operation*, p. 4.

72 Andrew Mepham, cited in Boyd, 'The Discourses of Bioethics', p. 487. For more background on Mepham's report, see L. A. Reynolds and E. M. Tansey (eds), *Medical Ethics Education in Britain, 1963–1993* (London: Wellcome Trust Centre for the History of Medicine, 2007) pp. 76–7.

73 For a first-hand account of the origins and early years of the LMG, see Edward Shotter, 'Appendix One: A Retrospective Study and Personal Reflection on the Influence of the Medical Groups', in Reynolds and Tansey (eds), *Medical Ethics Education in Britain*, pp. 71–121. See also Whong-Barr, 'Clinical Ethics Teaching in Britain', pp. 75–6.

74 The list of lecture topics is held at the LMG files at the Wellcome Library: GC/253/A/31/8. Shotter also organised annual LMG conferences from 1964 onwards. The first, a 'Conference for Medical and Theological Students', was held in February 1964 and consisted of seminars on medical ethics, human relations, preparing for death, and neurosis and Christian belief. 'Conference for Medical and Theological Students' (February 1964). Wellcome archives

GC/253/A/31/8. Subsequent annual conferences looked at the welfare state (1965), aspects of guilt (1966), the prolongation of life (1967), the pattern of medical care (1969) and the new poor (1970).

75 LMG lecture lists for 1963/64 and 1964/65. Wellcome archives GC/253/A/31/8.

76 Whong-Barr, 'Clinical Ethics Teaching in Britain', p. 76. Whong-Barr notes that to ensure topics were openly discussed, and in order to maintain confidentiality, clinicians never spoke at hospitals where they worked.

77 Edward Shotter, *Twenty-Five Years of Medical Ethics: A Report on the Development of the Institute of Medical Ethics* (London: Institute of Medical Ethics, 1988) pp. 6–7.

78 Dunstan, 'Authority of a Moral Claim', p. 192.

79 Whong-Barr, 'Clinical Ethics Teaching in Britain', p. 77.

80 Shotter, 'Retrospective Study and Personal Reflection on the Influence of the Medical Groups', p. 85.

81 Dunstan, 'Authority of a Moral Claim', p. 192.

82 Whong-Barr, 'Clinical Ethics Teaching in Britain', p. 76.

83 Edward F. Shotter to Ian T. Ramsey, 9 May 1968. See also Reynolds and Tansey (eds), *Medical Ethics Education in Britain*, pp. 197–8.

84 Ian T. Ramsey to Edward F. Shotter, 24 August 1966. Ramsey offered to resign from the LMG governing body after missing several meetings in 1969, but his resignation was rejected. Ian T. Ramsey to Edward F. Shotter, 18 February 1969. Ramsey archives.

85 Ian Ramsey (chair), 'First Meeting of a Group Convened Under the Provisional Title "Science and Personality" at the CIBA Foundation', 12 May 1967. Ramsey archives. The CIBA Foundation's preference for technical expertise was evident in its 1963 symposium on 'Man and His Future', where the vast majority of participants were doctors or scientists. It was also evident in the 1966 symposium on 'Ethics in Medical Progress, with Special Reference to Transplantation', which included greater numbers of 'outsiders' and was chaired by a lawyer, but was still dominated by doctors and medical scientists.

86 Ian Ramsey, 'Introduction', typescript notes (1968). Ramsey archives. See also Ian T. Ramsey, 'Introduction', in Ramsey and Porter (eds), *Personality and Science*, pp. 1–5.

87 Ramsey, 'A New Prospect for Theological Studies', p. 530.

88 Church of England Social Morality Council, 'Moral Education Project' (1968).

89 Ibid. Emphasis in original.

90 Ibid. Emphasis in original.

91 Ramsey was one of twenty-six bishops with seats in the House of Lords, and often used this position to 'take the lead in expressing

the Church of England's reactions to moral problems of the day'. See Edwards, 'Ramsey, Ian Thomas'.

92 Walter Hedgecoe, Secretary BMA Board of Science and Education, to Ian T. Ramsey, 30 October 1970. Ramsey archives.
93 Ramsey, 'Moral Problems Facing the Medical Profession'.
94 Ibid.
95 Ibid.
96 Ibid. For more on how moral pluralism necessitated new modes of debate, see also Ian Ramsey, 'Biology and Personality: Some Philosophical Reflections', in Ian T. Ramsey (ed), *Biology and Personality: Frontier Problems in Science, Philosophy and Religion* (Oxford: Basil Blackwell, 1965) pp. 174–95.
97 Ramsey, 'Christian Ethics in the 1960s and 1970s', p. 224.
98 Ibid, p. 226.
99 Ramsey, 'Moral Problems Facing the Medical Profession'. See also Ramsey, 'Christian Ethics in the 1960s and 1970s', p. 221.
100 James M. Gustafson, 'Context Versus Principles: A Misplaced Debate in Christian Ethics', *Harvard Theological Review*, Vol. 58 (1965) pp. 171–202 (p. 177). Gustafson was not endorsing a 'contextual' approach here, but was paraphrasing supporters such as Lehmann.
101 Jonsen, *The Birth of Bioethics*, p. 41. See also Ramsey, 'Christian Ethics in the 1960s and 1970s', p. 224.
102 One major example was Paul Ramsey. For an overview of Ramsey's critique of 'situational ethics', see Gustafson, 'Context Versus Principles', pp. 189–90.
103 Jonsen, *The Birth of Bioethics*, p. 43.
104 Joseph Fletcher, *Morals and Medicine* (London: Victor Gollancz, 1955) p. xii. Emphasis in original.
105 Fletcher, *Morals and Medicine*, p. 25.
106 Ibid, p. 10. Emphasis in original.
107 Ibid, p. 33.
108 Ibid, p. 26.
109 Ibid, p. 28.
110 Joseph Fletcher, *Situation Ethics: The New Morality* (London: SCM Press, 1966) p. 29.
111 Ibid, p. 37.
112 Ibid, p. 30.
113 Ibid, p. 33.
114 Ibid, p. 39.
115 Jonsen, *The Birth of Bioethics*, p. 42. Fletcher also acknowledged his debt to the work of theologians such as Paul Lehmann and the pragmatist philosophy of John Dewey, who argued that the morality of a specific issue hinged on the circumstances in which it arose,

and that ethics was 'not a catalog of acts or set of rules to be applied like drugstore prescriptions or cook-book recipes'. See John Dewey, *Reconstruction in Philosophy* (London: University of London Press, 1920) pp. 169–70. See also Fletcher, *Situation Ethics*, pp. 14–15, 40–2.

116 Ramsey, 'Christian Ethics in the 1960s and 1970s', p. 224.

117 Ian T. Ramsey to Joseph Fletcher, 22 November 1965. Joseph Francis Fletcher papers, University of Virginia Historical Collections, File number 006:054. Ian T. Ramsey to Revd David L. Edwards, SCM Press, 16 June 1966. Ramsey archives.

118 Fox and Swazey, *Observing Bioethics*, p. 65.

119 Ibid.

120 See, for example, Jonsen, *The Birth of Bioethics*, pp. 36–9.

121 Sandbrooke, *White Heat*, p. 438.

122 For a full discussion of Catholic attitudes to medical ethics, see John Wilson, 'Abortion, Reproductive Technology and Euthanasia: Post-Conciliar Responses from within the Roman Catholic Church in England and Wales, 1965–2000', PhD thesis, University of Durham, 2010. Available online at http://etheses.dur.ac.uk/3076 (accessed 7 February 2014).

123 Ian T. Ramsey to J. D. Newth, Messrs. A & C. Black Ltd, 2 February 1966. Ramsey archives.

124 Paul Ramsey, 'Lehmann's Contextual Ethics and the Problem of Truth-Telling', *Theology Today*, Vol. 21 (1965) pp. 466–76 (p. 473). See also Paul Ramsey, *The Just War: Force and Political Responsibility* (New York: Scribner's, 1968); *Deeds and Rules in Christian Ethics* (New York: Scribner's, 1967).

125 Paul Ramsey, *Fabricated Man: The Ethics of Genetic Control* (New Haven, CT, and London: Yale University Press, 1970) pp. 17–18. See also John H. Evans, *The History and Future of Bioethics: A Sociological View* (Oxford: Oxford University Press, 2012) pp. 17–18

126 Paul Ramsey, *The Patient as Person: Explorations in Medical Ethics* (New Haven, CT, and London: Yale University Press, 1970) p. 4.

127 Ramsey, *Fabricated Man*, p. 19. See also Albert Jonsen, 'The Structure of an Ethical Revolution: Paul Ramsey, the Beecher Lectures and the Birth of Bioethics', in Paul Ramsey, *The Patient as Person: Explorations in Medical Ethics* (New Haven, CT, and London: Yale University Press, 2nd edn, 2002) pp. xvi–xxix.

128 Ramsey, *Fabricated Man*, p. 18.

129 Ramsey, *The Patient as Person*, pp. xiv, 11–12. See also Paul Ramsey, *The Ethics of Fetal Research* (New Haven, CT: Yale University Press, 1975).

130 Ramsey, *The Patient as Person*, p. xviii.

131 Ibid, p. xvi.
132 Ibid, p. xviii.
133 Ibid, p. xvii.
134 Walter Mondale, cited in Rothman, *Strangers at the Bedside*, p. 169.
135 Jerald Brauer, cited in Jonsen, *The Birth of Bioethics*, p. 92.
136 Owen Wangansteen, cited in Rothman, *Strangers at the Bedside*, p. 173.
137 Owen Wangansteen, cited in Jonsen, *The Birth of Bioethics*, p. 93.
138 Ramsey, *The Patient as Person*, p. xi.
139 Ramsey, *Ethics of Fetal Research*, p. xxi.
140 Jay Katz, *Experimentation with Human Beings* (New York: Russell Sage Foundation, 1972) p. 606.
141 Jay Katz, 'Who is to Keep Guard over the Guardians Themselves?', *Fertility and Sterility*, Vol. 23 (1972) pp. 604–9.
142 Katz, *Experimentation with Human Beings*, p. 1.
143 For more on this controversy, see Susan Reverby (ed), *Tuskegee's Truths: Rethinking the Tuskegee Syphilis Study* (Chapel Hill, NC: University of North Carolina Press, 2000); James L. Jones, *Bad Blood: The Tuskegee Syphilis Study* (New York: The Free Press, 1992); Allan M. Brandt, 'Racism and Research: The Case of the Tuskegee Syphilis Study', *The Hastings Center Report*, Vol. 8, no. 6 (1978) pp. 21–9.
144 Jonsen, *The Birth of Bioethics*, p. 148.
145 On American discussion of foetal research in this period, see Wilson, *Tissue Culture in Science and Society*, pp. 93–9.
146 Fox and Swazey, *Observing Bioethics*, pp. 128–45; Jonsen *The Birth of Bioethics*, pp. 102–4.
147 Robert Edwards and Patrick Steptoe, *A Matter of Life: The Story of IVF – A Medical Breakthrough* (London: Finestride & Crownchime, 2nd edn, 2011) p. 113.
148 Paul Ramsey, 'Shall We "Reproduce"? The Ethics of In Vitro Fertilization I', *Journal of the American Medical Association*, Vol. 220, no. 10 (1972) pp. 1346–50 (p. 1346).
149 Culliton and Waterfall, 'Flowering of American Bioethics', p. 1270. See also Paul Ramsey, 'Manufacturing Our Offspring: Weighing the Risks', *Hastings Center Report*, Vol. 8 (1978) pp. 7–9.
150 See Wilson, *Tissue Culture in Science and Society*, pp. 94–5; Turney, *Frankenstein's Footsteps*, pp. 180–1.
151 Barbara Culliton, 'National Research Act: Restores Training, Bans Fetal Research', *Science*, Vol. 185 (1974) pp. 426–7.
152 L. W. Fussell, E. A. Barton, F. H. Stevens, Jack Ramsey, A. B. Dawes, F. Nickalls, C. Lawson-Tancred, W. R. Ford and D. Rodgers to Ian Ramsey, 22 December 1971. Ramsey archives.

153 Ian Ramsey to Edwin Barker, 14 January 1972. Ramsey archives.
154 Gordon Dunstan, *The Artifice of Ethics* (London: SCM Press, 1974) p. 67.
155 Ramsey, 'Introduction', p. 3.
156 Anon, 'General Discussion', in Ramsey and Porter (eds), *Personality and Science*, pp. 133–43 (p. 143).
157 Ramsey, 'Moral Problems Facing the Medical Profession'.
158 Edward F. Shotter to Ian T. Ramsey, 18 May 1971. Ramsey archives.
159 Ian T. Ramsey to Edward F. Shotter, 28 June 1971. Ramsey archives.
160 Alastair Campbell, telephone interview with the author (May 2009). See also Reynolds and Tansey (eds), *Medical Ethics Education in Britain*, pp. 64–5.
161 Alastair Campbell, *Moral Dilemmas in Medicine* (Edinburgh and London: Churchill Livingstone, 1972) p. 7.
162 Campbell, interview with the author (2009).
163 Alastair Campbell, 'Philosophy and Medical Ethics', *Journal of Medical Ethics*, Vol. 2 (1976) pp. 1–2 (p. 2).
164 C. Gordon Scorer and Douglas Hill, 'Continuing the Debate: The Role of the Medical Ethicist', *Journal of Medical Ethics*, Vol. 4 (1978) p. 157.
165 David Morley, *The Sensitive Scientist: Report of a British Association Study Group* (London: SCM Press, 1978) p. 128.
166 Robert G. Edwards, 'Fertilization of Human Eggs *In Vitro*: Morals, Ethics and the Law', *Quarterly Review of Biology*, Vol. 49 (1974) pp. 3–26 (p. 15). See also Robert G. Edwards, 'Aspects of Human Reproduction', in Fuller (ed), *The Social Impact of Modern Biology*, pp. 108–22.
167 Edwards, 'Fertilization of Human Eggs *In Vitro*', p. 19.
168 Ibid, p. 19. The MRC mainly refused to fund Edwards and Steptoe because they had not undertaken work on primates, which raised fears about potential developmental abnormalities in humans. For more on this decision, see Martin H. Johnson, Sarah B. Franklin, Matthew Cottingham and Nick Hopwood, 'Why the MRC Refused Robert Edwards and Patrick Steptoe Support for Research on Human Conception in 1971', *Human Reproduction*, Vol. 25 (2010) pp. 2157–74.
169 Edwards rejected Paul Ramsey's claim that IVF was unethical because embryos could not provide informed consent beforehand. 'Every medical treatment', he argued, 'from eating aspirin to open-heart surgery, carries a risk for each patient, and foetuses are not asked beforehand about their own conception or even their abortion.' See Edwards, 'Fertilization of Human Eggs *In Vitro*', pp. 14, 16.

170 Ibid, p. 11.
171 Baroness Shirley Williams, interview with the author (Millbank House, London, September 2010).
172 Alun Jones and Walter Bodmer, *Our Future Inheritance: Choice or Chance? A Study by a British Association Working Party* (Oxford: Oxford University Press, 1974) pp. 40, 132.
173 Ibid, p. 41.
174 Gordon Wolstenholme and David W. Fitzsimmons (eds), *The Law and Ethics of AID and Embryo Transfer* (London and New York: Elsevier, 1973).
175 On positive coverage of IVF in the 1970s, see Imogen Goold, 'Regulating Reproduction in the United Kingdom: Doctor's Voices, 1978–1985', in Imogen Goold and Catherine Kelly (eds), *Lawyers' Medicine: The Legislature, the Courts and Medical Practice, 1760–2000* (Oxford: Hart Publishing, 2009) pp. 167–95 (pp. 168–75).
176 Anthony Tucker, 'The Brave New World of Test Tube Babies', *Guardian*, 27 July 1978, p. 11.
177 Edwards and Steptoe, *A Matter of Life*, p. 105.
178 Goold, 'Regulating Reproduction in the United Kingdom'.
179 Shirley Williams, interview with the author (2010).
180 Williams, 'Responsibility of Science'.
181 Mold, 'Patient Groups and the Construction of the Patient-Consumer', p. 510.
182 Shirley Williams, *Climbing the Bookshelves: The Autobiography* (London: Virago, 2009) p. 234.
183 This working party was established following the recommendations of an earlier review of the potential risks and benefits of genetic manipulation, which was chaired by the botanist Lord Ashby. For more detail, see David Bennett, Peter Glasner and David Travis, *The Politics of Uncertainty: Regulating Recombinant DNA Research in Britain* (London: Routledge and Kegan Paul, 1986).
184 Ibid, p. 49.
185 Sir Robert Williams (chair), *Report of the Working Party on the Practice of Genetic Manipulation* (London: HMSO, 1976) p. 13. Most participants in the working party and GMAG attribute this change to Shirley Williams, although Bennett, Glasner and Travis note that she was not yet Secretary of State for Science and Education when the Williams report was published. See Bennett et al., *The Politics of Uncertainty*, pp. 57–8.
186 Anon, 'GMAG Source of Authority and Terms of Reference' (1977). National Archives, Kew Gardens, London (henceforth National Archives). Medical Research Council files: File number FD7/2523. See also Bennett et al., *The Politics of Uncertainty*, pp. 73–7, 185.

187 Anon, 'GMAG Source of Authority and Terms of Reference' (1977).
188 Bennett et al., *The Politics of Uncertainty*, p. 71. Two of the four trade union representatives were scientists with a direct interest in genetic manipulation.
189 Anon, 'Present Status of Genetic Manipulations in Your Country: Composition of GMAG' (1977). National Archives: FD7/2523.
190 Jerome Ravetz, interview with the author (Oxford, August 2010); Bennett et al., *The Politics of Uncertainty*, p. 164.
191 For more on the President's Commission, see Fox and Swazey, *Observing Bioethics*, pp. 133–4.
192 Dr Vickers, Medical Research Council, 'Background Note for Item 5 – Safeguards Against Genetic Manipulation' (February 1977). National Archives: FD7/2523.
193 Anonymous lay member of GMAG, cited in Bennett et al., *The Politics of Uncertainty*, p. 166; Ravetz, interview with the author (2010).
194 Bennett et al., *The Politics of Uncertainty*, p. 169.
195 Ravetz, interview with the author (2010).
196 Bennet et al., *The Politics of Uncertainty*, p. 169. On the public nature of American committees, see Alexander Capron, 'Looking Back at the President's Commission', *The Hastings Center Report*, Vol. 13, no. 5 (1983) pp. 7–10.
197 Ravetz, interview with the author (2010).
198 Ibid.
199 Jones and Bodmer, *Our Future Inheritance*, p. 117.
200 Ravetz, interview with the author (2010); Werskey, 'The Marxist Critique of Capitalist Science', p. 431.
201 Campbell, interview with the author (2009). Illich made his remarks to Campbell after an LMG talk, which was reprinted in the *Journal of Medical Ethics*. See Ivan Illich, 'The Medicalization of Life', *Journal of Medical Ethics*, Vol. 1 (1975) pp. 73–7. See also Edmund Pellegrino, 'Book Review: *Ethics in Medical Progress: With Special Reference to Transplantation*', *Quarterly Review of Biology*, Vol 43, no. 4 (1968) pp. 478–9 (p. 479). Pellegrino argued that in contrast to similar meetings in the United States, 'social scientists, philosophers and theologians' were absent from the CIBA symposia, and 'the philosophical substratum for the recommendations and actions of the participants are touched on only tangentially'.
202 Culliton and Waterfall, 'Flowering of American Bioethics', p. 1270.
203 Sir Douglas Black, 'Both Sides of a Public Face', *British Medical Journal*, Vol. 282 (1981) pp. 2044–5 (p. 2044).
204 H. Tristram Engelhardt Jr, *The Foundations of Bioethics* (New York and Oxford: Oxford University Press, 1986) p. ix.

205 Ramsey, 'Christian Ethics in the 1960s and 1970s', p. 224.
206 Richard Ashcroft, 'Emphasis Has Shifted from Medical Ethics to Bioethics', *British Medical Journal*, Vol. 332 (2001) p. 302.
207 Ramsey, 'Moral Problems Facing the Medical Profession'.

3

'Who's for bioethics?' Ian Kennedy, oversight and accountability in the 1980s

Bioethics ceased to be an 'American trend' during the 1980s, when growing numbers of British outsiders publicly demanded greater external involvement in the development of guidelines for medicine and biological science. Their arguments were certainly successful. By the beginning of the 1990s, when the *Guardian* described the growing 'ethics industry', supporters of this new approach were influential public figures. One of the earliest and most high profile of these supporters was the academic lawyer Ian Kennedy. Since the late 1960s, Kennedy has written on medical definitions of death and mental illness, euthanasia, the doctor–patient relationship and the rights of AIDS patients. In line with the 'hands-off' approach of lawyers, Kennedy's early work stressed that decisions should rest solely with the medical profession; but this stance changed after he encountered bioethics during a spell in the United States. In 1980 Kennedy used the prestigious BBC Reith Lectures to endorse the approach that he explicitly labelled 'bioethics', critiquing self-regulation and calling for external involvement in the development of professional standards. Kennedy's Reith Lectures, entitled *Unmasking Medicine*, are recognised as a pivotal moment in the history of British bioethics, with a senior doctor identifying them as 'one of the key events in the retreat from paternalism'.[1]

In addition to *Unmasking Medicine*, Kennedy endorsed bioethics in academic publications, newspaper columns and several other radio and television programmes during the 1980s. In this period, he also established a Centre of Medical Law and Ethics at King's College, London, and served on several professional and regulatory bodies. During the 1990s he was a founding member of the Nuffield Council on Bioethics and chaired a public inquiry into human–animal 'xenotransplants'. His contribution to British bioethics has led one lawyer to claim that he 'virtually invented the field in the United Kingdom'.[2] In 2002 the Labour government endorsed this view when it awarded him a knighthood for 'services to bioethics'.[3]

On the one hand, there was little particularly new in Kennedy's call for outside involvement. This was pointed out in 1981 by Dame Elizabeth Ackroyd, chair of the Patients Association, who claimed that 'the proposals which Mr Kennedy puts forward are certainly ones which I support, and which indeed the Patients Association have advocated for a long time'.[4] 'I do not think', Ackroyd continued, 'that Mr Kennedy would claim that he was putting forward new ideas.' Kennedy's proposals did indeed echo those made by the Patients Association and Maurice Pappworth during the 1960s. They also drew on the civil rights campaigns Kennedy encountered in the United States, on Ivan Illich's critique of professions and, perhaps most significantly, on the work of American bioethicists such as Paul Ramsey and Jay Katz.

But while there was little new in Kennedy's calls for external involvement, they were certainly more influential than earlier British proposals. This owed a great deal to the changing political climate in the 1980s. Kennedy's arguments dovetailed with a central belief of the Conservative government that was elected in 1979, which believed that professions should be exposed to outside scrutiny in order to render them accountable to their end-users. It is no coincidence that bioethics emerged as a recognised approach in Britain once the Conservatives promoted external oversight as a way of ensuring public accountability and consumer choice.

This analysis provides a framework for understanding the broad context in which British bioethics emerged and operated, connecting with major themes in contemporary history, such as declining trust in professions among neo-liberal politicians and the rise of measures designed to enforce public accountability, which Michael Power has characterised as the 'audit society'. Power details how the 1980s saw the growth of mechanisms designed to monitor professional actions, whose main ingredient was reliance on experts independent from the profession in question. The early history of British bioethics offers substantive evidence in support of Power's thesis. It also deepens our understanding of how the 'audit society' was shaped by the interaction between political ideologies and professional agendas. The new regimes of external oversight that emerged in the 1980s, such as bioethics, were not simply the top-down product of Conservative demands for public accountability, but also depended on the presence of individuals and professional groups willing to define themselves as the new arbiters of best practice.[5] We can thus see Kennedy's criticism of self-regulation and calls for outside scrutiny as a fundamental constituent of the audit society, which helped create the demand for bioethics.

Examining Kennedy's work also dispels presumptions that are often made about the nature and function of bioethics. Several historians, sociologists and anthropologists have criticised bioethics for failing to ask fundamental questions about the political economy of medicine, or of medical power and authority.[6] But Kennedy regularly drew attention to the ideological aspects of medical decisions, criticised the focus on high-tech practices and claimed that professional authority infantilised patients. His calls for outside input were attempts to redress this perceived imbalance of power, involving others in 'the countless decisions taken by doctors which are not medical, but involve questions of morality or philosophy or economics or politics'.[7]

Yet while Kennedy asked critical questions about professional authority, his work was not, as some claim, simply an 'iconoclastic attack on medical paternalism'.[8] Kennedy also echoed American bioethicists when he claimed that outside involvement would benefit doctors, by relieving them of difficult decisions and helping them overcome public and political mistrust. This is crucial to helping us understand why bioethics became an important approach in the 1980s. Rather than simply challenging the authority of the medical profession, then, Kennedy presented it with a new means of legitimacy in a changed political climate. This ensured that many senior figures endorsed his proposals and Kennedy was soon 'embraced by much of established medicine'.[9] We can thus appreciate the growth of bioethics in the 1980s by seeing figures such as Kennedy as crucial *intermediaries* between politicians and doctors, who promised to fulfil the neo-liberal demand for oversight while also safeguarding medicine.

From paternalism to patient empowerment

Ian McColl Kennedy was born in the West Midlands on 14 September 1941, into what he described as a 'poor working-class' family.[10] His parents, a teacher and an electrician, encouraged their three sons to make the most of the opportunities provided by the postwar welfare state. In 2003 Kennedy recalled that: 'My father in particular was anxious to inculcate in us the notion that we were getting what opportunities we were enjoying by virtue of the taxes and the welfare state, on the back of those who had gone to war … It was our duty to give something back, if we made it.'[11]

Kennedy also grew up in a postwar era in which professions were well regarded.[12] This was especially true of medicine, following the creation of the NHS, the development of antibiotics and the production of 'magic bullets' against diseases such as polio. This high esteem was reflected by the fact that two of Ian Kennedy's brothers studied medicine at university, while he went on to read law at University College London (UCL) before attaining a Master of Laws degree from the University of California, Berkeley. During his time in the United States, Kennedy recalls, the growing civil rights movement strengthened his existing 'sense of social justice, of entitlement of anybody, no matter where they're from, to have an even break, to have a chance'.[13]

Kennedy returned to Britain in 1965, when he was appointed lecturer in law at UCL. While teaching jurisprudence, he became interested in the longstanding issue of when someone began and ceased to be legally defined as a person. Much of this interest stemmed from contemporary debates prompted by new medical technologies. Prominent lawyers such as Glanville Williams had previously investigated how medical techniques such as resuscitation impacted on legal definitions of life and death; but questions surrounding exactly when a person died had increased during the 1960s, thanks to the development of artificial respirators for brain-damaged and seriously ill individuals, and the realisation these so-called 'twilight patients' were a source of organs for newly developed transplant techniques.[14] Since death was legally defined as 'absence of vital functions' such as circulation and breathing, and since a fundamental requirement in the crime of murder was that the killing must have been of a 'life in being', various groups questioned whether a patient dependent on a ventilator was alive or dead and, consequently, whether a doctor who turned a machine off could be charged with murder.[15]

These questions were highlighted by a 1963 coroner's inquest, *Re Potter*, which investigated the death of a man who had been seriously assaulted, stopped breathing and was then placed on an artificial respirator. Having decided that he would not recover, doctors removed a kidney for transplantation, pronounced him dead and turned the machine off. A neurosurgeon later admitted that the patient had no hope of recovery and was only placed on the respirator because another patient needed a kidney transplant. The case raised the question of when death occurred and whether it had been caused by the original assailant, the doctor who removed the kidney, or the doctor who turned the machine off. The doctors involved told the inquest that they believed the patient died when he originally stopped breathing and the coroner agreed, clearing them of any wrongdoing and charging the assailant with manslaughter. But according to existing legal criteria, the patient had not died until the machine had been turned off and 'vital functions' had permanently ceased. Although the coroner's inquest diverged from this view, he ventured no firm opinion on when death now occurred. This uncertainty was compounded following the advent of heart transplants in 1967, when surgeons who removed hearts from ventilated patients in Japan and the United States were charged with murder, and British newspapers portrayed transplant doctors as 'vultures' hovering over ill and vulnerable patients.[16]

Kennedy engaged with this issue in 1969, writing an article that outlined 'the legal problems surrounding the moment of death' as they related to transplant surgery.[17] He used a discussion of *Re Potter* to claim that 'the accepted legal definition of death seems no longer to fit the realities of modern medicine and proves unworkable in certain circumstances'.[18] As he would throughout his career, Kennedy condemned the 'very English reluctance to do anything about the situation until it has caused difficulty' and called for guidelines to forestall legal cases.[19] He warned that if the present uncertainty continued, 'techniques and practices which have come to be regarded as established must stop or forever be open to challenge as regards their legality'.[20]

But Kennedy notably endorsed the 'hands-off' approach that lawyers adopted when it came to medicine, claiming that 'the re-definition of death should be left wholly to the medical profession'.[21] Far from leading or guiding doctors, he argued, the law should only change 'once there is an established consensus in the medical world'.[22] Kennedy believed this would give legal recognition 'to what is now accepted as a matter of practice ... that the turning off of a machine seems not a positive act of killing'.[23] 'In other words', he suggested, 'since it is suggested that the law turns a blind eye to what doctors now do, the insecurity which dogs the doctor should be dispelled by the *gradual acceptance of agreed medical practice as lawful*.'[24] 'The law would be then,' he continued, 'that if the doctor could prove that what he has done was in good faith and was skilful there would be no further inquiry into the relative worth or propriety of his actions.'[25] Kennedy argued that this would help ensure 'security for the doctor', by fostering 'a realisation that the medical profession is a responsible body requiring a high standard of conduct of its members'.[26]

Kennedy reiterated this position in a 1972 article for the *Medico-Legal Journal*, written while he was adjunct professor of law at the University of California, Los Angeles (UCLA). Again claiming that the 'old legal definition of death needs modification in light of advances in medical science', he outlined growing support for the view that death occurred when destruction of the brain stem caused irreversible coma and dependence on a ventilator, as an *ad hoc* committee from Harvard had proposed in a 1968 report.[27] Kennedy claimed that in order to avoid 'the impression of haste by over-zealous surgeons' if the concept were adopted, two sets of doctors should employ a battery of standard tests to determine 'brain death' and the consent of relatives should always be sought for organ transplants.[28] He also argued that doctors should not support patients on ventilators once brain death had been confirmed and 'there was no hope of survival'.[29] As before, Kennedy concluded by stating that whatever criteria were adopted, the 'doctor's judgment must prevail' and the courts must 'follow the consensus of medical opinion'.[30]

These two articles illustrate how British lawyers continued to defer to the medical profession in the 1960s and 1970s. Like Kennedy, others endorsed the *Bolam* ruling and claimed that doctors should be left to determine their own standards of care. In his closing remarks to a 1966 CIBA symposium on ethics and organ transplantation, the judge Lord Charles Kilbrandon stated that a lawyer would never answer the question of 'what is death ... because that is a technical, professional medical matter. It is entrusted to medical men to say when a man is dead, and nobody but a doctor can decide that.'[31] At the same symposium, David Daube, professor of civil law at Oxford, similarly claimed that defining death was 'of a scientific character and *prima facie* not for us'.[32] Daube also echoed the *Hatcher* ruling when he warned that legal interference would 'frighten doctors into passivity' – preventing them from thinking about medical progress and the good of their patients.[33] He argued that when they considered specific practices, lawyers should always 'be generous and leave the verdict to the rectitude and good sense of the doctor'.[34]

But Kennedy began to criticise this position after his return from the United States and appointment to King's College, London, in the late 1970s. In several publications, lectures and radio talks he now argued that patients should have greater say in their treatment and, crucially, that outsiders should play a role in setting standards for the medical profession. This was first evident in a 1976 article for the *Criminal Law Review*, in which he claimed that patients had a fundamental right to self-determination that overrode the paternalistic view that 'decisions concerning a person's fate are better made *for* him than *by* him'.[35] This, he argued, included terminally ill or elderly patients who wished to discontinue treatment that was keeping them alive. Kennedy stated that once a patient declared a wish to have treatment discontinued, the doctor was 'obliged to respect it'.[36] This principle, he continued, should be 'guaranteed and safe-guarded' by consent law so that 'if a patient withholds consent, if he refuses to be touched by a doctor, any further touching will be unlawful and give rise to civil and criminal liability'.[37]

The same year, Kennedy published an article that departed from the usual line in the *Journal of Medical Ethics*, calling for outside involvement in the development of medical guidelines. This proposal arose in a discussion of issues raised by the case of Karen Quinlan, a young American woman who fell into a coma in April 1975 and was then attached to a ventilator 'without any prospect of regaining consciousness'.[38] Kennedy detailed how doctors and a county judge had refused a parental request for the ventilator to be turned off, as Quinlan showed evidence of residual brain activity and was therefore alive according to the 'brain-death' criteria.[39] He argued that the ongoing controversy and uncertainty surrounding the Quinlan case 'serves as a timely reminder of the need for a code of practice'.[40] Should a similar case arise in Britain, he continued, 'the unfortunate position exists whereby the doctor must make a decision which obviously could have grave legal ramifications without any legal guidance'. In contrast to his earlier work, which stated that decisions regarding ventilated patients should 'be left wholly to the medical profession', Kennedy chastised lawyers for 'saying that these are medical matters' and shifting 'the responsibility for decision [*sic*] back to the hapless doctor'.[41] He now believed that they were 'patently not merely medical matters' and asserted that 'doctors function within a framework of legal and social rules which go beyond the rules of their particular profession and must be observed'.[42] This led Kennedy to conclude that any code of practice should be 'worked out by the medical profession after consultation with lawyers, theologians and other interested parties'.[43]

After meeting a BBC radio producer, Kennedy had the chance to make these arguments in public. Between 1976 and 1978 he presented several radio programmes on the care of disabled babies, euthanasia and reform of the Mental Health Act.[44] In his 1977 documentary *The Check-Out*, Kennedy asserted that euthanasia was 'a matter on which not just doctors or lawyers, but all of us, must have our say and our way'. The only way to ensure this, he concluded, was to give 'all interested parties' a role in the development of regulatory codes.[45] Although the subject matter of Kennedy's documentaries varied, his underlying message remained the same. A *British Medical Journal* review of the 1978 programme *The Defect*, which debated screening for spina bifida during pregnancy, noted that Kennedy's core argument was that doctors' opinions 'should be challenged by other members of society'.[46]

What influenced Kennedy's retreat from paternalism? His work from 1976 onwards certainly incorporated elements from Ivan Illich's and Thomas Szasz's radical critiques of medical authority. In a 1979 lecture at the Middlesex Hospital medical school, which highlighted the moral, political and economic aspects of medical decisions, and reiterated that they were 'not for doctors alone to make', Kennedy acknowledged his debt to Illich's claim that 'the whole of medicine is a moral enterprise, since it defines what is normal and, in behavioural terms, what is proper'.[47] Later in the lecture, he endorsed Illich's 'description of the doctor's attitude to his patient as one of infantilization'. Kennedy also shared Illich's scepticism towards the current state of medical ethics, believing that groups such as the SSME and the LMG were 'inward looking' and did little to challenge professional authority.[48] This was evident in his 1976 *Criminal Law Review* paper, where he quoted Szasz's 1961 claim that 'much of what passes for medical ethics is a set of rules the net effect of which is the persistent infantilization and subjugation of the patient'.[49]

But while he endorsed their critiques of paternalism, Kennedy distanced himself from the more radical aspects of Szasz's and Illich's work. He was clear that he did 'not necessarily endorse' Illich's sweeping denunciation of the professions and belief that 'nemesis for the masses is now the inescapable backlash of industrial progress'.[50] And in a radio talk on the Mental Health Act, Kennedy ridiculed as 'preposterous' Szasz's claim that 'there is no such thing as mental illness'.[51] 'Most people', he countered, 'regard mental illness as a reality, not a myth', and there was little to be gained from believing that 'psychiatrists act as agents of a malevolent government intent on locking away or otherwise suppressing those who deviate from an accepted norm'.[52]

Kennedy also acknowledged his debt to more moderate critics in this period, including the doctors Thomas McKeown and Muir Gray. McKeown and Gray both argued that the major causes of illness were poverty, poor public health and nutritional problems, and called for a less interventionist, technocratic approach to medicine.[53] In his 1979 lecture at the Middlesex Hospital, Kennedy drew on their work to claim that:

> there seems little doubt that the single largest cause of illness, however defined, is poverty and what it brings in its wake ... Yet we continue to ride the same tired whirligig of disease identification, exchanging one problem for a new one. And we do so, notwithstanding the fact that, by comparison with the effects produced by sanitation and clean water, medicine's advances are really rather limited.[54]

In this and other talks, Kennedy used McKeown and Gray to endorse a broad 'reorientation' of medicine, arguing that doctors should focus more on 'promotion of health' rather than simply the treatment of disease.[55] This reflected his own enthusiasm for social fairness and 'the politics of welfarism'.[56] And it ultimately bolstered his calls for outside involvement in setting standards and priorities: for 'if we are to change the way medicine is thought of and practised, it is we who must take that action. It is our responsibility.'[57]

Yet the greatest influence on Kennedy's changing worldview appeared to be the 'brilliant insights' of the bioethicists he encountered while teaching in America during the early 1970s.[58] In his 1976 paper on self-determination, for instance, he cited Paul Ramsey's 'outrage' at the fact that it was 'possible to deprive many a patient of a fulfilment of the wish to have a death of one's own'.[59] At the beginning of *The Patient as Person*, Ramsey stated that a patient's interests would be better served by involving outsiders in medical ethics, and this became a constant theme in Kennedy's work from 1976. Kennedy also claimed to find 'much of value' in the work of Jay Katz and his young research associate Alexander Capron, who both also endorsed outside involvement with medical decision-making.[60]

Kennedy was struck by this 'seminal work on bioethics' at a time when he believed 'we were doing nothing in this country'.[61] On returning from the United States, he argued that Britain had 'no vehicle' for the public discussion of issues such as euthanasia, patient rights and medical decision-making. He dismissed the medical groups and the *Journal of Medical Ethics* as 'too narrow' and 'preaching only to the converted, namely the people who came to the lectures were the people you didn't need to have at the lectures, and the people who didn't come were the people you needed to reach'.[62] This frustration was apparent in Kennedy's regular calls for external involvement from 1976 onwards. As he stated in *The Check-Out*: 'It's a deplorable fact that for far too long lawyers and others have ignored this important area [medical ethics] and left doctors alone to wrestle with its complexities'.[63]

Seeing bioethics as the major influence on Kennedy's work also helps us determine why his calls for external involvement eventually became so influential. Like American figures such as Jay Katz, who promised not to 'indict science or stifle research', Kennedy stressed that involving outsiders would benefit medicine.[64] He spent most of his *Criminal Law Review* article, for example, assuring doctors that they would not be prosecuted for meeting a terminally ill patient's request to have their treatment discontinued. Indeed he argued, on the contrary, that meeting the growing demand for self-determination was less likely to prompt a legal challenge than the traditional approach of 'doctor knows best'.[65] And in his lecture at the Middlesex Hospital, he stressed that 'it is important at this point to make clear that I am not criticizing doctors or attacking them or purporting to sit in judgement over them'.[66] Instead, he sympathised that:

> I think it is unfair that responsibility in many areas of human concern has been improperly shifted onto doctors by the rest of us, simply

> because we are happy to have others bear this responsibility, and because the doctor, at least initially, seems prepared to take it on.[67]

Kennedy promised that a more active role for outsiders would help doctors resolve the 'many hard decisions which it is not really their job to make'.[68]

But this conciliatory approach initially made little headway. While doctors may have encouraged interdisciplinary debates during the 1970s, they were less enthusiastic about devolving power to outsiders. In 1977 the BMA argued that outside involvement in medical decisions would damage doctor–patient relations, 'endanger research, increase waiting-lists and threaten the health and morale of doctors'.[69] Their resistance was not lost on Kennedy, who admitted that 'the moment I offer guidance or suggest what should be done, I am met with a chorus of cries, all variations on the theme that I do not really understand, that these are medical matters after all, that I should not trespass on the professional competence of others'.[70]

But this attitude softened in the 1980s, when political changes fostered the 'audit society'.[71] Kennedy's arguments now carried greater weight amid a political emphasis on oversight and public accountability, and senior doctors conceded that traditional forms of self-regulation might be untenable. He consequently became central to a growing form of public debate and regulation, which newspapers and the medical press joined him in labelling '*bioethics*'.

'Who's for bioethics?' The Reith Lectures, the Conservatives and the 1980s[72]

Following Kennedy's radio lectures, which gained him a reputation as a skilled broadcaster, the BBC's director-general invited him to give the prestigious Reith Lectures on Radio Four.[73] The Reith Lectures were established in 1948 to honour Sir John Reith, the BBC's first director-general. They are delivered annually by public intellectuals, and speakers before Kennedy had included Bertrand Russell, the biologist John Z. Young and the anthropologist Edmund Leach.[74] When Kennedy was approached in spring 1979, the lecture themes were the only piece of programme content that the BBC board of directors had the power to select. After requesting several options from Kennedy, they chose a broad analysis of the state of modern medicine.[75]

The BBC officially announced Kennedy was its thirty-second Reith Lecturer in December 1979. The major focus of newspaper profiles in the build-up to his lectures was that he was talking on a subject which, the *Observer* stated, 'the great panjandrums of the medical profession like to reserve for themselves'.[76] Journalists also detailed how a major premise of the lectures was that 'the community should take back some of the control which it has ceded to the medics', with 'lawyers looking over doctor's shoulders ... and a vigilant public endorsing their large decisions'.[77] The *Guardian* notably described this outside perspective as '*bio-ethics*', which was a term that British newspapers had traditionally joined the *British Medical Journal* in attributing to 'Americans, with their unfortunate gift for inventing new specialisms'.[78]

Kennedy's Reith Lectures, entitled *Unmasking Medicine*, consisted of six thirty-minute talks that were broadcast during November and December 1980. Each lecture discussed aspects of his work since 1976. Kennedy began the first lecture by stating that when it came to issues such as the definition of death, the treatment of the mentally ill and care of disabled babies, 'medicine is in the hands of experts and sets its own path'.[79] He claimed that doctors had attained this power by portraying definitions of health and illness as 'terms of scientific exactitude'.[80] Kennedy then drew on Illich and Foucault to contend that 'the normal state against which to measure abnormality is a product of social and cultural values and expectations. It is not some static, objectively identifiable fact.'[81] He continued that in medicine generally, and psychiatry especially, there was in fact a 'relationship between calling someone ill and making a moral judgement about him'.[82] 'If illness is a judgement', he argued, 'the practice of medicine can be understood in terms of power. He who makes the judgements wields the power.'[83]

Drawing on McKeown and Gray, Kennedy then discussed broader determinants of health and disease and claimed that 'Very many of the people to whom we are readily prepared to ascribe the status "ill" find themselves ill because they are poor, grow up in bad housing, eat poor food, work, if at all, in depressing jobs, and generally exist on the margin of survival.'[84] This led Kennedy to the broad conclusion that permeated all his Reith Lectures:

> As long as it is accepted that health is the exclusive preserve of doctors, something only they have competence in, then this state of affairs will continue. It is a matter of balance; the power is now with the professional. Only when it is realised that health is far too important to be left entirely to doctors, that it is a matter for all of us, will conditions be created for the necessary redirection of effort and resources. Only then will any real movement towards health be achieved.[85]

In the second lecture, Kennedy revived his critiques of interventionist approaches to argue that medicine was 'pursued in ways that do not best serve the needs of society'.[86] He claimed that this led to disproportionate investment in fields such as transplant surgery, which treated relatively few patients, while fields such as geriatric medicine and mental health were largely ignored. Kennedy concluded that this emphasis led the public to believe 'in magic cures and the waving of magic wands', while the reality was a 'constant disappointment' where 'the promised or expected cures are not there'.[87]

In lecture three, he outlined 'a better path for the future' and stated that 'we must curb our predilection for medicine in the form of ever more complex technology' and 'direct more of our energy and resources towards the promotion of good health'.[88] The focus here was firmly on primary care and education, on preventing deaths through 'cigarette smoking, alcohol consumption, appalling dietary habits, dangerous workplaces and roads'.[89] Kennedy concluded that 'If GPs were more adequately prepared for the real health needs of their patients, which are as much to do with social problems as with particular diseases, then the beginnings of a better movement towards health could emerge.'[90]

While his first three lectures drew mainly on Illich, McKeown and Gray, Kennedy's fourth lecture owed a large debt to American bioethics. Here, he echoed Paul Ramsey's claim that 'medical ethics are not separate from but part of the general moral and ethical order by which we live'. Drawing on Jay Katz, he also called for 'a wholesale re-examination of the sphere of alleged competence of the doctor'.[91] Kennedy claimed that in choosing whether or not to treat severely disabled babies, doctors currently 'decide on the basis of some rough-and-ready calculus of the future quality of life'.[92] And this, he argued, led to uneven outcomes, 'where in figurative terms, the baby in Barnsley lives, the baby in Bradford dies'.[93]

Kennedy then pointed out that deciding issues such as quality of life was in fact 'profoundly difficult', involving not only medical but also legal, economic, philosophical and social considerations.[94] He claimed to find it striking that 'despite their significance they are not widely discussed. They are resolved in the consulting room and debated, if at all, in the medical journals.'[95] As before, he insisted that the solution lay in ensuring 'that doctors conform to standards set down by all of us'.[96] This, he continued, would foster 'regularity if not uniformity in the decisions arrived at but also some conformity between these decisions and those which the rest of us might take'.[97] Although he ventured no firm plan of how this might be achieved, Kennedy stated that a vital first step was ensuring that 'doctors have some educational grounding in ethical analysis'. And in a now familiar swipe at paternalism, he stressed that this 'must be taught not be some superannuated elder statesman nor by the latest medical star in the firmament, but by an outsider, someone who is not deafened by the rhetoric of medicine'.[98]

After a fifth lecture in which he discussed the categorisation of mental illness and questioned the appropriate norm for mental health, Kennedy again endorsed outside involvement in the sixth and final lecture.[99] He argued that viewing patients as consumers rather than passive recipients of healthcare gave them greater 'power to participate responsibly in decisions made about [their] life'.[100] Kennedy spent much of the lecture dismissing the suggestion that the best route to consumerism in Britain was an increase in private litigation. He argued that litigation was more justifiable in the United States because patients paid for their own healthcare through private insurance schemes and 'if someone suffers harm unexpectedly, he needs money to pay for additional medical care or to meet other costs'.[101] In Britain, by contrast, he claimed that there was 'less need for this form of consumerist litigation' thanks to 'a social welfare system and free health care ... which can serve as the basic source of funds for patients who complain of harm'.[102] Kennedy believed that litigation consequently had a 'more limited' role in Britain, with patients only being justified in suing doctors if they were detained without consent or treated without full disclosure of potential risks.

Kennedy nevertheless believed that this small number of cases might, if successful, 'ensure that standards of practice were established which met the approval of outsiders'.[103] But he also noted that British courts 'tend toward conservatism' and would be 'reluctant to break new ground' by departing from the *Bolam* ruling and judging medical conduct themselves.[104] He proposed that consumerism in Britain should therefore 'take another tack'. This involved the establishment of a supervisory 'board or committee' charged with 'establishing standards which doctors must meet in their practice, measuring the doctor's performance in the light of these standards, and creating means of redress for the patient and sanctions against the doctor if these standards are breached'.[105]

Kennedy was keen to distance his proposed body from 'paternalistic' organisations such as the GMC. 'Standards will have to be set and measured by others', he argued, and 'the principle of outside scrutiny, a key feature of consumerism, seems inevitable.'[106] He closed the Reith Lectures by proposing that the impetus for this 'separate method of supervision' would 'have to come from the consumer, and the consumer will have to be prominently represented on any Board or Committee which is set up'.[107]

Transcripts of the Reith Lectures appeared weekly in the BBC's *Listener* magazine, and all six were published as a book by Allen and Unwin in May 1981. As Kennedy wrote in a preface, the book provided 'the opportunity to put a bit more flesh on the bones of my arguments' and contained an additional chapter on the definition of death. It also included a detailed bibliography 'to show how wide is the range of materials which someone entering into this area of study needs to cover'.[108] This included books by Illich, McKeown and Szasz, by bioethicists such as Ramsey and Katz, and by practically minded philosophers such as Mary Warnock and Peter Singer. Kennedy then defined precisely what this 'area of study' entailed and firmly aligned his Reith Lectures with the approach he encountered in the United States. 'Fundamentally', he stated, 'it is the study of the practice of medicine today.' But this, critically, was 'not a field in which it is necessary to be trained in medicine. *Indeed, it could be said that only someone who is free from any claims which medical professional loyalty may make on his objectivity who can successfully examine the institution of medicine*.'[109] Kennedy outlined how this approach involved 'ethics and law, together with sprinklings of philosophy, sociology and politics ... as they relate to medicine'. While he admitted that there was no 'single label for it' in Britain, he noted that 'In the United States the area goes by the name of "Bioethics"'.[110]

In the book chapter based on his sixth Reith lecture, Kennedy argued that support for outside involvement was growing and that paternalistic attitudes were 'clearly out of line with the political tenor of the day'. 'Consumerism is with us', he stated, and 'the doctor has the choices only of accepting it willingly and co-operating, or of accepting it unwillingly.'[111] But Kennedy had to rely mainly on American examples to support this claim, including the 'series of ethics committees' that included a majority of non-doctors. 'We have much to learn', he concluded, 'from how this aspect of consumerism has developed in the United States.'[112] When it came to Britain, Kennedy cited the presence of lay members on GMAG as 'a good example of the sort of arrangement I envisage'.[113] But GMAG was not really comparable to the American committees that Kennedy praised. Scientists remained in the majority and were viewed as its 'backbone' by civil servants, while lay members lacked influence.

But this situation was to change during the 1980s, thanks to the election of a government that shared Kennedy's enthusiasm for outside involvement and 'empowered consumers'. Kennedy's call for the public to 'take back control of medicine' dovetailed with a central ideology of Margaret Thatcher's Conservative Party, which won the 1979 general election. While they lauded private enterprise, the Prime Minister and politicians on the right of the Conservatives, such as Keith Joseph and Nicholas Ridley, regarded state-supported and self-regulating professions as complacent, wasteful and unresponsive to the market forces they saw as vital to regenerating the economy.[114] Seeking a coherent strategy for revitalising Britain, they drew on neo-liberal theorists such as Milton Friedman and William Niskanen, who believed that welfare states had allowed professions to become overly bureaucratic and self-serving, and argued that the solution lay in remodelling them on market lines.[115]

The influence of this neo-liberal worldview was apparent in a 1980 speech by Nigel Lawson, who encouraged privatisation of the public sector during his time as Treasury Secretary, Secretary of State for Energy and Chancellor of the Exchequer. Lawson declared that the new government would 'break from the predominantly social democratic assumptions that have underlain policy in postwar Britain' by exposing many professions and public services to 'the disciplines of the market'.[116] As the 1980s progressed it became clear that this involved promoting outside scrutiny and involvement as a means of devolving power from professionals to end-users – to parents, patients, students etc. – and enabling them to make decisions that furthered their own interests.

Reflecting the Conservative commitment to 'rolling back the frontiers of the state', scrutiny was not to be performed directly by politicians but was entrusted to an array of consultants and agencies who acted on behalf of consumers, which Alex Mold defines as 'consumerism by proxy'.[117] Mold claims that when it came to medicine, 'consumerism by proxy' was evident in the Conservative belief that managers and fund-holding GPs were the best guardians of patients' interests.[118] But it also, crucially, dovetailed with Kennedy's belief that patient empowerment was best achieved through outside scrutiny of medical practices and decisions.

Throughout the 1980s, in professions such as teaching, medicine, academia, social services and local government, reliance on professional expertise subsequently gave way to new mechanisms of external audit that were designed to enforce value-for-money, public accountability and consumer choice.[119] Change was gradual and proceeded well into the 1990s, but Lawson's speech demonstrates that the Conservatives voiced their intentions early on. This was not lost on the medical profession, which linked Kennedy's Reith Lectures to this neo-liberal worldview. Writing in the *Lancet*, for example, John D. Swales, head of medicine at the University of Leicester, pointed out that 'Kennedy's views enjoy the enormous advantage of following the current political tide'. Swales claimed that doctors should 'therefore look a little more closely at what he is saying rather than succumbing to dismissive comments on his style'.[120] Sir Douglas Black, the president of the Royal College of Physicians, similarly believed that 'Kennedy's views have to be taken seriously, both for their own sake and because they are representative of the forces that seek to effect a radical change in the focus of medicine'.[121]

While the psychiatrist Stephen Little criticised Kennedy for a lack of concrete proposals, he also conceded that: 'To follow the rhetoric of the present government, the public must become more fully informed of the pressures on its medical practitioners and administrators, of the shortcomings as well as the advances.'[122] And Michael Thomas, chair of the BMA, endorsed Kennedy's call for a diverse committee that acted as proxy for patients and the public, as part of 'a situation where all doctors are willing to accept that the public has a right to take part in the decisions on major moral and ethical issues'. Such changes were needed, Thomas stated, because 'the era which required paternalism is past'.[123]

This complicates the 'origin myth' that bioethics was opposed by a recalcitrant medical profession. While some doctors dismissed Kennedy's lectures as 'doctor bashing', many senior figures saw the benefits, or inevitability, of external involvement with medicine.[124] These views were compounded between 1981 and 1984, when growing numbers of politicians and public figures echoed Kennedy's calls for external oversight and patient empowerment. In 1981 Margaret Thatcher appointed Normal Fowler as Secretary of State for Health and Social Services. Fowler explicitly viewed patients and the public as 'consumers', and believed that non-doctors should play a major role in designing policies that rendered medicine more transparent, competitive and publicly accountable.[125]

As the next chapter shows, this was evident when senior figures at the Department of Health and Social Security (DHSS) prioritised the appointment of an 'outside chair' to a public inquiry into IVF and embryo research in 1982. It was also clear in the 1983 decision to select the businessman Sir Roy Griffiths as chair of an inquiry into NHS management. Reflecting the government's enthusiasm for market-oriented reform, the other inquiry members were executives from British Telecom, United Biscuits and Television South West. Their report echoed Fowler's desire for consumer influence when it claimed that: 'Businessmen have a keen sense of how well they are looking after their customers. Whether the NHS is meeting the needs of the patient, and the community, and can prove that it is doing so, is open to question.'[126] In a further blow to paternalism, Griffiths's inquiry suggested that the NHS would be better run by general managers recruited from outside the medical profession.[127]

Further support for external involvement also came from public figures such as Mary Warnock, following her selection as chair of the government's IVF inquiry, and the Australian-born lawyer Geoffrey Robertson, who used a 1982 *Observer* column to claim that 'interdisciplinary co-operation and insistence on public participation' were vital to solving 'the present, not to mention the future, dilemmas of bio-ethics'.[128] Robertson argued it was no longer adequate for lawyers to 'wash their hands and leave decisions in the sterilized gloves of the medical profession'.[129] 'Workable and acceptable' rules for medicine, he stated, 'should not be developed behind a closed door marked "Medical Ethics – laymen and lawyers keep out".'[130]

Many speakers at a 1984 GMC conference on 'Teaching Medical Ethics' also endorsed outside involvement with medicine. The sociologist Margaret Stacey, for example, criticised 'the "closed system" in which the medical profession works wherein the greater part of social as well as professional time is spent with other members of the profession'.[131] Stacey argued that making decisions on a patient's behalf, without consulting other professionals or the patient themselves, derived from an outdated 'model of the [doctor–patient] relationship where the doctor is seen as active and the patient passive, as opposed to one of mutual activity, a partnership in healing or managing disease'.[132] She proposed that doctors should rectify this by opening their records to patients and, where applicable, the public. Stacey claimed that 'this would be not only in the interest of the public but also of the profession'. 'All doctors are aware how difficult such judgements are', she continued, 'and to make records more open would help the public share these problems too.'[133]

At the same conference, John Habgood, the Archbishop of York, argued that 'insights and values from another field of awareness should be fed into the practical business of decision-making'.[134] Habgood similarly presented outside involvement as beneficial to doctors, claiming that it would help them share the 'crushing burden' imposed by 'decisions to make which bear directly on the lives of individuals with whom you are personally involved'.[135]

Surveying this changing landscape for the *Hastings Center Report* in 1984, the doctor and philosopher Raanan Gillon argued that the 1980s marked the end of 'medicine's halcyon days when doctors – for the most part only senior doctors – discussed the dilemmas of medical ethics in privacy and leisure'. 'Today', he noted, 'everyone in Britain seems to be muscling in.' Gillon claimed that these changes were 'ably abetted by the lawyer whom doctors love-hate, Professor Ian Kennedy'.[136] While Gillon conceded that he was no longer the sole advocate of oversight, he nevertheless noted that Kennedy continued to 'vigorously stir the pot'.

Indeed, the regularity with which Kennedy continued to publicly 'stir the pot' led medical journals to dub him 'the ubiquitous Ian Kennedy'.[137] In a 1981 radio documentary, he proposed the establishment of outside 'inspectorates' that would 'ensure proper accountability across many professions'.[138] Writing for the *Journal of Medical Ethics* the same year, Kennedy justified this proposal on the grounds that: 'If a profession by definition exists to serve the public interest, then clearly it must ultimately be the public who judge what that interest is and whether it is being served.'[139] Kennedy's profile increased further in 1983, when he hosted the BBC television series *Doctors' Dilemmas*, in which actors presented a doctor with an ethical dilemma and their decision was scrutinised by a diverse studio panel. In a favourable review for the *British Medical Journal*, Raanan Gillon claimed the programme's message, like all Kennedy's work, was that 'doctors and medical students need far more interdisciplinary discussion and debate about medical ethics'.[140]

Kennedy used his high profile to reassert that outside involvement would benefit doctors. In his final Reith Lecture, he promised that if his proposals were implemented, 'it wouldn't only be the patient who would gain. The doctor too would benefit, as would the practice of medicine.'[141] He expanded on these benefits in a *Journal of Medical Ethics* article that rejected his portrayal as a 'doctor-basher'. Here, Kennedy criticised the tendency to label all non-doctors as 'laymen', which he believed rhetorically stripped them of any expertise. He argued that philosophers and lawyers were trained to analyse ethical or legal issues, and that when confronted with particular ethical dilemmas 'it may be the doctor who is the layman'.[142] Kennedy claimed that external input would thus offer 'great help to doctors if only they would understand that it offers a guide to what they need to do where none existed before'.[143]

In the preface to his book of the Reith Lectures, Kennedy also claimed that giving patients greater say in their treatment would 'reduce the burden of responsibility placed on doctors'. 'I am quite sure', he argued, 'that we do doctors a great disservice by shuffling off onto them a range of problems which they should not be expected to deal with.'[144] In an updated version of the book, published in 1983, Kennedy stressed that this would encourage 'a relationship of partners in the enterprise of health'.[145] The stress on 'partners' helped Kennedy frame bioethics as a collaborative endeavour, in which lawyers, philosophers, politicians and patients were 'not interfering, but trying to help'.[146] He concluded that giving patients greater responsibility and allowing outsiders to set standards would not impede medical practice, but would 'produce guidelines for future conduct, tools for analysis, which will forearm the doctor'.[147]

In the updated edition of *Unmasking Medicine* and a 1984 article for the *Criminal Law Review*, Kennedy also reassured doctors that he was not advocating outside involvement on a case-by-case basis in 'a ghastly on-site Committee'. He instead proposed that 'it is the guidelines for conduct, and the analytical tools, which will be worked out by the non-doctor, along with the doctor'.[148] As in his final Reith lecture, Kennedy recommended that these guidelines should be designed and issued by a 'permanent standing advisory committee' comparable to the President's Commission in the United States. In addition to drawing up codes of practice, he proposed that the committee's interdisciplinary staff would also keep 'developments in medicine under constant review, with a view to identifying and responding to ethical issues'.[149] And Kennedy again stressed that this committee would benefit doctors by aligning medicine with public expectations and thereby preventing 'a sense of bitterness and frustration, out of which grows further litigation'.[150]

By the mid 1980s growing numbers of doctors appeared to agree. Speaking at the GMC conference on 'Teaching Medical Ethics', the surgeon Ronald Welbourne argued that student doctors should be taught by individuals 'drawn from all relevant disciplines', including 'clinical practice, moral philosophy, theology, law, sociology and other branches of learning'.[151] Welbourne claimed that each of these disciplines 'is essential and none is adequate alone'.[152] He also shared Kennedy's belief that outside involvement in developing guidelines would benefit 'patients and doctors' by boosting public confidence and preventing excessive 'legislation and litigation'.[153]

We might expect Welbourne to have supported interdisciplinary approaches, as he served on the editorial board of the *Journal of Medical Ethics* and chaired the Institute of Medical Ethics (IME), which was the new name for the SSME. But support also came from other quarters. Although the *Lancet* was more guarded than Welbourne, identifying external involvement as 'an uneasy but necessary compromise', it nevertheless acknowledged that it had become vital to protecting the interests of 'the individual patient, those of the doctor, and those of scientific progress'.[154] In a review of *Doctor's Dilemmas*, it noted that if 'difficulties and decisions were aired more widely, decision-making might be more even and suspicions might be allayed'.[155] And in another article, entitled

'Who's for Bioethics Committees?', the *Lancet* reiterated that bioethics would safeguard 'not only patients but also doctors and the institutions in which they work'. Outside involvement, it concluded, would help doctors develop guidelines, prevent litigation and ration 'the available and now inadequate resources of the National Health Service'.[156]

This professional acceptance underpinned the increasing recruitment of philosophers, lawyers and other non-doctors to regulatory commissions and medical bodies during the 1980s. Thanks no doubt to his 'ubiquitous' profile, Ian Kennedy was especially popular. Between 1984 and 1988 he was appointed to the GMC, a parliamentary Commission on the Safety of Medicines, the government's Expert Advisory Group on AIDS and a parliamentary review of guidelines for research on foetuses and foetal tissues.[157] These appointments illustrate the political and medical utility of bioethics. Recruiting individuals such as Kennedy to professional bodies helped doctors appear publicly accountable, which safeguarded them from political criticism. Their presence on public inquiries and regulatory committees, meanwhile, helped politicians challenge vested professional interests and fulfilled the neo-liberal enthusiasm for oversight.

But this does not equate to the positivist accounts of 'moral progress' found in participant histories.[158] Despite growing support for bioethics, the government only convened 'broad-based' inquiries to look into contentious new procedures such as IVF and gene therapy during the 1980s. Non-doctors such as Kennedy remained firmly in the minority on bodies such as the GMC and had little influence in their meetings.[159] As before, they also had little say in the governance of clinical treatment. This offered a notable contrast to the United States, where hospital ethics committees that included bioethicists and 'community representatives' had the power to consider treatment and advise on individual cases.[160] Despite his very public lobbying, the permanent 'inspectorate' that Kennedy often endorsed remained conspicuous by its absence.[161] This led him to complain that Britain lagged behind countries with national ethics councils, and that 'apart from the odd *ad hoc* committee, we seem happy to stumble along; so doctors, patients, nurses, and their advisers often seek in vain for guidance'.[162]

But Kennedy appeared most frustrated by the fact that the courts still relied on the 'hands-off' philosophy embodied in the *Bolam* ruling. This was apparent in the 1984 case *Sidaway* v. *Board of Governors of the Royal Bethlem Hospital*, which arose when a woman sued her doctor and his hospital for not disclosing the full risks of a pain-relieving operation that left her partially paralysed.[163] Rejecting her claim for damages, the Court of Appeal and the House of Lords both ruled that the doctor was not negligent since most responsible neurosurgeons elected not to warn patients that the operation carried a small risk of paralysis.[164] This verdict distinguished Britain from the United States and Canada, where courts increasingly argued that disclosure of information should be judged against what a reasonable patient would want to know. And it also led Kennedy to bemoan the fact that instead of recognising a patient's right to control their own treatment, the British courts continued to endorse 'the "right" of doctors to decide for patients'.[165]

Indeed, they endorsed the *Bolam* ruling well into the 1990s, ruling against patients who sued their doctors for failing to disclose the failure rate of sterilisation procedures and the possible risks of contraceptive drugs, on the grounds that the doctors in question had conformed with professional norms.[166] If we are to read bioethics as a decisive shift in the location of biopower, then, it appears that figures such as Ian Kennedy only made inroads into regulatory committees and public debates. In the clinic and the courtroom, as before, doctors remained the arbiters of best practice.

Conclusion

This chapter has detailed why bioethics ceased to be an 'American trend' during the 1980s. Calls for outside involvement with science and medicine became increasingly influential in Britain during this period thanks to the interaction between personal, political and professional agendas. Figures such as Ian Kennedy drew on the work of American bioethicists, among others, to endorse an approach that the medical lawyer Jonathan Montgomery calls 'ethical consumerism', proposing the introduction of mechanisms that redressed paternalism and gave 'outsiders' greater say in the development of professional standards.[167]

Kennedy claimed that this approach resulted from 'a changed attitude among the products of the welfare state towards the medical profession, whereby the doctor is expected to see his patients as partners in the enterprise of healthcare'.[168] Like many 'products of the welfare state', Kennedy was influenced by the leftist politics of the 1960s and 1970s. In addition to American bioethicists, his calls for outside involvement drew on Ivan Illich's critique of paternalism and reiterated the civil rights belief that 'we should respect each person's autonomy, his power to reach his own decisions and act on them'.[169]

This political background and his enthusiasm for the welfare state ensured that Kennedy was no fan of the Conservative government. Indeed, he often criticised its belief that many aspects of public life could be 'regulated (if that is the right word) entirely by market forces'.[170] But while his demands for outside involvement and patient autonomy were influenced by a markedly different 'sense of social justice' to that of Margaret Thatcher's cabinet, they nevertheless became influential thanks to the way they mapped on to the government's neo-liberal desire for publicly accountable and 'customer-focused' professions.[171] This overlap is crucial to understanding why Kennedy's calls for outside involvement were more influential than those of earlier figures such as Maurice Pappworth. While doctors resisted these earlier proposals, they had little choice but to accept them once the Conservatives came to power and it became clear that 'the era which required paternalism is past'.[172]

This latter point highlights that the demand for oversight did not emanate solely from Kennedy or politicians. While there were disgruntled voices at the outset, doctors were certainly willing partners in the emergence of bioethics. This stemmed partly from their sensitivity to the 'political tide' and a desire to align medicine with the growing demand for oversight. But it also stemmed from the way in which Kennedy drew on American bioethicists and framed outside involvement as beneficial to medicine. This undermines the 'origin myth' that portrays bioethics as a radical critique of a conservative and reluctant medical profession. Indeed, Kennedy acknowledged this in 2007, telling the *Guardian* that politicians and doctors would have both ignored him had he been nothing more than 'a pain in the neck'.[173]

This helps us identify what bioethics is and why it became influential. As Charles Rosenberg states, bioethics is best viewed as a 'mediating element' between politicians, the public and health professionals.[174] But the form it takes varies between different locations, thanks to the specific contexts in which it emerges and the individuals who position themselves as bioethicists. In contrast to the United States, where theologians and then philosophers dominated, Ian Kennedy's work ensured that lawyers were integral to the emergence of British bioethics.[175] His Reith Lectures, in particular, engendered a public debate on the law relating to medical practices and the position the courts should adopt *vis-à-vis* doctors. This gave a greater profile to lawyers who already looked at medicine, such as Margaret Brazier and Sheila MacLean, and encouraged others to do likewise. The focus of much writing in this burgeoning area of 'medical law' had more in common with work in American bioethics than traditional legal fields such as tort, family and contract law, and focused on the moral aspects of medical practices and the ethical values that underpinned patient rights.[176]

Specific national factors also ensured that Kennedy's vision of bioethics was more limited in Britain that in the United States. Judges were reluctant to overturn the longstanding *Bolam* ruling and decide the appropriate standards for medicine, while his calls for a national ethics committee were ignored. Those lawyers interested in medical law instead exerted their greatest influence as members of *ad hoc* inquiries into new biomedical technologies, which included greater numbers of 'non-experts' from the 1980s onwards. Yet despite the central role that lawyers played in the emergence of British bioethics, and to the surprise of many, the government chose a philosopher to chair its high-profile inquiry into IVF and embryo experiments in 1982.[177] The next chapter demonstrates how Mary Warnock's appointment fostered a debate on the place of philosophy in bioethics and, more contentiously, on how interdisciplinary committees formulated acceptable rules for science and medicine.

Notes

1 Professor Sir Christopher Booth, cited in Reynolds and Tansey (eds), *Medical Ethics Education in Britain*, p. 51.

2 Laurence Gostin, 'Honoring Ian McColl Kennedy', *Journal of Contemporary Health Law and Policy*, Vol. 14 (1997) pp. vi–xi (p. vi).

3 Anon, 'Professor Loses Job But Gains Knighthood', *Times Higher Education Supplement*, 4 January 2002, p. 11.

4 Dame Elizabeth Ackroyd, 'Mr Kennedy and Consumerism', *Journal of Medical Ethics*, Vol. 7 (1981) pp. 180–1 (p. 180).

5 Power's original work was criticised for a 'top-down' approach and lack of specific case studies, but this has been rectified by recent

studies on the history of oversight in social services, teaching and local government, while Power has also expanded and refined his thesis. See Michael Power, 'Evaluating the Audit Explosion', *Law and Policy*, Vol. 25 (2005) pp. 185–202. For professional case studies, see Duncan Campbell-Smith, *Follow the Money: The Audit Commission, Public Money and the Management of Public Services* (London: Allen Lane, 2008); Roy Lowe, *The Death of Progressive Education: How Teachers Lost Control of the Classroom* (London: Routledge, 2007); Eileen Munro, 'The Impact of Audit on Social Work Practice', LSE Research Online (2004). Available online at http://eprints.lse.ac.uk/523/1/Audit-SocialWork_05.pdf (accessed 13 February 2014).

6 For example, Rose, *The Politics of Life Itself*; Jasanoff, *Designs on Nature*; Evans, *Playing God*; Cooter, 'The Ethical Body'.

7 Ian Kennedy, 'What is a Medical Decision?', in Ian Kennedy, *Treat Me Right: Essays in Medical Law and Ethics* (Oxford: Clarendon Press, 1988) pp. 19–31 (p. 24). This chapter is the text of the Astor Memorial lecture, which Kennedy gave at the Middlesex Hospital medical school on 3 July 1979.

8 Boyd, 'The Discourses of Bioethics in the United Kingdom', p. 489.

9 Gostin, 'Honoring Ian McColl Kennedy', p. x.

10 Claire Donnelly, 'Inquiring Mind. The HSJ Interview: Sir Ian Kennedy', *Health Services Journal*, Vol. 113 (2003), pp. 22–3 (p. 22).

11 Ibid, p. 22.

12 Judt, *Ill Fares the Land*; Perkin, *The Rise of Professional Society*.

13 Sir Ian Kennedy, interview with the author (Portland House, London, July 2010).

14 See Glanville Williams, *The Sanctity of Life and the Criminal Law* (London: Faber and Faber, 1958). On ventilated patients as 'twilight' individuals, see Gordon Wolstenholme (ed), *Law and Ethics of Transplantation* (London: CIBA Foundation, 1968).

15 Edward Shotter (ed), *Matters of Life and Death* (London: Darton, Longman & Todd, 1970); Wolstenholme (ed), *Law and Ethics of Transplantation*; Church Assembly Board for Social Responsibility, *Decisions About Life and Death* (1965).

16 See Nathoo, *Hearts Exposed*; Margaret Lock, *Twice Dead: Organ Transplants and the Reinvention of Death* (Berkeley and London: University of California Press, 2001).

17 Ian Kennedy, 'Alive or Dead? The Lawyer's View', *Current Legal Problems*, Vol. 22 (1969) pp. 102–28 (p. 103). While his article concentrated primarily on organ transplants, Kennedy also highlighted other implications of cases such as *Re Potter*. What, he wondered, were the legal rights and liabilities where X makes a gift *inter vivos* and then six years and eleven months later becomes irreversibly

unconscious after suffering extensive brain injury, but is 'kept alive' on a ventilator for another month to avoid estate duty? Kennedy recounted that 'more than one hospital physician' assured him that 'this practice is not unknown'. Ibid, pp. 110–11

18 Ibid, pp. 108, 106.
19 Ibid, p. 106.
20 Ibid, p. 109.
21 Ibid, pp. 111, 115.
22 Ibid, pp. 115, 113.
23 Ibid, pp. 116, 124–5.
24 Ibid, p. 125. Emphasis in original.
25 Ibid, p. 126.
26 Ibid, pp. 116, 126.
27 Ian Kennedy, 'The Legal Definition of Death', *Medico-Legal Journal*, Vol. 99 (1972) pp. 36–41 (p. 38). For more background on the Harvard committee's report and the history of the 'brain-death' concept, see Lock, *Twice Dead*, pp. 89–91, 111–12; Rothman, *Strangers at the Bedside*, pp. 160–5. See also Ad Hoc Committee of Harvard Medical School, 'A Definition of Irreversible Coma', in Peter Singer and Helga Kushe (eds), *Bioethics: An Anthology* (London: Blackwell Publishers, 1999) pp. 287–92.
28 Kennedy 'Legal Definition of Death', p. 41.
29 Ibid.
30 Ibid, pp. 39, 40.
31 Lord Kilbrandon, 'Chairman's Closing Remarks', in Wolstenholme (ed) *Law and Ethics of Transplantation*, pp. 212–16 (p. 213).
32 David Daube, 'Transplantation: Acceptability of Procedures and the Required Legal Sanctions', in Wolstenholme (ed), *Law and Ethics of Transplantation*, pp. 188–201 (p. 190).
33 Ibid, p. 189.
34 Ibid, p. 196.
35 Ian Kennedy, 'The Legal Effect of Requests by the Terminally Ill and Aged not to Receive Further Treatment from Doctors', *Criminal Law Review* (1976) pp. 217–32 (p. 219). Emphasis in original.
36 Ibid, p. 229.
37 Ibid, pp. 231, 217.
38 Ian Kennedy, 'The Karen Quinlan Case: Problems and Proposals', *Journal of Medical Ethics*, Vol. 2 (1976) pp. 3–7 (p. 3).
39 Ibid, p. 3. Following a successful appeal, doctors removed Karen Quinlan from the ventilator in March 1976; but she continued to breathe unaided until her death from pneumonia in 1985. For more on the Quinlan case, see Stevens, *Bioethics in America*, pp. 109–49; Rothman, *Strangers at the Bedside*, pp. 222–46.

40 Kennedy, 'The Karen Quinlan Case', p. 7.
41 Ibid, p. 3.
42 Ibid.
43 Ibid, p. 4.
44 Kennedy, interview with the author (2010).
45 Ian Kennedy, 'The Check Out: A Humane Death', in Kennedy, *Treat Me Right*, pp. 300–14. This chapter was based on a BBC radio talk originally broadcast in August 1977.
46 Anon, 'Medicine and the Media', *British Medical Journal*, Vol. 2, issue 6148 (11 November 1978) p. 1361.
47 Ian Kennedy, 'What is a Medical Decision?'
48 Ian Kennedy, quoted in Reynolds and Tansey (eds), *Medical Ethics Education in Britain*, p. 46.
49 Kennedy, 'The Legal Effects of Requests by the Terminally Ill', p. 219, n 4. See also Thomas Szasz, *The Myth of Mental Illness: Foundations of a Theory of Personal Conduct* (London: Paladin, 1972).
50 Kennedy, 'What is a Medical Decision?', p. 23. See also Illich, *Medical Nemesis*, p. 154.
51 Ian Kennedy, 'The Mental Health Act: A Model Response that Failed', *World Medicine*, Vol. 37 (1978) pp. 37–8, 75–6, 81–2 (pp. 37, 75). This article was based on a BBC radio talk broadcast in February 1978.
52 Ibid, pp. 37, 75.
53 Kennedy, interview with the author (2010); Kennedy, 'What is a Medical Decision?', p. 22. See also J. A. Muir Gray, *Man Against Disease: Preventive Medicine* (Oxford: Oxford University Press, 1979); Thomas McKeown, *The Role of Medicine: Dream, Mirage or Nemesis?* (Oxford: Blackwell, 1979). For an overview of McKeown's work and influence, see Bill Bynum, 'The McKeown Thesis', *Lancet*, Vol. 371 (2008) pp. 644–5.
54 Kennedy, 'What is a Medical Decision?', p. 22.
55 Ian Kennedy, 'Tinkering with or Retooling the NHS is Not What is Needed', *Listener*, 20 November 1980, pp. 677–9. See also Kennedy in Reynolds and Tansey (eds), *Medical Ethics Education in Britain*, p. 48.
56 Kennedy, interview with the author (2010).
57 Kennedy 'Tinkering with or Retooling the NHS', p. 679.
58 Kennedy, 'What is a Medical Decision?', p. 27.
59 Kennedy, 'The Legal Effect of Requests by the Terminally Ill', p. 220. See also Ramsey, *The Patient as Person*, p. 116.
60 Kennedy, interview with the author (2010).
61 Ibid.
62 Ibid.

63 Kennedy, 'The Check-Out', p. 313.
64 Katz, *Experimentation with Human Beings*, p. 5.
65 Kennedy, 'The Effect of Requests by the Terminally Ill', p. 226.
66 Kennedy, 'What is a Medical Decision?', p. 23.
67 Ibid, pp. 23–4.
68 Ibid, p. 24.
69 Anon, 'Complaints by Patients', *Lancet*, Vol. 310 (1977) p. 1238.
70 Kennedy, 'What is a Medical Decision?', p. 24.
71 Power, *The Audit Society*. See also Marilyn Strathern (ed), *Audit Cultures: Anthropological Studies in Accountability, Ethics and the Academy* (London: Routledge, 2000).
72 Anon, 'Who's for Bioethics Committees?', *Lancet*, Vol. 327 (1986) p. 1016.
73 In 1978 the BBC entered *The Check-Out* as its documentary entry for the prestigious Italia Prize.
74 Nathoo, *Hearts Exposed*, p. 214, n.16. The BBC has archived recordings and transcripts of all the Reith Lectures, which are available online at http://www.bbc.co.uk/radio4/features/the-reith-lectures/archive/ (accessed 13 February 2014).
75 Reynolds and Tansey (eds), *Medical Ethics Education in Britain*, p. 47.
76 Nigel Hawkes, 'Tough-Talking Lawyer who Wants to Cure the Ills of Medicine', *Observer*, 2 November 1980, p. 5.
77 John Cunningham, 'Ritual of the Medicine Man', *Guardian*, 3 November 1980, p. 11.
78 Paul Keel, 'The BBC's Quest, in the Name of Reith', *Guardian*, 15 December 1979, p. 17. On press portrayals of 'bioethics' as an American approach, see Nigel Hawkes, 'The Making of Baby Brown', *Observer*, 30 July 1978, p. 9.
79 Ian Kennedy, 'We Must Become the Masters of Medicine, Not Its Servants', *Listener*, 6 November 1980, pp. 600–4 (p. 600).
80 Ibid, p. 600.
81 Ibid. There are traces here of Foucault's mentor, Georges Canguilhem, although Kennedy did not acknowledge him as he did Illich and Foucault. See Georges Canguilhem, *The Normal and the Pathological* (New York: Zone Books, 1991).
82 Ibid. Kennedy here discussed the controversy surrounding the American Psychiatric Association's definition of homosexuality as a mental illness, which was overturned during the 1970s following protests by campaign groups.
83 Ibid, p. 601.
84 Ibid, p. 602.
85 Ibid, p. 603.

86 Ian Kennedy, 'Now, More than Ever, Wealthier Means Healthier', *Listener*, 13 November 1980, pp. 641–4 (p. 641).
87 Ibid, p. 643.
88 Ian Kennedy, 'Tinkering with or Retooling the NHS', p. 677.
89 Kennedy, 'Now, More than Ever, Wealthier Means Healthier', p. 677.
90 Ibid.
91 Ian Kennedy, 'Medical Ethics are not Separate from but Part of Other Ethics', *Listener*, 27 November 1980, pp. 713–15 (p. 713).
92 Ibid, p. 713.
93 Ibid, p. 713.
94 Ibid, pp. 714, 715.
95 Ibid, p. 715.
96 Ibid, p. 713.
97 Ibid, p. 715.
98 Ibid.
99 Ian Kennedy, 'Great Caution Must be Exercised in Visiting the Status of Mentally Ill on Anyone', *Listener*, 4 December 1980, pp. 745–8. This lecture essentially repeated claims made in the first three Reith Lectures, with Kennedy claiming that diagnosing mental illness was a 'moral, social and political' enterprise, and calling for a shift away from pharmaceutical intervention to tackling the economic and social causes of mental illness.
100 Ian Kennedy, 'Consumerism in the Doctor–Patient Relationship', *Listener*, 11 December 1980, pp. 777–80 (p. 777).
101 Ibid, p. 778.
102 Ibid.
103 Ibid, pp. 778–9.
104 Ibid, p. 780.
105 Ibid, pp. 780, 777.
106 Ibid, p. 777.
107 Ibid, p. 780.
108 Ian Kennedy, *Unmasking Medicine* (London: Allen and Unwin, 1981) p. vii.
109 Ibid, p. vii. Emphasis added.
110 While he may have approved of the approach, Kennedy admitted that he never found the term bioethics 'terribly appealing'. See Ibid, p. vii.
111 Ibid, p. 128.
112 Ibid, pp. 122–3.
113 Ibid, p. 122.
114 On Conservative distrust of public services and state-supported professions, see Eric J. Evans, *Thatcher and Thatcherism* (London: Routledge, 2004) pp. 65–78; Dennis Kavanagh, *Thatcherism and British Politics: The End of Consensus?* (Oxford: Oxford University

Press, 1989) pp. 85–7; Perkin, *The Rise of Professional Society*, pp. 472–80. On distrust of medicine in particular, see Stephen Harrison and Waqar I. U. Ahmad, 'Medical Autonomy and the State 1975 to 2025', *Sociology*, Vol. 34, no. 1 (2000) pp. 129–46.

115 William Niskanen, *Bureaucracy: Servant or Master?* (London: Institute for Economic Affairs, 1973). Conservative enthusiasm for this neo-liberal view was evident in Nicholas Ridley's glowing foreword to Niskanen's book. I am grateful to Raymond Plant for pointing me towards Niskanen's work.

116 Nigel Lawson, *The New Conservatism* (London: Centre for Policy Studies, 1980) pp. 6–7. Several writers stress that the various tactics and policies enacted by Thatcher's Conservative government were not a pre-formed ideology or political philosophy. What commonly became known as 'Thatcherism' was instead a mixture of different ideas, applied in order to think about and act upon specific problems in a rather *ad hoc* way. As Rose details, however, the neo-liberal rationalism of the Chicago School nevertheless came to provide a way of linking up these various tactics so that 'they appeared to partake in a coherent logic'. See Nikolas Rose, *Powers of Freedom: Reframing Political Thought* (Cambridge: Cambridge University Press, 1999) p. 27. See also Evans, *Thatcher and Thatcherism*.

117 Lawson, *The New Conservatism*, p. 5. On the neo-liberal belief that external audit ensured competitiveness, transparency and public accountability, see Power, *The Audit Society*; Rose, *Powers of Freedom*; Rose, 'Government, Authority and Expertise in Advanced Liberalism', pp. 283–99. On 'consumerism by proxy', see Mold, 'Patient Groups and the Construction of the Patient-Consumer', pp. 512–15.

118 Mold, 'Patient Groups and the Construction of the Patient-Consumer'. See also Alex Mold, 'Making the Patient-Consumer in Margaret Thatcher's Britain', *Historical Journal*, Vol. 54, no. 9 (2011) pp. 509–28.

119 I have written elsewhere on how the Conservative distrust of the professions led British universities to make fields such as biology more competitive and managerial during the 1980s. See Duncan Wilson and Gaël Lancelot, 'Making Way for Molecular Biology: Institutionalizing and Managing Reform of Biological Science in a UK University during the 1980s and 1990s', *Studies in the History and Philosophy of the Biological and Biomedical Sciences*, Vol. 39 (2008) pp. 93–108.

120 J. D. Swales, 'Thoughts on the Reith Lectures', *Lancet*, Vol. 316 (1980) pp. 1348–50 (p. 1348).

121 Black, 'Both Sides of a Public Face', p. 2044.

122 Stephen Little, 'Consumerism in the Doctor–Patient Relationship', *Journal of Medical Ethics*, Vol. 7 (1981) pp. 187–90 (p. 190).

123 Michael Thomas, 'Should the Public Decide?', *Journal of Medical Ethics*, Vol. 7 (1981) pp. 182–3 (p. 182).
124 Boyd, 'The Discourses of Medical Ethics in the United Kingdom', p. 489.
125 Norman Fowler, *Ministers Decide: A Memoir of the Thatcher Years* (London: Chapmans, 1991) p. 197.
126 Roy Griffiths (chair), *NHS Management Inquiry* (London: Department of Health and Social Security, 1983) p. 2.
127 For more background, see Klein, *The New Politics of the NHS*, pp. 117–23.
128 Geoffrey Robertson, 'The Law and Test Tube Babies', *Observer*, 7 February 1982, p. 8.
129 Ibid.
130 Ibid.
131 Margaret Stacey, 'The View of a Social Scientist', in 'Report of a Conference on the Teaching of Medical Ethics, held on Thursday 16 February 1984' (General Medical Council) pp. 10–15. File held at National Archives FD7/3268. A revised version of this talk was reprinted as Margaret Stacey, 'Medical Ethics and Medical Practice: A Social Science View', *Journal of Medical Ethics*, Vol. 11 (1985) pp. 14–18.
132 Stacey, 'The View of a Social Scientist', p. 12.
133 Ibid, p. 12.
134 The Most Revd and Rt Hon. J. S. Habgood, Archbishop of York, 'The View of a Theologian', in 'Report of a Conference on the Teaching of Medical Ethics, held on Thursday 16 February 1984' (General Medical Council) pp. 8–10 (p. 8). National Archives FD7/3268. This talk was reprinted as J. S. Habgood, 'Medical Ethics – A Christian View', *Journal of Medical Ethics*, Vol. 11 (1985) pp. 12–13.
135 Habgood, 'The View of a Theologian', p. 10.
136 Raanan Gillon, 'Britain: The Public Gets Involved', *Hastings Center Report*, Vol. 14 (December 1984) pp. 16–17 (p. 16).
137 Anon, 'Research Ethics Committees', *Lancet*, Vol. 321 (1983) p. 1026.
138 Ian Kennedy, 'Response to the Critics', *Journal of Medical Ethics*, Vol. 7 (1981) pp. 202–11 (p. 206).
139 Ibid, p. 206.
140 Raanan Gillon, 'Medicine and the Media', *British Medical Journal*, Vol. 286 (1983) p. 715.
141 Kennedy, 'Consumerism in the Doctor–Patient Relationship', p. 777.
142 Kennedy, 'Response to the Critics', p. 207.
143 Ibid, p. 204.
144 Kennedy, *Unmasking Medicine*, p. ix.
145 Ian Kennedy, *The Unmasking of Medicine* (London: Paladin Press,

rev. and updated edn, 1983) p. 124. Nikolas Rose has described how seeing individuals as empowered 'partners' in a relationship with professionals and service providers entails a distinctly neo-liberal conception of the human actor as 'an entrepreneur of his or her self', who is '*active* in making choices in order to further their own interests'. In this view, it is the individual rather than the state who is answerable for their own health, security, productivity, etc. See Rose, *Powers of Freedom*, p. 142. Emphasis in original. See also Foucault, *The Birth of Biopolitics*, pp. 225–6.

146 Kennedy, *The Unmasking of Medicine*, p. 115.

147 Ibid, p. 118.

148 Ibid p. 119.

149 Ibid, p. 129.

150 Ian Kennedy, 'The Patient on the Clapham Omnibus', *Modern Law Review*, Vol. 47, no. 4 (1984) pp. 454–71 (p. 468).

151 R. B. Welbourne, 'The View of an Editor of *The Journal of Medical Ethics*', in 'Report of a Conference on the Teaching of Medical Ethics, held on Thursday 16 February 1984' (General Medical Council) pp. 18–20 (p. 20). National Archives FD7/3268. This talk was reprinted as R. B. Welbourne, 'A Model for Teaching Medical Ethics', *Journal of Medical Ethics*, Vol. 11 (1985) pp. 29–31.

152 Welbourne, 'A Model for Teaching Medical Ethics', p. 29.

153 Ibid.

154 Anon, 'Research Ethics Committees', p. 1026.

155 Ibid.

156 Anon, 'Who's for Bioethics Committees?', p. 1016.

157 Kennedy, interview with the author (2010). A full list of Kennedy's appointments can be found in his *Who's Who* entry. See *Who's Who and Who Was Who* (Oxford University Press, 2010), http://www.ukwhoswho.com (accessed 13 February 2014).

158 Cooter, 'The Ethical Body', p. 435.

159 Kennedy, interview with the author (2010).

160 Rothman, *Strangers at the Bedside*, pp. 255–6.

161 Kennedy, 'Consumerism in the Doctor–Patient Relationship', p. 780.

162 Ian Kennedy, 'The Patient on the Clapham Omnibus: Updated and with a new Postscript', in Kennedy, *Treat Me Right*, pp. 175–213 (p. 176).

163 *Sidaway* v. *Board of Governors of the Royal Bethlem Hospital and the Maudsley Hospital* [1984] QB 493, [1984] 1 All ER 1018. For more background to the case, see Brazier, *Medicine, Patients and the Law*, pp. 103–5.

164 Brazier, *Medicine, Patients and the Law*, p. 104.

165 Kennedy, 'The Patient on the Clapham Omnibus: Updated', p. 210.

166 The first case was *Gold* v. *Haringey Health Authority* [1987] 2 All ER 888, where a woman sued her doctors when she became pregnant after a failed sterilisation operation. The second was *Blyth* v. *Bloomsbury Health Authority* [1993] 4 Med LR 151 CA, where a woman suffered long-term bleeding after the injection of a contraceptive drug. On the continued reliance on the *Bolam* ruling during the 1990s and beyond, see Brazier, *Medicine, Patients and the Law*; Jonathan Montgomery, 'Time for a Paradigm Shift? Medical Law in Transition', *Current Legal Problems*, Vol. 53 (2000) pp. 363–408. This attitude was questioned in 2001 by the senior judge Lord Woolf, who argued that the courts had long been excessively deferential to the medical profession and that this was beginning to change. For a discussion of Woolf's speech and an assessment of possible changes, see Jonathan Montgomery, 'Law and the Demoralisation of Medicine', *Legal Studies*, Vol. 26, no. 2 (2006) pp. 185–210.

167 See Montgomery, 'Time for a Paradigm Shift?'; Jonathan Montgomery, 'Medical Law in the Shadow of Hippocrates', *Medical Law Review*, Vol. 52 (1989) pp. 566–76.

168 Kennedy, 'Patient on the Clapham Omnibus', p. 454.

169 Ibid, p. 456. On how civil rights politics encouraged the postwar generation to concentrate on individual autonomy and critique professions, see Judt, *Ill Fares the Land*, pp. 86–9.

170 Ian Kennedy, 'Preface', in Kennedy, *Treat me Right*, pp. vii–viii (p. viii).

171 Kennedy, interview with the author (2010). Several writers have outlined how the civil rights belief in individual autonomy mapped on to the neo-liberal enthusiasm for 'empowered consumers'. See, for example, Mitchell Dean, *Governmentality: Power and Rule in Modern Society* (London: Sage, 2nd edn, 2010) pp. 180–2; Rose, *Powers of Freedom*, pp. 141–2.

172 Thomas, 'Should the Public Decide?', p. 182.

173 John Carvel, 'Duty Bound', *Guardian*, 17 October 2007, p. 18.

174 C. Rosenberg, 'Meanings, Policies and Medicine: On the Bioethical Enterprise and History', *Daedalus*, Vol. 128, no. 4 (1999) pp. 27–47 (p. 38).

175 Jonathan Montgomery, 'Lawyers and the Future of UK Bioethics', *Verdict – The Magazine of the Oxford Law Society* (2012). Available online at http://eprints.soton.ac.uk/341657/ (accessed 13 February 2014); Sheila MacLean, interview with the author (University of Glasgow, July 2009).

176 Montgomery, 'Law and the Demoralisation of Medicine', p. 207. As Roger Brownsword notes, 'there is no older or more deeply contested jurisprudential question than how we should understand the relation-

ship between law and morality', and medical lawyers often confronted the longstanding issue of what links, if any, should exist between the law and morals. Many principal figures in the field, such as Ian Kennedy, Kenneth Mason and Alexander McCall Smith, shared Lord Chief Justice Coleridge's 1893 view that 'It would not be correct to say that every moral obligation involves a legal duty; but every legal duty is founded on a moral obligation'. See, for example, J. K. Mason and G. T. Laurie, *Law and Medical Ethics* (Oxford: Oxford University Press, 8th edn, 2011) pp. 1–26; Ian Kennedy and Andrew Grubb, *Medical Law: Text and Materials* (London: Butterworths, 1989). For further discussion of this approach, see Roger Brownsword, 'Bioethics: Bridging from Morality to Law?', in Michael Freeman (ed), *Law and Bioethics* (Oxford: Oxford University Press, 2008) pp. 12–30; Jonathan Montgomery, 'The Legitimacy of Medical Law', in Sheila MacLean (ed), *First Do No Harm: Law, Ethics and Healthcare* (Aldershot: Ashgate, 2006) pp. 1–17.

177 MacLean, interview with the author (2009).

4

'Where to draw the line?' Mary Warnock, embryos and moral expertise

The political enthusiasm for external oversight was made clear in 1982 when officials at the DHSS broke from the longstanding reliance on scientific and medical expertise and prioritised 'an outside chairman' for their public inquiry into IVF and embryo experiments. After a brief discussion about possible chairs, politicians chose the moral philosopher Mary Warnock to chair an inquiry in which, for the first time, individuals from other professions outnumbered doctors and scientists. Warnock's involvement with IVF highlights the British emergence of what Jasanoff calls 'official bioethics', in which philosophers, lawyers and others serve on government committees and assist in policymaking.[1]

Once appointed, Warnock became a vocal supporter of external oversight. In language reminiscent of Conservative politicians and Ian Kennedy, she regularly argued that the public were 'entitled to know, and even to control' professional practices.[2] Like Kennedy, she also claimed that this would benefit researchers by safeguarding them from declining public and political trust. Many clinicians and researchers agreed that oversight would make their work 'socially palatable' and supported Warnock's calls for a 'monitoring body' to scrutinise IVF and embryo research.[3] Like Kennedy, then, Warnock both responded to and helped to generate the demand for bioethics, contributing to the public and political construction of the 'audit society'.

Despite the similarity in their arguments, Kennedy and Warnock promoted bioethics for different reasons. While Kennedy's endorsement drew on his encounters with civil rights politics and American bioethicists, Warnock was motivated by changing trends in philosophy. She believed that the refusal to discuss practical issues had rendered philosophy trivial and boring, and joined a growing number of philosophers who began to comment 'on the rightness or wrongness of particular issues' during the 1970s and 1980s.[4] Warnock's appointment as chair of the IVF inquiry provided her with the chance to engage with practical affairs and led other philosophers to view bioethics as the most profitable branch of what Peter Singer called 'applied ethics'.

But many of the philosophers who engaged with ethical issues could not shake off the belief that morality was a set of subjective and often incompatible views and premises.[5] Warnock was confronted with this problem when her committee disagreed over embryo research and she was unable to reconcile those 'who said "Look at the benefits" and those who, at the other extreme, said "I don't care what the benefits are: I feel it to be wrong"'.[6] Warnock recognised that there was no way of uniting these opposing views or of reasonably showing that one was more valid than the other. First, scientific evidence offered no resolution, as both sides used data on embryological development to justify their particular standpoint. The interpretation of scientific 'facts' here was not a neutral activity, since the question of which facts mattered was clearly shaped by an individual's moral preferences.

Secondly, Warnock drew on figures such as A. J. Ayer to claim that opposition to research was valid even if an individual simply felt it to be wrong, as 'morality cannot be divorced from sentiment'.[7] She publicly argued that this limited the role that philosophers had to play in practical affairs, where 'there is no such thing as authority. There is only a set of different opinions.'[8] Warnock's argument here further aligned bioethics with the sociopolitical climate of the 1980s. It dovetailed with the neo-liberal emphasis on individual autonomy and echoed Margaret Thatcher's belief that 'choice is the essence of ethics'.

Warnock's belief that 'there cannot be moral experts' also set her against figures such as Richard Hare and Peter Singer, who argued that philosophers could provide authoritative answers to moral dilemmas.[9] Their differences of opinion demonstrate that bioethicists held no consensus on what bioethics was or how it should function. While Hare and Singer believed that bioethics provided a vehicle for philosophers to act as 'ethics experts', Warnock saw it as a form of 'corporate decision-making' in which representatives of different groups and professions sought 'a middle way' between competing interests.[10]

When it came to Warnock's committee of inquiry, this 'middle way' involved using scientific data to try and reconcile supporters and opponents of embryo research. Warnock argued that permitting experiments up to fourteen days after fertilisation, when antecedents of the nervous system began to form, would retain many utilitarian benefits while offending as few people as possible. But this decision was heavily criticised by other bioethicists, in addition to supporters and opponents of research. Despite the emergence of 'official bioethics', then, the question of 'where to draw the line', and *who* exactly should draw it, remained publicly contentious.

From meta-ethics to 'applied ethics': British philosophy in the 1960–80s

Mary Warnock was born Mary Wilson in Winchester on 14 April 1924, seven months after her father had died from diphtheria. Despite being one of six children in a single-parent family, she enjoyed a comfortable childhood. Her family remained wealthy thanks to her maternal grandfather, the German-born banker Sir Felix Schuster, and she was educated at the prestigious St Swithin's school in Winchester.[11] After leaving this school in 1940, she spent three terms at St Prior's school in Surrey, which counted Julian and Aldous Huxley among its former pupils. In 1942 she won a scholarship to Lady Margaret Hall, Oxford, to study Classics. It was here that she met a fellow student, Geoffrey Warnock, who went on to become a well-known philosopher and Vice-Chancellor of Oxford. They married in 1949, and that same year the new Mrs Warnock was appointed lecturer in moral philosophy at St Hugh's College, Oxford.

Warnock recalls that 'philosophy in Oxford was then in the high point of success', with large student numbers and over thirty members of staff.[12] The dominant figures were Gilbert Ryle and J. L. Austin, who encouraged meta-ethical work on the meaning and classification of language. Although A. J. Ayer had recently left for London, Warnock noted that his influence 'seemed most difficult to shake off'.[13] Most Oxford staff believed that the focus on logic and language ensured that 'moral philosophy, as a subject, was over and done with'.[14] The only exception was Richard Hare, who argued that moral judgements were different from factual statements on account of their being prescriptive and universalisable, where claiming 'you ought to do X' at once commits me to doing so and instructs others to do likewise.[15] But Hare's work was still concerned with the nature of moral language, not with concrete questions of what ought to happen in specific situations. Even when he spoke at meetings on practical subjects, such as Ian Ramsey's symposium on 'Personality and Science', Hare simply clarified the use of words and concepts such as 'personality'.[16] Like their colleagues elsewhere, Oxford philosophers firmly believed they 'had no more right to pontificate about morals than anyone else'.[17]

This standpoint clearly frustrated Mary Warnock. In a 1960 book on *Ethics since 1900*, she complained that 'the concentration upon the most general kind of evaluative language, combined with the fear of committing the naturalistic fallacy, has led too often to discussions of grading fruit, or choosing fictitious games equipment, and ethics as a serious subject has been left further and further behind'.[18] But she closed the book on a cautiously optimistic note and claimed that 'the most boring days are over'.[19] Warnock drew encouragement from the work of the Oxford philosopher Philippa Foot, who published several papers in the late 1950s that criticised Hare's belief that a moral argument was both prescriptive and universalisable. Foot claimed, by contrast, that an argument could only be shown to be moral on the grounds that following or ignoring it entailed concrete benefits or harms to people.[20]

Foot also countered G. E. Moore's naturalistic fallacy by arguing that descriptive premises counted as evidence for normative conclusions, since notions such as 'good' could not be separated from the benefits or harms they produced in specific contexts. She believed morality was not simply a matter of choice and that 'man can no more decide for himself what is evidence of rightness or wrongness than he can decide what is evidence for monetary inflation or a tumour on the brain'.[21] Warnock recalled that Foot's conclusions freed philosophers 'from the restrictions of the so-called naturalistic fallacy … At last, the absolute barrier erected between fact and value had been breached, and moral realism began to be sniffed in the air.'[22]

Warnock believed that moral philosophy could now incorporate 'both description of the complexities of actual choices and actual decisions, and also discussion of what would count as reasons for making this or that decision'.[23] Her optimism was vindicated in the 1960s and 1970s, as a growing number of philosophers began to 'look at real moral problems, rather than the words or forms in which these problems are discussed'.[24] The subject matter of this work highlighted the influence of contemporary events. For example, when Harold Wilson's Labour government passed its Abortion Act in 1967, Foot wrote an article that considered different instances when abortion might be considered permissible.[25]

But this paper was something of a novelty. While issues such as IVF, euthanasia and organ transplantation were increasingly discussed in 'trans-disciplinary' groups during the 1960s, most of the philosophers who looked at practical issues did not consider medical ethics to be an important topic.[26] They were more concerned with political issues, including the ethics of the Vietnam war, student protests in the United States, France and Britain, and the ongoing campaigns against nuclear and chemical weapons. This was certainly the case in Oxford, which became central to the growth of 'applied ethics' despite the scepticism of some senior staff.[27] Here, young fellows such as Jonathan Glover worked on the morality of arguments relating to acts and omissions, investigating whether the belief that 'it makes no difference whether or not I do it' could justify developing weapons or selling arms to South Africa.[28] Under the supervision of Hare and Ronald Dworkin, students such as Peter Singer and John Harris wrote PhDs on political violence and civil disobedience. And the increasingly practical interests of PhD and undergraduate students, in turn, also encouraged senior figures such as Hare to write on the morality of subjects such as war and slavery.[29]

Despite her enthusiasm for practical philosophy, Mary Warnock left Oxford just as this approach was making inroads. While she enjoyed teaching philosophy, Warnock considered herself an 'entirely unoriginal thinker' and 'not much good at the subject'.[30] Having come to believe that her 'natural habitat was school, not university', she accepted the position of headmistress at the private Oxford School for Girls in 1966.[31] Although she enjoyed working at the school, Warnock returned to the University of Oxford in the summer of 1972, after Geoffrey Warnock was elected Principal of Hertford College. She noticed that medicine now featured more prominently in applied ethics, with students on a philosophy and theology degree encouraged to discuss issues such as euthanasia and abortion.[32]

By this point, philosophers such as Glover, Singer, Harris and Hare had also begun to write on medical and scientific ethics, extending their prior work on acts and omissions and the moral implications of violence and killing. Singer looked at animal experiments in essays and in his book *Animal Liberation*, in which he drew on civil rights campaigns to propose that 'we extend to other species the basic principle of equality that most of us recognise should be extended to members of our own species'.[33] After claiming that harming animals on account of their presumed inferiority was a form of discrimination known as 'speciesism', Singer argued that animals should be given equal consideration to humans on account of their capacity to suffer. Harris's work on whether people were causally responsible for harm they could have prevented, meanwhile, led him to write on a hypothetical 'survival lottery' and to ask if it was morally permissible to let two patients in need of organs die when one healthy person could be killed to save both their lives.[34]

Richard Hare's 1975 paper on 'Abortion and the Golden Rule' set out his own ambitions for applied ethics, which underpinned his later criticism of Mary Warnock. Hare argued that philosophers wasted time discussing the 'rights' of the foetus and its mother, or whether or not the foetus was a person, since these were not 'fully determinate' concepts and only served to complicate debates.[35] He believed the central question hinged instead on the 'golden rule' whereby: 'If we are glad that no-one terminated the pregnancy that resulted in *our* birth, then we are enjoined not, *ceteris paribus*, to terminate any pregnancy which will result in the birth of a person having a life like ours.'[36]

Hare did not support a pro- or anti-abortion stance here, but was instead appealing for public debates to be grounded in a thorough 'study of moral concepts and their logical properties'. Seeking to distance himself from the subjective view of ethics promoted by Ayer, he declared that appeals to intuition or sentiment were as fruitless as appeals to rights and notions of personhood. All they did, he argued, was highlight an individual's or group's prejudices without 'telling which prejudices ought to be abandoned'.[37] Hare believed that philosophers should instead enable scientists, politicians and the public to reach clear answers by using moral frameworks such as utilitarianism to show 'which are good and bad arguments'.[38]

Hare also promoted the benefits of philosophy in the growing number of interdisciplinary publications and symposia concerned with medical ethics. In a 1977 book on *Philosophical Medical Ethics*, he argued that if a philosopher could not help doctors to understand and resolve ethical dilemmas, 'then he ought to shut up shop'.[39] 'A failure here', he continued, 'really would be a sign of either the uselessness of the discipline or the failure of the particular practitioner.' During 1981 Hare joined Jonathan Glover in teaching an LMG course on 'An Introduction to Ethics', while John Harris regularly spoke to the Manchester Medical Group following his appointment as lecturer in the philosophy of education.[40] From its first edition, the *Journal of Medical Ethics* also contained a series of regular articles on the 'introduction to ethical concepts', where philosophers such as Robin Downie, from the University of Glasgow, claimed that doctors would be better equipped to deal with an ethical issue 'if they have some theoretical grasp of the principles underlying it'.[41]

Mary Warnock clearly took heart from the growth of what Singer and others now called 'applied ethics'. In the afterword to a 1978 edition of *Ethics since 1900*, she wrote that philosophy was now becoming 'a practical subject, and therefore more urgent and interesting'.[42] By this point Warnock was engaging with practical issues herself. In 1974 Margaret Thatcher, then Secretary of State for Education, asked her to chair an inquiry into teaching children with special educational needs.[43] And in 1977 Warnock was given the chance to engage with a more contentious issue when the Labour Home Secretary, Merlyn Rees, asked her to join a newly reconstituted Home Office advisory committee on the administration of the 1876 Cruelty to Animals Act.

Like many aspects of biomedical research, animal experiments became increasingly controversial during the 1970s. This stemmed largely from a public outcry over the use of dogs in smoking research, a growing belief that animal tests should be replaced by 'humane' alternatives such as tissue culture, and a focus on the 'rights' of laboratory animals.[44] While they had barely criticised animal experiments for most of the twentieth century, newspapers and some politicians now called for stricter legislation and condemned scientists for performing vivisection when alternatives existed. James Callaghan's Labour government responded to this controversy by issuing a charter for animal protection, entitled *Living Without Cruelty*, and pledging to reduce the number of animal tests.[45] In line with its belief that different stakeholders should have a say in the development of public policies, it also recruited greater numbers of lay people to the Home

Office advisory committee that had comprised one lawyer and ten scientists since 1913.[46]

Warnock believed that this appointment allowed her to tackle the longstanding philosophical question of 'how we ought to behave toward the natural world', including when people were justified in utilising animals for their own ends.[47] Her first task was to help the committee investigate the LD50 test, which scientists used to evaluate the single dose of a compound needed to kill 50 per cent of a given animal population (predominantly rats and mice). Animal welfare groups claimed that this test was crude, wasteful and unnecessarily cruel, since there was no limit to the maximum dose scientists could administer.[48]

When the committee's report was issued in 1979, it discussed broad questions of when it was ethically acceptable for humans to use or inflict pain on animals. The committee argued that inflicting pain on animals only amounted to cruelty when it was 'not compensated by the consequential good'.[49] To the disappointment of animal rights campaigners, they continued that: 'In applying this criterion, there must be assumed a presumption in favour of humans over animals. We believe that while it is not legitimate to use one human being, without his consent, as a means to an end, it is, within limits, legitimate to use animals for human ends.'[50] The committee argued that the use of animals was acceptable where the envisaged human benefit was 'a serious and necessary one, not a frivolous or dispensable one'.[51] While they acknowledged that the LD50 test caused 'appreciable harm' to a large proportion of experimental animals, the committee argued it was nevertheless essential 'for the proper testing of new substances'.[52]

These recommendations reflected and were partly influenced by Warnock's own views. Although she believed that people should treat animals humanely, she argued that: 'Speciesism is not the name of a prejudice which we should try to wipe out. It is not a kind of injustice. It is a natural consequence of the way that we and our ancestors have established the institution of society.'[53] Shortly after the Conservatives won the 1979 election, the new Home Secretary, William Whitelaw, asked Warnock to chair a restructured Home Office advisory committee on animal experiments. In addition to Warnock, the new committee included five scientists, two veterinarians, one experimental psychologist, two laypeople and two representatives of animal welfare groups.[54] It also had a broader remit than previous committees, and was charged with recommending new legislation for animal experiments. The committee's report, which was published in 1981, reflected Warnock's belief that oversight of science was vital to both safeguarding research and maintaining public confidence.[55] Its main recommendation involved the formation of an advisory committee to complement the Home Office inspectorate, which would act as 'a detached observer in touch with both the concerns of the public and the legitimate requirements of science and industry'.[56]

In a 1980 article for the *Journal of Medical Ethics*, the Oxford philosopher Michael Lockwood surveyed the increasing engagement with practical matters and claimed that 'what were once questions largely for philosophical debate' now turned 'with bewildering rapidity into matters of widespread concern'.[57] But despite the growth of applied ethics, philosophers were not yet considered public authorities on medical ethics to the same extent as doctors, scientists or legal 'outsiders' such as Ian Kennedy. Their discussion of issues such as abortion and vivisection took place in academic journals, conferences or government committees, and not in newspapers, on the radio or television. Although Mary Warnock was the most high-profile 'applied' philosopher at the start of the 1980s, this reveals what Stefan Collini calls an 'intriguing disjunction between professional and public standings'.[58] Warnock was publicly known for chairing the inquiry into special educational needs and writing for publications such as the *Times Literary Supplement*; but no newspapers covered her appointment as chair of the advisory committee on animal experiments, and she did not comment publicly on its work or recommendations.[59]

Philosophers first entered public debates on medical ethics in Britain following the 1981 trial of Dr Leonard Arthur, who had ordered 'nursing care only' and prescribed a course of strong sedatives to a newborn baby with Down's Syndrome after his parents had indicated that they did not wish him to survive. The baby contracted pneumonia and died shortly afterwards, prompting a member of hospital staff to contact the anti-abortion group LIFE, which then informed the police. The prosecution argued that Arthur had used drugs to intentionally kill the baby, while the defence argued that he had conformed with standard practice for severely disabled babies of waiting to see if they would survive or whether 'nature would take its course'.[60] In summing up the evidence for the jury, the judge distinguished between doing something active to kill a child and electing not to follow a particular course of action that might have saved it. He reminded them that the former was unlawful, and the latter lawful. The jury, who also learned that the child suffered from serious heart and lung problems, acquitted Arthur on 5 November 1981.[61]

Although LIFE attacked the verdict, many journalists and doctors greeted it as vindication for a conscientious doctor who had acted 'within the professionally accepted limits of paediatric practice'.[62] Arthur also received public support from Jonathan Glover, who wrote in the *London Review of Books* that 'a verdict of guilty would have been a morally undeserved calamity'.[63] Glover used the Arthur case to reiterate the main points of his 1977 book *Causing Death and Saving Lives*, exploring the moral implications of non-treatment and promoting the benefits of 'applied ethics'. He stressed that deciding whether or not to treat disabled babies was 'not simply a legal or medical matter', but was firmly linked to philosophical discussions on when it was acceptable to kill or let someone die. 'The conventional view that philosophical discussions are quite remote from having any practical upshot', Glover stated, 'has very little to be said for it.'[64]

Glover argued that philosophers provided vital clarity to debates on the non-treatment of disabled babies by helping identify the relative strengths and weaknesses of the moral principles that various groups evoked to support their position. While he acknowledged there were 'unresolved problems' surrounding arguments in favour of non-treatment, he believed that they nevertheless outweighed the arguments of groups such as LIFE.[65] Glover stated that if a 'right to life' entailed a right to treatment, then opponents of non-treatment would also have to take a stand on the deaths of children that resulted from a shortage of organs or government cuts to hospital services. Until they did, he continued, it was difficult to frame the right to life as superior to the belief that non-treatment was a humane course of action for parents and 'the baby facing a terrible life'.[66] This led Glover to endorse a situation in which the authorities indicated that 'they will not bring prosecutions where parents and doctors allow severely handicapped newborn babies to die'.[67]

Leonard Arthur also received public support from a more unlikely figure. In a long piece for *The Times*, A. J. Ayer argued that severely handicapped babies who had been rejected by their parents should 'not be suffered to live', and that doctors who discontinued treatment 'ought not to be morally or legally condemned'.[68] Ayer argued that he was not simply giving his personal views on current affairs, as he often did in newspapers and on television, but was using his professional capacity as a philosopher to analyse the 'moral questions which the trial of Dr Arthur and similar cases pose'.[69] This reflected his changing opinion on philosophy's relevance to public affairs. Following his retirement from UCL in 1978, Ayer had taken up a visiting professorship at the philosophy department at the University of Surrey, which was run by a former student, Brenda Cohen, and concentrated on applied ethics. Perhaps because of his new surroundings, Ayer now criticised the 'rather insular position' he had adopted for much of the twentieth century and claimed that philosophers had a role to play in public affairs.[70] Ayer's changed attitude was illustrated by his support for a new Society for Applied Philosophy, which Cohen and Anthony O'Hear founded in 1982 to 'provide a focus for philosophical research with a direct bearing on areas of practical concern'.[71] His encouragement for this new venture was rewarded when Cohen made Ayer the society's inaugural president – to the consternation of Richard Hare, who protested that in 'Ayer's rare incursions into practical ethics, he does no more than tell us what he thinks without reasons beyond the appeal to the authority of B. Russell'.[72]

But Glover and Ayer also conceded that there were limits to what philosophy could contribute to ethical debates. In the *London Review of Books*, Glover claimed that this stemmed from the fact that disputes on issues such as treating disabled babies 'reflect much wider disagreements about morality'.[73] He characterised British society as split between a declining but still powerful 'morality derived from religious commands and prohibitions' and 'consequentialist views, such as utilitarianism, which also vie with rights theories, and with agent-centred views stressing purity of motive and character'. Deriving clear answers was often difficult, Glover argued, because 'many people have an eclectic mixture of these, with no general agreement on the criteria to be used in moral debate'.[74] Glover stated that even if philosophers highlighted what were good or bad arguments, as Hare had recommended, 'no claim is made that moral argument can establish general principles which any rational and informed person can accept'.[75] Since different people had deeply held views that were often incompatible with opposing beliefs, 'it would be possible for someone to accept all the arguments put forward [by a philosopher] and yet legitimately to reject many of the conclusions because they do not accept the premises'. This meant, Glover concluded, that 'there is always the possibility, and sometimes the reality, of ultimate disagreement'.[76]

Ayer also doubted whether applied ethics could provide clear answers. Although he moderated his subjective view of ethics during the 1980s, Ayer still denied that morality could be reduced to a set of universally binding principles.[77] This was evident in his *Times* article on the Arthur case, where he contended that 'no moral judgement can be founded on authority'.[78] While he recognised that religion or secular frameworks such as utilitarianism were legitimate sources of morality, Ayer claimed that 'the fact remains that one still has to make the independent judgement that what the authority enjoins in this case is right'.[79] He believed that philosophers could not hope to provide widely accepted answers, since individuals who held religious premises would reject arguments made on utilitarian grounds, whatever their validity, and vice versa. This meant there were no obviously 'correct' solutions to issues such as abortion or euthanasia, and that when it came to making recommendations it was 'to some extent arbitrary where one draws the line'.[80]

By 1982, then, 'applied ethics' had become a recognised branch of philosophy in a relatively short space of time. It had a dedicated society, and philosophers now discussed practical issues in academic journals, symposia, government committees and increasingly in public. But while many philosophers agreed that 'applied ethics' had a role to play in public affairs, they disagreed on its scope and importance. Some, such as Richard Hare, believed that they could reconcile differing groups by highlighting 'good or bad arguments' and evoking principles 'which we ought all to try to preserve'.[81] Supporters of this view believed that the philosopher was a 'specialist in ethics' or a 'moral expert', who provided clear answers that appealed to any rational person.[82]

But others, including A. J. Ayer, argued that while philosophers could help clarify the values that underpinned a particular standpoint, there was no rational way of reconciling them with opposing standpoints. While this was a largely academic debate during the 1970s, it played out in public to a small extent following the trial of Leonard Arthur in 1981. However, it soon received far greater coverage thanks to the controversy surrounding IVF and embryo experiments, which emerged in 1982 and lasted for the rest of the decade. This controversy gave 'applied ethicists' an unprecedented opportunity to publicly engage with topics that many now included under the heading of 'bioethics', and prompted questions about exactly what form this new approach should take.[83]

'We must ALL have a say on test tube babies': outside involvement with IVF and embryo research[84]

Despite a brief flurry of newspaper articles that discussed their eugenic implications, IVF and embryo research received little criticism in Britain during the 1970s. Religious figures claimed to see 'no problem', provided that scientists treated embryos responsibly, while politicians supported the use of IVF by married couples 'who could not have a baby any other way'.[85] The main concerns were raised by fellow scientists and hinged on the possibility that IVF might produce abnormalities in developing embryos.[86] There was certainly nothing like the ethical scrutiny and criticism found in the United States, where bioethicists such as Paul Ramsey criticised Edwards and Steptoe and urged Congress to ban IVF. As the *British Medical Journal* noted, thanks to the 'flowering of American bioethics' the major question here was not whether a baby conceived *in vitro* 'would be a girl or a boy, but whether its presumably unprecedented manner of coming into being is ethical'.[87]

British newspapers also highlighted the differing transatlantic attitudes to IVF. When Louise Brown, the first 'test-tube baby', was born in Oldham, Greater Manchester, on 25 July 1978, the *Guardian* noted how Britain lacked the 'moral and ethical outrage' that characterised American debates.[88] Newspapers greeted Louise Brown as the 'Baby of the Century' and claimed, like Edwards, that IVF was a valuable medical technique.[89] A long *Observer* report, which again contrasted British and American attitudes, argued that any misgivings were likely to have been 'softened somewhat by the pictures of Mrs Brown's obviously normal baby'. Claiming that IVF was no more 'unnatural' than using hormones to stimulate ovulation, the paper predicted that if it 'can be proven to be safe, reliable and free of complications, then it will join those other medical techniques which have helped thousands of women become mothers – and men to become fathers'.[90]

But the British enthusiasm for IVF soon evaporated. By 1982

newspapers claimed it raised troubling questions that the *Daily Express* called 'the aberrations of the baby revolution'.[91] Journalists now dwelt less on the benefits to infertile couples and criticised the possible use of IVF by single women or unmarried couples, implanting multiple embryos in a single pregnancy, paying women to act as commercial surrogates and using embryos in research. The *Daily Mail*, which had greeted Louise Brown's birth as a 'miracle', withdrew the money it had pledged for the private IVF clinic that Edwards and Steptoe were building at Bourne Hall, Cambridge.[92] And political figures, such as the Conservative peer Lord Campbell, now predicted that IVF would 'imperil the dignity of the human race, threaten the welfare of children, and destroy the sanctity of family life'.[93]

Michael Mulkay argues that this change can be partly explained by the renewed emphasis on 'traditional' morals that followed the 1979 election of Margaret Thatcher's Conservatives.[94] Keen for a 'return to Victorian values', members of her government spoke regularly about the need to reaffirm social principles undermined by the 'permissive' Bills on homosexuality, abortion and capital punishment that were passed during the 1960s. Anti-abortion organisations such as LIFE and the SPUC, which remained marginal while the Conservatives were in opposition, now had the opportunity to alter the 'rhetoric of British political life'.[95]

These changes certainly helped transform attitudes to IVF. Whereas Edwards and Steptoe presented the recipients of IVF as married couples during the 1970s, supporters of 'traditional' morals now framed IVF as a potent threat to the nuclear family. They claimed that there was nothing to prevent single women from having multiple embryos implanted in one IVF cycle, or homosexual couples from paying third parties to act as commercial surrogates.[96] Some even foretold an 'Oedipus tragedy', in which a fertilised embryo was implanted into a surrogate and, years later, grew into an adult who unwittingly married their biological mother.[97] Although the lawyer Geoffrey Robertson was not an advocate of 'traditional' morals, he highlighted this as a potential problem in the *Observer* and argued that such children should be entitled to discover the identity of their genetic parents 'for the better avoidance of incest'.[98]

But research on embryos *in vitro* undoubtedly became the most controversial issue. In a television documentary screened in

February 1982, Robert Edwards admitted to experimenting on embryos that he had no intention of implanting into patients and claimed 'these spare embryos can be very useful … they can teach us things about early human life'.[99] The immensely critical response to this admission indicated just how much opinion had swung against IVF – especially when we consider that embryo experiments had prompted little controversy when either Edwards or Gordon Dunstan discussed them in the 1970s. Within a week, the Labour politicians Leo Abse and Gwyneth Dunwoody had urged the Prime Minister to establish an 'urgent' inquiry.[100] A *Times* editorial argued that embryos 'ought not to be regarded as dispensable matter' and joined calls for an inquiry into 'which of the many strange possibilities now opening up are acceptable, which need controls, and which are unacceptable'.[101] When Edwards again admitted that he had experimented on fifteen 'spare' embryos, representatives from LIFE demanded that he be immediately prosecuted and warned that 'unless test tube technology is brought under immediate control, we will find that manipulation of life on a horrifying scale has overtaken us'.[102]

While it was clearly a significant factor, the resurgence of 'traditional' morals only partly explains why IVF and embryo research became so contentious in the 1980s. Criticism of these practices also reflected, and bolstered, growing calls for external involvement with scientific and medical ethics. While Ian Kennedy used a range of examples to endorse 'bioethics' during his Reith Lectures, calls for outside regulation centred almost exclusively on IVF from 1982 onwards. Leo Abse, for one, claimed that the establishment of inquiries into IVF by the MRC, the BMA, and the Royal College of Obstetricians and Gynaecologists was insufficient as the procedure raised ethical questions 'too enormous to be left to doctors'.[103] The only adequate solution, he argued in Parliament, was to convene an 'inter-departmental and inter-disciplinary inquiry'.[104]

In her *Observer* column, the journalist Katherine Whitehorn similarly criticised Edwards for protesting that 'he didn't understand what the fuss was about'. She argued that his indifference highlighted the inadequacies of self-regulation and showed that 'if ever there was a case where it shouldn't be left to doctors and scientists alone, where society ought to have a say, as Ian Kennedy insisted in last year's Reith Lectures, this is it'.[105] In a letter to *The Times*, Kennedy used the controversy over embryo experiments to reiterate that decisions 'cannot be left simply to one professional group, whether doctors, lawyers or whatever'.[106] Geoffrey Robertson also claimed that the backlash against IVF 'shows the threadbareness of the claim "leave it to the professionals"', and called for an inquiry whose members were 'representative of the entire people'.[107]

These demands resonated with the Conservative enthusiasm for outside scrutiny and regulation of professions. It is no surprise, then, that when ministers responded to 'repeated calls [from] both inside and outside Parliament' and decided to hold an inquiry into IVF in April 1982, they stressed that it should involve members of several professions and at least 'four or five non-experts'.[108] As civil servants at the Department of Education and Science (DES) noted in correspondence to the MRC, ministers felt that there was 'a strong case' for a diverse committee, because none of the professional inquiries were 'sufficiently broadly based or sufficiently representative to be regarded as a source of authoritative advice to Government'.[109]

Civil servants prioritised the appointment of 'an outside chairman' once they began to organise the inquiry.[110] During April and May, officials at the DES and DHSS, which co-sponsored the inquiry, suggested possible chairs with no connection to IVF or reproductive medicine. Although their initial preference was for a legal chair, and 'perhaps a judge from the family division', they eventually settled on four candidates.[111] These were Sir Norman Lindop, an osteopath; James Sutherland, a specialist in commercial law; Lady Gillian Wagner, head of the children's charity Dr Barnardo's; and Mary Warnock, then a senior research fellow at St Hugh's College, Oxford, and still chair of the Home Office advisory committee on animal experiments.[112]

Warnock was the favoured candidate from the outset, with civil servants identifying her as 'very well qualified for the job'.[113] But these qualifications only stemmed partly from her 'non-expert' status. Warnock was a typical member of the so-called 'Great and Good' whom politicians and civil servants regularly turned to when selecting public inquiries; she was Oxbridge educated, well-connected and had proved her reliability on previous inquiries. In some respects, the decision to appoint Warnock as chair embodied the longstanding Whitehall belief that eminent and non-partisan figures were critical sources of policy advice, which persisted during the 1980s despite Margaret Thatcher's disdain for the 'Great and Good'.[114] Newspapers that profiled Warnock following her appointment certainly highlighted this background, portraying her as 'a well-seasoned member of the great and good who fight honourable battles in the committees and quangos of British public life'.[115]

It was also nothing new for politicians to select an educationalist or philosopher to chair an inquiry into a moral issue, as Kenneth Wolfenden and Bernard Williams had led inquiries into the legalisation of homosexuality and the regulation of pornography during the 1950s and 1970s.[116] But the undoubted novelty of Warnock's appointment lay in the fact that a philosopher now led a public inquiry into science and medicine, where doctors and scientists had long been recognised 'as key holders of expertise'.[117] This represents a subtle but important change in British politics. Although the government still looked to Establishment figures for regulatory advice, they now sought individuals who had no connection to the profession or field under scrutiny.

This preference for 'non-experts' provides an example of how the Conservatives sought to break from the form of government associated with the welfare state, in which politicians had believed that professional expertise was vital to the development of public policy.[118] Thanks to their distrust of self-regulation and enthusiasm for public accountability, the Conservatives believed that policy should now be shaped by 'outsiders' who functioned as a proxy for different stakeholder and consumer interests. Warnock acknowledged this herself when she claimed that her appointment demonstrated how 'politics had entered medical ethics' during the 1980s. She argued that declining trust in professional expertise transformed what were once 'matters of professional behaviour' into 'questions of public policy, which merit public discussion and therefore, because we are a democratic society, ultimate discussion in Parliament'.[119]

In June 1982 Warnock received a letter from Norman Fowler asking her to chair the government's inquiry into human fertilisation and embryology. She was initially hesitant, because of a heavy teaching load at Oxford, her workload as chair of the animals advisory committee and fears that the inquiry would bring unwelcome publicity.[120] But she accepted the invitation after realising that it presented her with the chance to engage with two philosophical issues. The first involved questions surrounding the relationship between morality and the law, which had interested her following Wolfenden's report on homosexuality and a 1978 book on *Public and Private Morality*.[121] The second was the longstanding question of when in development a human embryo attained moral status, which dated back to Aristotle's work on embryogenesis and 'seemed to many people to raise fundamental questions about the nature and value of human life'.[122]

After agreeing to chair the inquiry, Warnock met Norman Fowler and civil servants at the DHSS in order to select the other members. This resulted in the appointment of seven doctors and scientists, with different religious backgrounds, and eight individuals from other professions, including a solicitor, a court recorder, two social workers, two managers of a healthcare trust, a theologian and the vice-president of the UK Immigrants Advice Service. Conservative politicians dwelt on the fact that members of other professions outnumbered doctors and scientists when they discussed the inquiry. During a Commons statement that announced its formation, in July 1982, Norman Fowler distinguished it from the 'examinations already underway by medical bodies', dwelling on its 'broad-based' membership and stressing that it would 'hear from many lay and religious viewpoints'.[123]

Like Fowler, the committee also viewed their diverse backgrounds as a means of ensuring public accountability. During their second meeting, in December 1982, they criticised representatives from the MRC for only having one non-scientist on their inquiry and noted that the individual in question, the Bishop of Durham, used to be a scientist anyway. They claimed that this would simply increase distrust since 'it might be seen by the public as a situation when scientists who had an interest in this research quite naturally gave it their approval'.[124] In these early meetings, committee members also endorsed Fowler's claim that opinions on IVF and embryo research should be sought from a 'wide range of interested bodies'.[125] They invited written or oral evidence from over three hundred organisations and individuals, including scientists such as Robert Edwards; anti-abortion, family planning and feminist groups; marriage counsellors and adoption agencies; university departments such as law, theology and medicine; and representatives of all the major religious denominations.[126] They also sought the views of Ian Kennedy, who praised Warnock's appointment as 'evidence that progress along the lines I advocate has recently been made'.[127]

The amount of evidence that it was due to hear and the range of issues it considered, including IVF, donor insemination, egg donation, surrogacy and embryo experiments, meant that Warnock's committee was not due to publish its recommendations until 1984. In the meantime, newspaper reports continued to demand greater public scrutiny and influence over IVF. An *Observer* editorial stated that test-tube babies were 'now a public subject' and claimed that if scientists were allowed to proceed unchecked, 'then we can hardly complain at the lack of faith shown by the public [in science]'.[128] And a *Mail on Sunday* editorial, entitled 'why we must ALL have a say on test tube babies', similarly argued that 'the time has come for the public to be involved in the decisions which are being made in the laboratory'.[129]

At the same time Warnock became a public advocate of external oversight herself. In a 1983 edition of the *Philosophical Quarterly*, she argued that her committee's main priority was to ensure that discussion and even regulation of IVF 'be taken, not in the private, but in the public sphere'.[130] The only way to render IVF publicly accountable, she claimed, was to establish a 'system of surveillance' that ensured that it was 'constantly watched, not merely by the medical profession and the research biologists, but the lay as well'.[131]

In several newspaper interviews published before the committee's report was issued in July 1984, Warnock claimed that their key recommendation hinged on the establishment of 'a monitoring body to keep all innovations and technical developments under constant review'.[132] These articles and interviews clearly highlighted Warnock's enthusiasm for external oversight, which she had acquired during her spell on the animals advisory committee.[133] They also demonstrate how she became publicly synonymous with 'applied ethics' following her appointment as chair of the IVF inquiry, with one journalist detailing how 'her influence runs deeper than the usual philosophy don's, as a moulder of moral policy for generations yet unborn by methods as yet not fully explored or developed'.[134]

Warnock's promotion of oversight, and public demands for an external 'watchdog', clearly influenced the committee's thinking.[135] The minutes from a 1983 meeting show that members voiced 'scepticism ... about the effectiveness of self-monitoring by doctors and a strong desire that, especially on sensitive issues, there should be a mechanism for reflecting the views of the general public'.[136] This belief underpinned large sections of the committee's report when it was published in July 1984. The committee claimed that the formation of a supervisory body was the 'most urgent' of their sixty-four recommendations and stressed that it must not be 'exclusively, or even primarily, a medical or scientific body'. In order to ensure that it would not be 'unduly influenced' by professional interests, they proposed that it should have a wide-ranging membership and, crucially, 'that the chairman must be a layperson'.[137] Warnock justified this proposal in the *New Scientist* by framing the public as increasingly empowered stakeholders in science and medicine. She claimed that when research raised a moral dilemma:

> there is no reason why scientists should be responsible by themselves for solving it ... A society in which what might or might not be done was decided solely by those committed to the advance of knowledge would not be acceptable to those of us who are not scientists. There are other values to be considered. *Increasingly, and rightly, people who are not experts expect, as of right, to help determine what is or is not a tolerable society to live in.*[138]

Her calls for outside involvement distinguished Warnock from other British philosophers involved with applied ethics, who clarified the moral aspects of specific practices but rarely, if ever, claimed that 'outsiders' should help establish professional standards. But this did not mean that she was a radical critic of science. Like Ian Kennedy, Warnock regularly stressed that outside involvement would benefit scientists and doctors. She argued that it would safeguard public and political trust by ensuring 'that no nameless horrors were going on, hidden away in laboratories', and that this would allow scientists 'to get on with their work, without the fear of private prosecutions, or disruption by those who object to what they are doing'.[139]

Biomedical journals endorsed this positive view of external oversight. A 1983 editorial in *Nature* argued that it would help make IVF 'socially palatable' and supported Warnock's call for a statutory body that would 'exert a supervisory influence, consider difficult questions as they arise, and keep the general public informed'.[140] Following the publication of her committee's report, the *British Medical Journal* similarly claimed that scientists 'will welcome the suggestion that a new licensing authority should be set up to regulate infertility services, monitor new developments, and vet individual research projects'.[141] And a *New Scientist* editorial also welcomed the prospect of oversight when it asserted that: 'Science policy should not be left entirely in the hands of scientists and matters as important as those addressed by the Warnock committee *should* remain under public scrutiny and regulation.'[142]

While Warnock and Kennedy both echoed the government's enthusiasm for 'non-experts', their differing political affiliations show how support for bioethics had broad social and political origins. While Kennedy was influenced by civil rights campaigns and a leftist critique of authority, Warnock confessed to being a 'dripping wet Conservative'.[143] She recalled that

> from the age of about fifteen, I knew I was a natural Tory. All my instincts and loves were Trollopian. I loved the thought of a landed aristocracy … I loved hunting; I loved time-honoured hierarchies; I loved cathedrals. I wanted to become an old-fashioned scholar. Nothing could be further from the politics of the Left.[144]

Perhaps more strikingly, Warnock may also have played a role in encouraging the Conservatives to adopt neo-liberal ideologies. Like many members of the 'Great and Good', she moved in the same social circles as senior politicians. When they were in opposition during the late 1970s, Keith Joseph asked Warnock for help in finding a worldview that would help the Conservatives 'justify the overriding value of individual choice and minimise the power of the state'.[145] She recommended that Joseph and colleagues read Robert Nozick's 1974 book *Anarchy, State and Utopia*, which argued that governments should only concern themselves with protecting citizens against force, theft and fraud, leaving everything else to individual choice and market forces. Warnock recalls that Joseph 'certainly read this book' and passed it on to other senior Conservatives.[146] Although there is no evidence that Margaret Thatcher read Nozick's book herself, debates on embryo research demonstrate that she and Warnock had similar views on ethics and individual choice.

Embryo research and 'moral experts'

While the Warnock committee, the press, scientists and politicians all agreed on the need for external oversight, there was less consensus when it came to deciding specific policies for embryo research.

By 1983 Warnock and her committee had identified embryo experimentation as 'the most significant of the moral problems posed by *in vitro* fertilization techniques'.[147] This was partly due to the fact that it touched on difficult questions about life before birth, including when in development embryos began to deserve legal protection. But as Warnock was finding, it also stemmed from the fact that supporters and opponents of research both mobilised equally valid but incompatible claims to support their case.

As the *Daily Telegraph* outlined in 1983, the Warnock committee faced an 'ethical log jam of conflicting evidence' when they came to consider embryo research.[148] They heard strong support from the Royal Society, the BMA and the MRC, who argued that embryo experiments were vital to understanding development and improving IVF techniques. These groups agreed that work should be permitted up to an agreed cut-off point, corresponding either to the beginning of implantation *in utero* at around five days, the end of implantation at eleven days, the point where cell differentiation began at around fourteen days, or the point where the nervous system began to form between seventeen and twenty-three days. Supporters of research argued that experiments were justified before these points as the embryo equated to little more than a bundle of cells and was 'not recognizably human'.[149]

However, a committee memo noted that this 'essentially pragmatic and utilitarian' stance ran counter to the 'substantial body of opinion which is opposed to the use of embryos in any circumstance'.[150] Individuals such as the Chief Rabbi, Immanuel Jacobovits, told the committee that 'upholding the sanctity of life' from conception onwards was 'an overriding moral imperative' that outweighed utilitarian considerations.[151] A delegation of Catholic bishops similarly argued that permitting embryo research 'involved people sitting in judgement on another's life and treating that life as a mere means to an end, which undermined the basic dignity of human beings'.[152] Opponents of research, which also included anti-abortion groups and the Women's Institute, notably stressed that their stance was not anti-science or based simply on religious dogma. For the Guild of Catholic Doctors, it was supported by the fact that 'as any microgeneticist will tell you, whether or not more individuals result, the genetic coding is laid down on fertilization and [is] discernable as human on the first mitosis'.[153]

The seemingly irreconcilable nature of this division was captured by representatives from the Family Planning Association, who told the committee that 'debate continues without advancing, largely due to the fact that we lack any common basis for resolving these ethical issues of what can be done to the embryo, and we do not know who should decide, and on what grounds, where the ethical line should be drawn'.[154] When the committee began to draft their recommendations later in 1983, Warnock tried to bypass this problem by persuading members that the 'central issue' was not 'when does life begin', which she argued was 'a matter of belief as much as of science'.[155] It hinged instead, she claimed, on using scientific evidence to determine when in development embryos should be accorded moral status and legal protection.[156]

Yet this move offered little resolution, and remained as much a 'matter of belief as of science'. The fact that scientists disagreed on when to implement any cut-off indicated that there were several potential stages when an embryo could be afforded moral status and legal protection.[157] No one stage appeared objectively more significant than another, and a person's moral preferences still conditioned where they believed the line should be drawn.

Perhaps unsurprisingly, given the lack of scientific clarity and their 'broad-based' membership, there was little consensus when the committee attempted to draw up proposals for embryo experiments. Three members argued that since scientific observation and philosophical reflection 'cannot answer' the question of when an embryo attained moral status, it should be protected from fertilisation onwards 'because of its potential for development to a stage at which everyone would accord it the status of a human being'.[158] While the majority believed that experiments were essential for understanding development and overcoming infertility, they disagreed on where to draw the cut-off and whether embryos should be created specifically for research.

This division became public early in 1984, when an unnamed committee member broke ranks and spoke to journalists.[159] A lengthy *Times* report subsequently detailed 'growing concern among members that they will be unable to produce a unanimous view at a critical time'. It argued that these differences had placed the committee 'in a quandary' and ensured that 'the Government is likely to face serious difficulties in deciding on controls over test tube baby developments and research on human embryos'.[160]

Following these leaks, Warnock admitted to newspapers that

there was 'no compromise between the view that no research at all should be done on embryos and the scientists' views that it should and must'.[161] Like Glover and Ayer, she argued that the dispute hinged on irreconcilable premises that could not be settled by either philosophical argument or scientific evidence. 'Neither side', she wrote in the *New Scientist*, 'can be disregarded. Neither party can say of the other that it is mad or stupid or frivolous ... No accusations of ignorance or prejudice will dissipate such strongly held beliefs.'[162]

While the Warnock committee was disagreeing over embryo research, Robert Edwards used the prospect of moral disagreement to revive his opposition to outside involvement with science and medicine. In a *Horizon* lecture for the BBC, he argued that it was unrealistic to expect 'philosophers, theologians and lawyers' to provide guidance for IVF and embryo experiments, since this 'search for advice, for leadership from the traditional purveyors of moral standards, usually ends in frustration'.[163] Edwards outlined how there was 'great confusion between the religions of the world. The Roman Catholics are absolutists, stressing that fertilization begins life and that embryos must have full moral protection.' He contrasted this to the more liberal view of Anglican figures such as Gordon Dunstan, who searched instead 'for the stage of life where moral protection must be given to the embryo' and supported 'conclusions remarkably similar to those that we as scientists have reached'.[164] Edwards then stated that there was similar confusion among philosophers, where 'different schools define good and harm in their own way'. And he regarded lawyers as even more ineffectual, as 'they demand a clear lead from the moralists, whoever they are, before any law is written. But a clear answer will not be forthcoming if the attitudes of different religions or philosophers are any guide.'[165] Despite the public clamour for oversight and external involvement, Edwards again concluded that: 'It would be far better to leave standards of practice primarily to the scientific and medical societies, especially those well-versed in those affairs. In this way, new procedures can be adapted smoothly to the public needs, and unacceptable methods can be suppressed'.[166]

Warnock, by contrast, believed the divisions within her committee proved that they were fulfilling their role as a proxy for the public. She told the *Observer* that ministers would be more worried if members had 'presented a united front, because the whole country is split on this'.[167] In another interview, she reiterated that had members agreed on embryo experiments, they 'would have been a very strange, unrepresentative committee'.[168]

But Warnock also recognised that the committee had been charged with making policy recommendations, and acknowledged that there was no way of arriving at a proposal that satisfied all of the social and professional groups with an interest or a stake in IVF. She admitted that committee members were 'no longer in the business of working out exactly what is right or wrong', and argued that any solution therefore lay 'in the messier, less tidy business of compromise ... of attempting to come up with a solution which, while retaining as many of the calculated benefits to society as possible, will nevertheless offend and horrify people as little as possible'.[169]

For all the emphasis on its 'broad-based' membership, the committee fell back here on the expertise of the developmental biologist Anne McLaren, who Warnock later identified as 'indispensable'.[170] McLaren advised the committee to adopt fourteen days after fertilisation as a cut-off for embryo experiments. Around this point in development, cells in one pole of the rudimentary embryo condense to form the so-called 'primitive streak', which differentiates into the antecedents of the spinal cord and nervous system.[171] McLaren argued that scientists could legitimately experiment on embryos before this stage in development, as there was no possibility of them experiencing pain. And she also claimed that the primitive streak could be framed as the beginning of individual development, since it marked the last point where the embryo could cleave to form twins. McLaren claimed that it was only once the primitive streak had formed that:

> we can for the first time recognise and delineate the boundaries of a discrete human entity, an individual, that can become transformed through growth and differentiation into an adult human being. If I had to point to a stage and say "This was when I began being me", I would think it would have to be here.[172]

McLaren argued that the term 'pre-embryo' should be used before the primitive streak formed.[173] This new term had biological and ethical significance – portraying specimens younger than fourteen days as 'different in kind from the later, more complex, and ontologically distinguishable organism known as the embryo'.[174] It ensured that the 'pre-embryo' was 'safely bounded off from personhood, and hence could be an object for research, as opposed to the embryo proper, the authentic precursor of human life'.[175]

McLaren's arguments satisfied the majority of the committee, including Warnock, who sought to permit research up to a specific stage in development. Their report subsequently recommended fourteen days as the cut-off for embryo experiments. In doing so, it portrayed the primitive streak as a significant biological and ontological landmark, where a 'loosely packed configuration of cells' attained the 'first features of the embryo proper'.[176] Warnock even claimed that the primitive streak settled longstanding questions of when a human individual began and asserted that:

> Up to the [primitive streak] it is difficult to think of the embryo as an individual, because it might still become two individuals. None of the criteria that apply to me, or Tom or Dick or Harry, and distinguish us from the others, are satisfied by the embryo at this early stage. The collection of cells, though loosely strung together, is hardly yet one thing, nor is it several ... But from the fourteenth or fifteenth day onwards, there is no doubt that it is Tom or Dick or Harry that is developing.[177]

But these arguments did not appease those individuals and groups who believed that embryos attained moral status at fertilisation and should never be used in research. The three committee members who opposed any experiments refused to endorse the fourteen day cut-off and set out their objections at the end of the report.[178] They received support from the Christian gynaecologist Ian Donald, a founding member of the SPUC and longstanding opponent of embryo research. In several publications and the television programme *Credo*, Donald claimed that divisions within the committee highlighted the 'fatuous inconsistency in seeking to differentiate the rights of a 13 day embryo from those of a 15 day old one'.[179] He attacked the committee's proposals as 'atheistic and amoral' and called for 'a halt to experimentation until the public as a whole, and Parliament in particular, do not find themselves on a slippery slope they cannot hope to remount'.[180] He was joined by members of LIFE, who argued that embryo experiments were 'not in keeping with the respect due to human life' and called for legislation banning all research.[181]

For a while it appeared that Parliament would do just this. After debates in which many MPs and Lords criticised the fourteen-day limit, and after strenuous lobbying by LIFE and the SPUC, the Ulster Unionist and former Conservative politician Enoch Powell introduced a Private Member's Bill late in 1984 which sought to prohibit all research on embryos. This Unborn Children (Protection) Bill was only defeated after a pro-research lobby, including Mary Warnock, distinguished the early 'pre-embryo' from an 'unborn child' and warned politicians that a total ban would stifle essential research.[182]

At the same time, others attacked the fourteen-day limit as too restrictive. Articles in *Nature* and the *New Scientist* urged the government to 'devise more liberal legislation' to ensure that important research was not halted prematurely.[183] In a letter to the MRC, meanwhile, Robert Edwards claimed that he saw 'no reason' why the primitive streak gave embryos any particular moral status, and warned that 'many fundamental studies on differentiation, human anomalies and other major advances may require more days *in vitro*'.[184]

Describing the fourteen-day limit as 'particularly unfortunate' for the same reasons, Michael Lockwood suggested that research should be permitted on embryos up to six weeks after fertilisation, as before this point there was 'not the remotest possibility of the brain structures that are a *sine qua non* of human existence having developed'.[185] Another philosopher, John Gray, went even further on *Credo* and suggested that experiments should continue until the foetus had attained elements of consciousness, even if this meant allowing research up to birth.[186] And John Harris claimed that if scientists developed the technology to maintain foetuses to term in the laboratory, then research should be permitted up to and including the third trimester. Experimentation was justified until this late stage, he argued, because the embryo or foetus was not yet capable of valuing its own existence and was therefore not a person whose interests outweighed the potential beneficiaries of research.[187]

The growing number of philosophers who publicly commented on 'where to draw the line' shows how Warnock's engagement with IVF pushed medicine and science to the forefront of applied ethics. This was clear in May 1984, when the Society for Applied Philosophy announced that its next annual conference would be on 'bio-ethics'. Conference organisers drew up a list of speakers who they considered to be leading figures in this new field, including Ian Kennedy, John Harris, Jonathan Glover, Gordon Dunstan and

Mary Warnock.[188] When the Society for Applied Philosophy began to arrange this 'bio-ethics' conference, it claimed that attendees 'showed a strong bias' for discussing IVF and set aside a whole day to discuss the Warnock report.[189] In its autumn 1985 newsletter, it reported that most of the conference 'was directed to the issues generated by new reproductive technologies' and argued that this focus 'justified the Society's major interest in issues in the field of medicine'.[190]

In his 1986 book on *Applied Ethics*, Peter Singer claimed that Warnock's engagement with IVF showed how 'the broader community has willingly accepted the relevance and value of philosophers to practical issues', and argued that this was 'particularly noticeable in bioethics'.[191] Michael Lockwood also believed that Warnock had raised the profile of moral philosophers, and claimed that they should thank her for helping 'the powers that be recognise and value the philosopher's peculiar kind of expertise'.[192] But he also noted that her spell as chair had been viewed 'with somewhat mixed feelings', since philosophers often disagreed 'at a rather fundamental level' and 'a fellow philosopher may wonder, therefore, whether the philosophically trained chairman of a government committee is providing the right sort of guidance'.[193] These 'mixed feelings' were illustrated, of course, by the various criticisms of the fourteen-day cut-off. But Warnock's role as chair also prompted more specific discussion about the nature of ethical expertise and exactly what form bioethics should take.

Her major critic here was Richard Hare, who used discussion of the Warnock report to voice his frustration at much work in bioethics. This was already clear in 1982, when Hare complained to the Society for Applied Philosophy that 'the increase in philosophical interest in medical ethics is generating what I call a "garbage explosion", because the philosophers who jump on the bandwagon have either given no attention to problems in theoretical ethics or do not understand them'.[194] Although his target here was A. J. Ayer, Hare turned his attention to Mary Warnock during a 1985 lecture at Oxford. He argued that she had failed 'to do a lot of hard *philosophical* work' when it came to embryo experiments by not getting her fellow committee members to scrutinise the 'good and harm that would have come from allowing or forbidding such research'.[195] Hare believed that by not persuading members to think in utilitarian terms, Warnock was simply

> content with the second best alternative, which was perhaps all she could manage. This was to find some conclusions, which the members of the committee, or as large a majority as possible, would sign, and not bother about finding defensible reasons for them. Since the members were fairly typical in their moral attitudes or prejudices, it might be hoped that conclusions to which they would agree would also be acceptable to the public.[196]

Hare continued that by allowing committee members to reach decisions based on 'expressions of moral conviction without any support', Warnock had ensured that the committee failed to provide firm guidance to politicians and the public.[197] And this, he concluded, meant that opponents of research had 'a field day and the public is still floundering'.[198]

In 'abjuring utilitarianism', Warnock fell below Hare's estimation of how philosophers should behave when engaging with practical affairs. He believed that a philosopher was a 'specialist in ethics', who had a duty to provide a 'sound and generally accepted method of argumentation, with which those who start with different views can ... in the end, we hope, reach agreement'.[199] This was through adhering to a broadly utilitarian framework, which yielded 'clear answers that would commend themselves to anyone who had a firm understanding of the questions he was asking and of the facts'.[200] Hare's argument here drew on Peter Singer and Deanne Wells's 1984 book *The Reproduction Revolution*, which claimed that governments would only get clear policy advice if bioethics functioned as a vehicle for 'ethical experts', who answered questions through 'reason and logical argument' and employed 'general principles which depend on no sectarian allegiances'.[201]

Warnock responded to this criticism by stating that Hare and Singer were mistaken in presuming that 'morality, like logic, is a matter of reason only'.[202] Echoing A. J. Ayer, and citing David Hume's claim that morality was 'more properly felt than reasoned', she argued that moral opinions could not be divorced from sentiment and 'that such a divorce, if attempted, would spell the end of morality itself'.[203] 'Ethical decisions', she claimed, 'cannot be taken without the examination of ethical feelings', and utilitarianism alone could not answer the question of whether eight-cell embryos were morally significant enough to be included in the calculus of benefits and harms.[204] Warnock countered that when it came to her committee, it was little help to 'say that the utilitarian party were reasonable, their opponents irrational' since 'it is the nature of morality to be at least partly irrational'.[205] This meant that while a philosopher could help to clarify the properties and consequences of a particular moral standpoint, they could not 'prove or otherwise show conclusively that one view is to be preferred to another'.[206]

Warnock claimed that such disagreement was 'unavoidable' as pluralistic societies lacked 'an agreed set of principles which everyone, or the majority, or any representative person believes to be absolutely binding'.[207] It followed from this, she argued, that no one field or approach should dominate ethical oversight and decision-making. Warnock encapsulated this position in the afterword to a popular edition of her committee's report, where she argued that:

> In matters of life and death, of birth and the family, no-one is prepared to defer to judgements made on the basis of a superior ability in philosophy. For these are areas that are central to morality, and everyone has a right to judge for himself. Such issues indeed lie at the heart of society; everyone not only wants to make their own choices but are bound to do so. And this is why there cannot be moral experts. Everyone's choice is his own.[208]

In line with the broader distrust of experts and the emphasis on public accountability, she believed that simply replacing the expertise of doctors and scientists with that of philosophers was 'not only out of place, but simply unacceptable'.[209] Warnock argued that government inquiries should instead provide a form of 'corporate decision-making', where various professions and interest groups sought a 'middle way' between competing claims and moral worldviews.[210]

Warnock's dismissal of 'moral experts' reflected several underlying concerns in applied ethics. Many philosophers believed that engagement with practical issues was limited by the presence of equally valid but incompatible moral premises. Jonathan Glover, A. J. Ayer, Bernard Williams, Alasdair MacIntyre and Mary Warnock, among others, argued that philosophers could not provide 'correct' answers, since there 'was no rational way of securing agreement' in pluralist societies.[211] The best they could do, in this view, was to help clarify the differing moral viewpoints associated with a particular issue. As Warnock told the BBC programme *Talk of the '80s*, which selected her as one of the most influential people of the decade, philosophy was simply a 'useful analytical tool' that helped different groups examine the consequences of their own and rival viewpoints.[212]

The belief that philosophy was 'a clarifying rather than a critical activity' set many of the British philosophers who engaged with bioethics apart from their American counterparts, who claimed that ethical issues could be resolved by applying universal principles such as respect for persons, beneficence and justice.[213] This view, which underpinned the *Belmont Report* and Tom Beauchamp and James Childress's influential *Principles of Biomedical Ethics*, ensured that Hare and Singer's view of 'moral experts' carried more weight in the United States than Britain. It also ensured that the strongest critics of principle-based methods in the United States were British philosophers such as Stephen Toulmin and Alasdair MacIntyre, who both emigrated in the 1970s. Toulmin, for example, criticised the 'tyranny of principles' in the *Hastings Center Report*, while MacIntyre claimed that the 'character of moral debate in our liberal, secular, pluralist culture' meant there was 'no rational method' to resolve ethical problems.[214] Warnock also criticised this principles-based approach in the 1990s, when she dismissed 'the suggestion that bioethics has a tool box, ready to hand, out of which it extracts a finite number of tools to hack at … certain problems involving cases of life or death'. She instead hoped that 'serious practitioners of philosophy, even perhaps the numerous class of bioethicists, do not feel themselves limited to certain preformed off-the-shelf tools, but are able to survey new problems with their usual weapons of common sense and their ability to draw distinctions'.[215]

Warnock's criticism of 'moral experts' also drew on an individualistic view of ethics, where the only authority that moral viewpoints possessed was that which particular agents chose to give them. This subjective viewpoint permeated *Language, Truth and Logic*, and underpinned Ayer's public discussion of the Arthur case. During the 1980s, moreover, it resonated with the neo-liberal conviction that the individual consumer was the prime locus of decision-making.[216] This was embodied by Margaret Thatcher's famous belief that 'choice is the essence of ethics: without choice there would be no ethics'.[217] In a 1979 conference speech Thatcher had argued that 'morality is personal' and denied there 'was such a thing as a collective conscience'.[218] Her conviction that individual choice was the best route to social change dovetailed with Mary Warnock's dismissal of 'moral experts' and claim that 'everyone's choice is his own'.[219] By the mid 1980s, when Warnock became a 'moulder of public policy' and the Conservatives won a landslide second election, their rhetoric helped produce a sociopolitical climate that defined people primarily in terms of their ability to make autonomous choices.[220]

The synergy between Warnock and the government's view of ethics, their shared distrust of experts and belief in oversight, all help explain why she became 'synonymous with British bioethics' during the 1980s.[221] After being appointed to the House of Lords as a cross-bench peer in 1985, Warnock contributed the first two articles to a new journal of *Bioethics*, continued to publicly discuss the ethics of IVF, gene therapy and animal experiments, endorsed the formation of a national bioethics committee, and was the first British representative on a new European Council on Bioethics.[222] By 1990, when her committee's recommendations were passed into law in the Human Fertilisation and Embryology Act, she was undoubtedly the most recognised figure in the growing 'ethics industry'.

Conclusion

There is no doubt that Mary Warnock's appointment as chair of the IVF inquiry was a significant moment in the history of British bioethics. Alastair Campbell recalls that 'a lot of us were very encouraged by the composition of that committee and felt that the government actually was taking notice of something other than the medical profession in defining its legislation and its policies for emerging medical technologies'.[223] The composition of Warnock's committee illustrates how the demand for external oversight found expression in neo-liberal forms of government, with Conservative politicians recruiting 'non-experts' in order to make procedures such as embryo experiments publicly accountable. Warnock also became a figurehead for bioethics during and after her spell as chair, promoting external oversight in her committee's report, in academic journals, lectures and popular media. The various spheres in which her arguments resonated demonstrates how bioethics is a multi-sited activity, functioning as the basis for policymaking and a new form of public discourse on the morality of medical and scientific practices.

Warnock's inquiry also provided a model for later committees.

When they assembled inquiries into gene therapy and human-animal transplants during the 1980s and 1990s, Conservative and New Labour governments appointed Cecil Clothier and Ian Kennedy to chair groups that again included several 'non-experts'.[224] But Warnock remains the most influential of these outside chairs. Nearly thirty years after her committee's report was published, scientists continue to work to rules that it suggested for IVF and embryo research. The former head of the Human Fertilisation and Embryology Authority (HFEA), the 'monitoring body' that Warnock proposed, argues that 'rarely can an individual have had so much influence on public policy'.[225] Even those who disputed her proposals agree that Warnock was invaluable for raising the profile of bioethics and asserting the value of outside involvement with medicine and science.[226]

Examining why Warnock became so influential highlights the value of actor-centred studies of bioethics. As this and the previous chapter have demonstrated, bioethics did not simply emerge thanks to the reforming ambitions of neo-liberal politicians during the 1980s. It arose thanks to the way in which certain individuals engaged with the political demand for oversight and framed bioethics as beneficial to the public, politicians, doctors and scientists. Like Ian Kennedy, Warnock echoed broader criticism of self-regulation and simultaneously promised that oversight would safeguard research. This ensured that her calls for a 'monitoring body' received support from politicians, the press and biomedical journals, and that Robert Edwards was now firmly in the minority when he endorsed self-regulation in his *Horizon* lectures.

At the same time, Warnock also positioned herself as an intermediary between different groups when she argued that philosophers and other 'lay' members had to 'work with scientists' in order to develop recommendations for a specific issue.[227] In working with scientists to determine 'where to draw the line' for embryo experiments, she engaged in what Jasanoff terms 'ontological surgery'. By criticising her proposals as either too lax or severe, other bioethicists, doctors and pro-life groups demonstrated how this 'dual work of biological classification and moral clarification' is an ongoing and often public process.[228] It is, crucially, also a historically contingent one. By prioritising questions of when in development embryos began to deserve legal protection, instead of questions over costs and equal access to IVF, these public debates both reflected and helped shape a sociopolitical climate that was preoccupied with notions of individual personhood and rights.[229]

But perhaps Warnock's greatest engagement with the broader climate, and the reason for her enduring influence, lay in her dismissal of 'moral experts'. This chimed with the contemporary distrust of expertise, where many people believed 'no profession should become a law unto themselves'.[230] Echoing the Conservative distaste for experts, and Margaret Thatcher's claim that 'there is no such thing as society', Warnock argued that when it came to ethics 'there is no such thing as authority'.[231] She believed that the presence of incompatible viewpoints on issues such as embryo research meant that bioethics should function as a proxy for the different views that existed in pluralistic societies, 'where we are compelled to accept that "common morality" is a myth'.[232] In doing so, Warnock framed moral pluralism and disagreement not as a problem for bioethics, but as the source of its utility, in that it provided various professions and groups with the chance to facilitate what Roger Brownsword calls 'the process of practical decision-making'.[233]

This line of thought was clearly influential, and continues to determine how many people view bioethics today. To the political philosopher Onora O'Neill, a former student of Mary Warnock and founding member of the Nuffield Council on Bioethics, 'bioethics is not a discipline' but has instead 'become a meeting ground for a number of different disciplines, discourses and organisations'.[234] The philosopher David Archard, a member of the HFEA who shares Warnock's distrust of 'moral experts', similarly believes that disagreements between individuals and groups can be read as a sign of 'robust good health' in bioethics rather than 'evidence of systematic ignorance'.[235]

This is crucial to understanding the growth of the 'ethics industry' during the 1980s and 1990s. Bioethics became a valued enterprise in this period not because it provided a vehicle for experts to provide obviously 'correct' answers, but precisely because 'the problems multiply and the proffered solutions are disputed'.[236] Bioethics emerged thanks to a broader distrust of experts and demand for public accountability, both in government and beyond, and its continued growth stemmed from the way in which figures such as Warnock endorsed the neo-liberal conviction that '*everyone's choice is his own*'.[237]

Notes

1 Jasanoff, *Designs on Nature*, p. 173.
2 Mary Warnock, *A Question of Life: The Warnock Report on Human Fertilisation and Embryology* (London: Basil Blackwell, 1985) p. xiii. Emphasis added.
3 Anon, 'Embryology Needs Rules, Not New Laws', *Nature*, Vol. 302 (1983) pp. 735–44 (p. 735).
4 Anthony Kenny, *A New History of Western Philosophy* (Oxford: Oxford University Press, 2012) p. 944.
5 For more background, see MacIntyre, *Short History of Ethics*; Isaiah Berlin, *The Crooked Timber of Humanity: Chapters in the History of Ideas* (London: Pimlico, 2003); Mary Warnock, *Ethics since 1900* (Oxford: Oxford University Press, 3rd edn, 1978).
6 Mary Warnock, 'Government Commissions', in U. Bertazzoni, P. Fasella, A. Klepsch and P. Lange (eds), *Human Embryos and Research: Proceedings of the European Bioethics Conference, Mainz 1988* (Frankfurt and New York: Campus Verlag, 1988) pp. 159–68 (p. 165).
7 Mary Warnock, 'Moral Thinking and Government Policy: The Warnock Committee on Human Embryology', *The Millbank Memorial Fund Quarterly. Health and Society*, Vol. 63, no. 3 (1985) pp. 504–22 (p. 518); Warnock, 'Government Commissions', p. 166.
8 Warnock, 'Government Commissions', p. 167.
9 Warnock, *A Question of Life*, p. 97.
10 Mary Warnock, interview with the author (House of Lords, London, March 2009). On bioethics as 'corporate decision-making' see Mary Warnock, 'Embryo Therapy: The Philosopher's Role in Public Debate', in D. R. Bromham, M. E. Dalton and P. J. R. Millican (eds), *Ethics in Reproductive Medicine* (London: Springer Verlag, 1992) pp. 21–31 (p. 31).
11 On Warnock's early life and education, see Mary Warnock, *People and Places: A Memoir* (London: Duckworth, 2000) especially pp. 1–15.
12 Ibid, p. 18.
13 Ibid, p. 49.
14 Ibid.
15 R. M. Hare, 'Universalisability', *Proceedings of the Aristotelian Society*, Vol. 55 (1955) pp. 295–312. See also A. W. Price, 'Richard Mervyn Hare (1919–2002)', *Oxford Dictionary of National Biography*, www.oxforddnb.com/view/article/76706 (accessed 21 February 2014); Peter Singer, 'R. M. Hare's Achievements in Moral Philosophy', *Utilitas*, Vol. 14, no. 3 (2002) pp. 309–17.
16 R. M. Hare, 'Man's Interests', in Ruth Porter and Ian Ramsey

(eds), *Personality and Science* (London: CIBA Foundation, 1971) pp. 97–101.

17 Warnock, interview with the author (2009).

18 Warnock, *Ethics since 1900*, p. 204.

19 Ibid, p. 204.

20 Warnock, *People and Places*, pp. 53–4.

21 Philippa Foot, 'Moral Arguments', *Mind*, Vol. 67 (1958) pp. 502–13 (pp. 504–5). See also Philippa Foot, 'Moral Beliefs', *Proceedings of the Royal Aristotelian Society*, Vol. 59 (1958-59) pp. 83–104.

22 Warnock, *People and Places.*

23 Warnock, *Ethics since 1900*, p. 206.

24 Warnock, *People and Places*, p. 54.

25 Philippa Foot, 'The Problem of Abortion and the Doctrine of Double Effect', reprinted in Phillipa Foot, *Virtues and Vices* (Oxford: Oxford University Press, 2002) pp. 19–31.

26 Onora O'Neill, interview with the author (Royal Society, London, June 2009); Warnock, interview with the author (2009).

27 The philosophy faculty initially rejected John Harris's PhD on political violence, on the grounds that it was a non-philosophical matter. They only relented after his supervisor, Ronald Dworkin, persuaded them that it was a legitimate topic. John Harris, interview with the author (University of Manchester, November 2011). See also Singer, 'R. M. Hare's Contribution to Moral Philosophy', p. 316.

28 See Jonathan Glover and M. J. Scott-Taggart, 'It Makes No Difference Whether or Not I Do It', *Proceedings of the Aristotelian Society*, Vol. 49 (1975) pp. 171–90.

29 R. M. Hare, 'What's Wrong with Slavery?' *Philosophy and Public Affairs*, Vol. 8 (1978) pp. 103–21; R. M. Hare, 'Rules of War and Moral Reasoning', *Philosophy and Public Affairs*, Vol. 1 (1972) pp. 166–81. Although these were his first articles dedicated to applied ethics, Hare devoted the final chapter of his 1963 book *Freedom and Reason* to a practical discussion of racism, contending that for a racist argument to truly carry weight, the perpetrator must be prepared to universalise their own beliefs and be subjected to racism themselves. See R. M. Hare, *Freedom and Reason* (Oxford: Oxford University Press, 1963) pp. 202–24.

30 Mary Warnock, *Nature and Morality: Recollections of a Philosopher in Public Life* (London: Continuum, 2004) p. 14; Warnock, *People and Places*, p. 23.

31 Warnock, *People and Places*, p. 23.

32 Warnock, interview with the author (2009).

33 Peter Singer, 'All Animals are Equal', *Philosophic Exchange* (1974) pp. 10–16; reprinted in Peter Singer and Helga Kushe (eds), *Bioethics:*

An Anthology (Oxford: Blackwell, 2000) pp. 461–70 (p. 461). See also Peter Singer, *Animal Liberation: Toward an End to Man's Inhumanity to Animals* (London: Cape, 1976).

34 John Harris, 'The Survival Lottery', *Philosophy*, Vol. 50 (1975) pp. 81–7; John Harris, 'The Marxist Conception of Violence', *Philosophy and Public Affairs*, Vol. 3, no. 2 (1974) pp. 192–220.

35 R. M. Hare, 'Abortion and the Golden Rule', *Philosophy and Public Affairs*, Vol. 4 (1975) pp. 201–22 (p. 205).

36 Ibid, p. 208. Applying this rule led Hare to conclude that abortion was prima facie wrong 'in default of sufficient countervailing circumstances'. While this initially seemed to support the anti-abortion lobby, he continued that the prevalence of often serious illness and disability which did real harm to the foetus ensured that 'such countervailing circumstances are not too hard to find in many cases'. See Ibid, p. 221.

37 Ibid, p. 202.

38 Ibid, p. 201.

39 R. M. Hare, 'Medical Ethics: Can the Moral Philosopher Help?', in Stuart Spicker and H. Tristram Engelhardt (eds), *Philosophical Medical Ethics: Its Nature and Significance* (Dordrecht: Riedel Publishing, 1977) pp. 49–63 (p. 49).

40 Harris, interview with the author (2011). Details of the 1981 course on ethics are held in the files relating to the London Medical Group at the Wellcome Trust Library for the History of Medicine: file GC/253/A/31/8.

41 R. S. Downie, 'Ethics, Morals and Moral Philosophy', *Journal of Medical Ethics*, Vol. 6 (1980) pp. 33–4 (p. 34). See also Colin Honey, 'Acts and Omissions', *Journal of Medical Ethics*, Vol. 5 (1979) pp. 143–4; Patrick J. McGrath, 'Is and Ought', *Journal of Medical Ethics*, Vol. 1 (1975) pp. 150–1.

42 Warnock, *Ethics since 1900* (3rd edn) p. 139.

43 We should be wary of seeing these appointments as evidence of a political enthusiasm for non-professional involvement. As I noted in chapters 2 and 3, this first emerged with Callaghan's Labour government and was fully implemented by the Conservatives during the 1980s. Warnock believes that she was appointed chair of the education committee in 1974 because she had been headmistress of a reputable school and had also served on the Oxford County Education Authority – i.e., she was an 'insider'. Warnock, interview with the author (2009).

44 For more detail on debates surrounding the replacement of animals with tissue culture, see Wilson, *Tissue Culture in Science and Society*, pp. 81–91.

45 Ibid, pp. 87–8.
46 Judith Hampson, 'Animal Welfare – A Century of Conflict', *New Scientist*, 25 October 1979, pp. 280–2.
47 Warnock, *Nature and Morality*, p. 149.
48 Ibid, pp. 152–3. For criticism of the LD50 test, see Ryder, *Victims of Science*, pp. 42–3.
49 Lord Geoffrey Cross (chair), *Report on the LD50 Test* (London: Home Office, 1979) p. 12.
50 Cross, *Report on the LD50 Test*, p. 12. For criticism of the report, see Ryder, *Victims of Science*, p. 153; Anon, 'Bitter Dose for Laboratory Animals', *New Scientist*, 12 July 1979, p. 24.
51 Cross, *Report on the LD50 Test*, p. 12.
52 Ibid, pp. 12, 19.
53 Warnock, *People and Places*, pp. 174–5.
54 Despite this broad composition, animal rights campaigners criticised the government for 'inadequately' representing their interests. See Ryder, *Victims of Science*, p. 157.
55 Warnock, interview with the author (2009).
56 Advisory Committee on Animal Experiments, *Report to the Secretary of State on the Framework of Legislation to Replace the Cruelty to Animals Act 1876* (London: Home Office, 1981) p. 33.
57 Michael Lockwood, 'Ethical Dilemmas in Surgery: Some Philosophical Reflections', *Journal of Medical Ethics*, Vol. 6 (1980) pp. 82–4 (p. 84).
58 Collini, *Absent Minds*, p. 403.
59 Following her spell as chair of the education inquiry, Warnock wrote a popular book and several newspaper columns on education. See Mary Warnock, *Education: A Way Ahead* (London: Blackwell, 1979). She also reviewed children's books and encyclopedias for the *Times Literary Supplement*. See, for example, Mary Warnock, 'Looking It Up', *Times Literary Supplement*, 19 September 1980, p. 1034.
60 The original charge of murder was reduced to attempted murder after it became clear that the drug could not be proved to have caused the baby's death. For a useful summary, see Brazier, *Medicine, Patients and the Law*, pp. 344–5.
61 Anon, 'The Vindication of Dr Arthur', *The Times*, 6 November 1981, p. 12.
62 Anon, 'When a Child is Born', *The Times*, 6 November 1981, p. 15. See also Anon, 'Vindication of Dr Arthur'.
63 Jonathan Glover, 'Letting People Die', *London Review of Books*, 4 March 1982, pp. 3–5 (p. 3).
64 Jonathan Glover, *Causing Death and Saving Lives* (Harmondsworth: Penguin, 1977) p. 16.
65 Glover, 'Letting People Die', p. 5.

66 Ibid, p. 4.
67 Ibid, pp. 3, 5. For a longer dismissal of the 'right to life' argument, see Glover, *Causing Death and Saving Lives*, pp. 39–58, 150–68.
68 A. J. Ayer, 'Why the Dr Arthur Verdict is Right', *The Times*, 16 November 1981, p. 14.
69 Ibid.
70 Ayer, quoted in Rogers, *A. J. Ayer*, pp. 319–20.
71 Society of Applied Philosophy, 'Mission Statement' (November 1981). Uncatalogued archive, held at the University of Lancaster at the time of writing. See also Rogers, *A. J. Ayer*, p. 319.
72 R. M. Hare to Brenda Cohen, 17 February 1982. Society of Applied Philosophy archive, University of Lancaster. Cohen informed Hare that Ayer had been selected as president due to his ongoing support for the society, and in the hope that his public profile would be of 'enormous value' in promoting their work. Brenda Cohen to R. M. Hare, 23 February 1982.
73 Glover, 'Letting People Die', p. 5.
74 Ibid.
75 Glover, *Causing Death and Saving Lives*, p. 35.
76 Ibid.
77 For more detail, see Rogers, *A. J. Ayer*, p. 329.
78 Ayer, 'Why the Dr Arthur Verdict is Right'. See also Rogers, *A. J. Ayer*, pp. 328–9.
79 Ayer, 'Why the Dr Arthur Verdict is Right'. Emphasis added.
80 Ibid.
81 Hare, 'Medical Ethics: Can the Moral Philosopher Help?', p. 55.
82 R. M. Hare, '*In Vitro* Fertilization and the Warnock Report', in Ruth Chadwick (ed), *Ethics, Reproduction and Genetic Control* (London: Croom Helm, 1987) pp. 71–90 (pp. 71, 77). See also Peter Singer, 'Moral Experts', *Analysis*, Vol. 32, no. 4 (1972) pp. 115–17.
83 Society for Applied Philosophy, 'Spring Newsletter' (1985) p. 1. University of Lancaster archive.
84 Anon, 'Why We Must ALL Have a Say on Test Tube Babies', *Mail on Sunday*, 20 May 1984, p. 16.
85 Edwards and Steptoe, *A Matter of Life*, p. 106.
86 In their article assessing Edwards's funding application to the MRC in the early 1970s, Johnson et al. note that 'In marked contrast to the debate during the 1980s, which came to focus on the moral status of the embryo, only Glenister [an MRC referee] addressed this point … However, other referees were more concerned that those embryos to be placed *in utero* might be abnormal.' See Johnson et al., 'Why the MRC Refused Robert Edwards and Patrick Steptoe Support', p. 2166. See also Michael Mulkay, *The Embryo Research Debate: Science and*

the Politics of Reproduction (Cambridge: Cambridge University Press) pp. 11–12.

87 Culliton and Waterfall, 'Flowering of American Bioethics', p. 1270.

88 Anthony Tucker, 'Brave New World of Test Tube Babies', *Guardian*, 27 July 1978, p. 11.

89 For more on British and American responses to the birth of Louise Brown, see Turney, *Frankenstein's Footsteps*, pp. 182–5.

90 Nigel Hawkes, 'The Making of Baby Brown', *Observer*, 30 July 1978, p. 9

91 Katherine Hadley, 'Tinkering With Life', *Daily Express*, 21 June 1982, p. 10.

92 Warnock, *Nature and Morality*, p. 74.

93 Baron Alan Campbell of Alloway, cited in *Parliamentary Debates: House of Lords*, Vol. 432 (9 July 1982) col. 1001.

94 Mulkay, *The Embryo Research Debate*, pp. 15–16.

95 Ibid, p. 15. See also Edward Yoxen, 'Conflicting Concerns: The Political Context of Recent Embryo Research Policy in Britain', in Ian Varcoe, Maureen McNeill and Steve Yearley (eds), *The New Reproductive Technologies* (London: Macmillan, 1990) pp. 173–200. On unsuccessful pro-life attempts to reverse the 1967 Abortion Act, see Gayle Davis, 'The Medical Community and Abortion Law Reform: Scotland in National Context, c. 1960–1980', in Goold and Kelly (eds), *Lawyer's Medicine*, pp. 143–65 (p. 161).

96 Hadley, 'Tinkering with Life'.

97 George Clark, 'Test-Tube Babies "Could Bring Oedipus Tragedy"', *The Times*, 9 February 1982, p. 3.

98 Geoffrey Robertson, 'The Law and Test Tube Babies', *Observer*, 7 February 1982, p. 8.

99 Peter Williams and Gordon Stevens, 'What Now for Test Tube Babies?', *New Scientist*, 4 February 1982, pp. 311–16 (p. 313). This article was based on the television programme 'Test Tube Explosion', which the authors directed and which was screened on ITV on 2 February 1982.

100 Anon, 'Debate Call over Test Tube Babies', *Guardian*, 10 February 1982, p. 2.

101 Anon, 'A Matter of Origins', *The Times*, 10 February 1982, p. 11.

102 Murray Davies and Ronald Badford, 'Test Tube Baby Doctor Warned Off', *Daily Mirror*, 28 September 1982, p. 7.

103 On these professional inquiries, see Goold, 'Regulating Reproduction in the United Kingdom'.

104 Anon, 'Debate Call Over Test Tube Babies', p. 2.

105 Katherine Whitehorn, 'Embryonic Problems', *Observer*, 3 October 1982, p. 27.

106 Ian Kennedy, 'Ethical Guidelines on Fertilization', *The Times*, 11 February 1982, p. 17.
107 Robertson, 'The Law and Test Tube Babies'.
108 Richard Norton, Department of Education and Science, to David Noble, Medical Research Council, 13 April 1982. National Archives: FD7/2307.
109 Ibid.
110 Ibid.
111 Ibid.
112 Keith Newton to Dr Vickers, Medical Research Council (24 May 1982). National Archives: FD7/2307.
113 Ibid.
114 The 'Great and Good' have constituted an important part of the British Establishment since the eleventh century. Their role in policymaking became increasingly contested during the late twentieth century, however, and Margaret Thatcher portrayed herself as 'an anti Great and Good' Prime Minister (although her government still selected esteemed figures such as Sir Oliver Franks, Cecil Clothier and Mary Warnock as chairs of public inquiries). For a comprehensive account of the 'Great and Good' in postwar Britain, see Peter Hennessy, *Whitehall* (London: Fontana, 1990) pp. 540–86.
115 Anon, 'The Good Woman of Oxbridge', *Observer*, 24 June 1984, p. 7.
116 See Hennessy, *Whitehall*, pp. 551, 578–9.
117 Moran, *British Regulatory State*, p. 61. See also Fisher, *Risk Regulation*; Gummett, *Scientists in Whitehall*, pp. 91–116.
118 See Moran, *British Regulatory State*, pp. 61–2; Rose, *Powers of Freedom* pp. 150–4.
119 Mary Warnock, 'The Politicisation of Medical Ethics', *Journal of the Royal College of Physicians of London*, Vol. 33, no. 5 (1999) pp. 474–8 (p. 474).
120 Warnock, *Nature and Morality*, p. 73.
121 Warnock, interview with the author (2009). See also Stuart Hampshire, T. M. Scanlon, Bernard Williams, Thomas Nagel and Ronald Dworkin (eds), *Public and Private Morality* (Cambridge: Cambridge University Press, 1978).
122 Warnock, *Nature and Morality*, pp. 72.
123 Norman Fowler, cited in *Parliamentary Debates: House of Commons*, Vol. 28 (23 July 1982) col. 329.
124 Warnock committee meeting minutes (14 December 1982). National Archives: FD7/2307.
125 DHSS correspondence to University Grants Committee (1982). National Archives: FD7/2307.

126 Warnock committee meeting minutes (14 December 1982); DHSS to University Grants Committee (1982). National Archives: FD7/2307.
127 Kennedy, *The Unmasking of Medicine*, p. 152.
128 Anon, 'Test Tube Babies Are Now a Public Subject', *Observer*, 20 May 1984, p. 18.
129 Anon, 'Why We Must ALL Have a Say on Test Tube Babies'.
130 Mary Warnock, '*In Vitro* Fertilization: The Ethical Issues (II)', *The Philosophical Quarterly*, Vol. 33 (1983) pp. 238–49 (p. 249).
131 Ibid, p. 249.
132 Suzanne Lowry, 'Birth of a New Ethic', newspaper clipping (*circa* July 1984) held at Archives and Manuscripts, Wellcome Trust Library for the History of Medicine: file PP/MLV/C23/1/6.
133 Warnock, interview with the author (2009).
134 Lowry, 'Birth of a New Ethic'.
135 Anon 'Why We Must ALL Have a Say on Test Tube Babies'; Warnock, interview with the author (2009).
136 Warnock committee meeting minutes, 'Other Questions' (October 1983). National Archives: FD7/2307.
137 Warnock, *A Question of Life*, p. 79.
138 Mary Warnock, 'Scientific Research Must have a Moral Basis', *New Scientist*, 15 November 1984, p. 36. Emphasis added.
139 Warnock, 'Moral Thinking and Government Policy', p. 514.
140 Anon, 'Embryology Needs Rules', p. 735.
141 Anon, 'A Welcome Report', *British Medical Journal*, Vol. 249 (1984), pp. 207–8 (p. 207).
142 Anon, 'Embryonic Research', *New Scientist*, 27 September 1984, p. 2. Emphasis in original.
143 Anon, 'The Good Woman of Oxbridge', p. 7.
144 Warnock, *People and Places*, p. 135. Warnock's support for the Conservatives wavered during the 1950s and early 1960s, when she joined the Labour Party. She returned to the Conservatives after Harold Wilson became Labour leader in 1963. See Ibid, pp. 135–69.
145 Ibid, p. 182.
146 Ibid.
147 Warnock, '*In Vitro* Fertilization', p. 238.
148 Peter Pallot, 'Towards a Brave New World', *Daily Telegraph*, 27 April 1983, p. 11.
149 Royal Society submission to DHSS Committee of Inquiry (1983). National Archives: FD7/2308.
150 Warnock committee memo, 'Experiments on Embryos: Key Questions' (October 1983). National Archives: FD7/2307.
151 Immanuel Jacobovits, quoted in Anon, 'Are the Baby-Makers Going too Far?', *The Sunday Times*, 27 May 1984, p. 13.

152 Anon, 'Embryo Experiments Ban Urged', *Guardian*, 12 May 1983, p. 3.
153 S. J. G. Spencer, 'Human *In Vitro* Fertilization and Embryo Replacement and Transfer', *British Medical Journal*, Vol. 286 (1983) pp. 1822–3 (p. 1823).
154 Family Planning Association, statement to the Warnock committee, 16 February 1983. National Archives: FD7/2307.
155 Warnock committee, 'Experiments on Embryos: Key Questions' (1983).
156 Ibid.
157 Warnock committee, 'Other Questions' (1983).
158 See Madeline Carriline, John Marshall and Jean Walker, 'Expression of Dissent: B. Use of Human Embryos in Research', in Warnock, *A Question of Life*, pp. 90–3 (p. 90).
159 Warnock later identified this 'mole' as the gynaecologist Dame Josephine Barnes. See Warnock, *Nature and Morality*, pp. 104–5.
160 Thomas Prentice and Nicholas Timmins, 'Surrogate Mothers and Embryo Research Put Committee in a Quandary', *The Times*, 6 February 1984, p. 3. See also Warnock, *Nature and Morality*, pp. 78–9.
161 Mary Warnock, quoted in Lowry, 'Birth of a New Ethic'.
162 Warnock, 'Scientific Research Must Have a Moral Basis', p. 36.
163 R. G. Edwards, 'The Horizon Lecture. Test Tube Babies: The Ethical Debate', *Listener*, 27 October 1983, pp. 11–19 (p. 12).
164 Ibid, p. 12.
165 Ibid.
166 Ibid, p. 19.
167 Anon, 'Good Woman of Oxbridge'.
168 Lowry, 'Birth of a New Ethic'.
169 Warnock, 'Do Human Cells have Rights?', *Bioethics*, Vol. 1 (1987) pp. 1–14 (p. 8). This article is the text of a 1986 lecture that Warnock gave at Ormond College, Melbourne.
170 Warnock, *Nature and Morality*, pp. 80–3.
171 Anne McLaren, 'Where to Draw the Line?' *Journal of the Royal Institution*, Vol. 56 (1984) pp. 101–21 (p. 117). This paper was based on McLaren's lectures to the committee and reprinted evidence given by supporters and opponents of embryo experiments.
172 Ibid, pp. 111–12.
173 Ibid, p. 106.
174 Jasanoff, 'Making the Facts of Life', p. 65.
175 Jasanoff, *Designs on Nature*, p. 152.
176 Warnock, *A Question of Life*, pp. 58–9.
177 Warnock, 'Do Cells Have Rights?', p. 11.

178 See Warnock, *A Question of Life*, pp. 90–3.
179 Ian Donald, 'Introduction', in Ian Donald (ed), *Test Tube Babies: A Christian View* (Oxford: Order of Christian Unity, 1984) pp. 1–15 (p. 6).
180 Ibid, pp. 8, 12. In a special edition of *Credo* dedicated to the Warnock report, Donald again called for a ban on research and declared that 'every human baby has a beginning and conception is it'. *Credo: The Warnock Committee* (screened July 1984), held at the British Film Institute, London.
181 Margaret M. Heley, LIFE Doctors' Group, 'Mrs Warnock's Brave New World', *Lancet*, Vol. 324 (1984) p. 290.
182 For more on these parliamentary debates, see Jasanoff, *Designs on Nature*, pp. 150–2; Michael Mulkay, 'Galileo and the Embryos: Religion and Science in Parliamentary Debate over Research on Human Embryos', *Social Studies of Science*, Vol. 25, no. 3 (1995) pp. 499–532; Rannan Gillon, 'In Britain, the Debate after the Warnock Report', *Hastings Center Report*, Vol. 17, no. 3 (1987) pp. 16–18. On Warnock's support for research, see Mary Warnock, 'Totally Wrong', *The Times*, 30 May 1984, p. 12.
183 Anon, 'Confused Comment on Warnock', *Nature*, Vol. 312 (1984) p. 389; Omar Sattaur, 'New Conception Threatened by Old Morality', *New Scientist*, 27 September 1984, pp. 12–17.
184 Robert Edwards correspondence to Joan Box, MRC, 13 August 1984. National Archives: FD7/2332.
185 Michael Lockwood, 'The Warnock Report: A Philosophical Appraisal', in Michael Lockwood (ed), *Moral Dilemmas in Modern Medicine* (Oxford: Oxford University Press, 1985) pp. 155–87 (p. 187).
186 John Gray, quoted in *Credo: The Warnock Report* (screened July 1984). Recording held at the British Film Institute, London.
187 John Harris, *The Value of Life: An Introduction to Medical Ethics* (London: Routledge and Kegan Paul, 1985) pp. 16–27, 120–35.
188 Minutes of the Fifth Meeting of the Society for Applied Philosophy, 25 May 1984. Society of Applied Philosophy archives.
189 Minutes of the Sixth Meeting of the Executive Committee of the Society for Applied Philosophy, 30 November 1984. Society of Applied Philosophy archives.
190 Society for Applied Philosophy Newsletter (Autumn 1985). Society of Applied Philosophy archives.
191 Singer 'Introduction', p. 4.
192 Lockwood, 'The Warnock Report', p. 155.
193 Ibid.
194 R. M. Hare to Brenda Cohen, 17 February 1982. Society of Applied Philosophy archive, University of Lancaster.

195 Hare, '*In Vitro* Fertilization and the Warnock Report', pp. 83, 88. Emphasis in original. This paper was initially delivered as a 1985 lecture at the University of Oxford.
196 Ibid, p. 82.
197 Ibid, p. 88.
198 Ibid.
199 Ibid, p. 71.
200 Ibid, p. 77.
201 Peter Singer and Deanne Wells, *The Reproduction Revolution: New Ways of Making Babies* (Oxford: Oxford University Press, 1984) pp. 199, 202.
202 Warnock, 'Government Commissions', p. 167.
203 Warnock, 'Moral Thinking and Government Policy', p. 518.
204 Ibid, p. 520.
205 Warnock, 'Government Commissions', p. 166.
206 Warnock, 'Embryo Therapy', p. 21.
207 Warnock, *A Question of Life*, p. xi.
208 Ibid, p. 96.
209 Ibid.
210 Warnock, 'Embryo Therapy', p. 31. Laura Stark's work on IRBs in the United States shows how ethicists and scientists similarly negotiate differing views to fashion policies for biomedical research, then have to justify their recommendations to particular stakeholders, e.g., researchers, funding bodies and politicians. In a claim that evokes Warnock's endorsement of the fourteen-day limit, Stark notes that '"good" recommendations are, by my account, social achievements and not reflections of the inherent qualities of the recommendations themselves'. See Stark, *Behind Closed Doors*, p. 39.
211 Alasdair MacIntyre, *After Virtue* (Notre Dame: University of Notre Dame Press, 3rd edn, 2007) p. 6. See also Bernard Williams, *Ethics and the Limits of Philosophy* (London: Fontana, 1985); Warnock, *A Question of Life*; Glover, *Causing Death and Saving Lives*. On the historical background to this belief, see MacIntyre, *A Short History of Ethics*.
212 *Talk of the '80s: Mary Warnock* (screened BBC 1, 7 December 1989). Recording held at the British Film Institute, London.
213 MacIntyre, *A Short History of Ethics*, p. 235. On the growth of principles-based approaches in the United States, see Evans, 'Sociological Account of the Growth of Principalism'.
214 Stephen Toulmin, 'The Tyranny of Principles', *Hastings Center Report*, Vol. 11, no. 6 (1981) pp. 31–9; Alasdair MacIntyre, 'How Virtues Became Vices: Values, Medicine and Social Context', in H. Tristam Engelhardt, Jr, and Stuart Spicker (eds), *Evaluation and Explanation*

in the Biomedical Sciences (Dordrecht: Riedel Publishing, 1975) pp. 97–111 (pp. 97–8); Alasdair MacIntyre, 'Why is the Search for the Foundations of Ethics so Frustrating?', *Hastings Center Report*, Vol. 9, no. 4 (1979) pp. 16–22. MacIntyre served as a consultant to the National Commission and became interested in the sources of medical error during the 1970s. He nevertheless claimed that his 'encounters with bioethics were by and large frustrating' because 'its practitioners did not take seriously the deep-cutting moral disagreements of our culture'. Alasdair MacIntyre, correspondence with the author (April 2009).

215 Mary Warnock, 'Commentary on "Suicide, Euthanasia and the Psychiatrist"', *Philosophy, Psychiatry and Psychology*, Vol. 5 (1998) pp. 127–30 (p. 128).

216 See Foucault, *Birth of Biopolitics*, pp. 225–7; Rose, *Powers of Freedom*, p. 142.

217 Margaret Thatcher, quoted in John Campbell, *Margaret Thatcher, Volume One: The Grocer's Daughter* (London: Vintage, 2007) p. 377.

218 Margaret Thatcher, notes for 1979 Conservative Party conference speech, 'Thoughts on the Moral Case 1', 3 October 1979. These notes have been archived as part of the Margaret Thatcher Foundation and are available online at www.margaretthatcher.org (accessed 14 February 2014).

219 Warnock, *A Question of Life*, p. 96.

220 Cooter, 'The Ethical Body', pp. 460–2; Rose, *Powers of Freedom*; Lowry, 'Birth of a New Ethic'.

221 Jasanoff, *Designs on Nature*, p. 152.

222 Ibid. See also Warnock, 'Do Human Cells Have Rights?'; Mary Warnock, 'The Good of the Child', *Bioethics*, Vol. 1 (1987) pp. 141–55.

223 Campbell, interview with the author (2009).

224 See, for example, Cecil Clothier (chair), *Report of the Committee on the Ethics of Gene Therapy* (London: HMSO, 1992).

225 Suzi Leather, quoted in Andrew Brown, 'The Practical Philosopher', *Guardian*, 19 July 2003.

226 Harris, interview with the author (2011); Singer, 'Introduction'; Lockwood, 'The Warnock Report'.

227 Warnock, quoted in *Talk of the '80s*. See also Warnock, 'Embryo Therapy', pp. 28–30.

228 Jasanoff, 'Making the Facts of Life', p. 61.

229 On how public and ethical debates overlooked issues such as patient welfare and equal access to IVF, which ultimately ensured that NHS coverage was patchy and forced most patients to seek treatment at private clinics, see Naomi Pfeffer, *The Stork and the Syringe: A History of Reproductive Medicine* (London: Polity Press, 1993) pp. 165–75.

230 Michael Lockwood, 'Introduction', in Lockwood (ed), *Moral Dilemmas in Modern Medicine*, pp. 1–9 (p. 3).
231 Warnock, 'Government Commissions', p. 67.
232 Warnock, *A Question of Life*, p. xi.
233 Brownsword, 'Bioethics: Bridging from Morality to Law?', p. 29.
234 O. O'Neill, *Autonomy and Trust in Bioethics* (Cambridge: Cambridge University Press, 2002) p. 1.
235 David Archard, 'Why Moral Philosophers Are Not and Should Not Be Moral Experts', *Bioethics*, Vol. 25, no. 3 (2011) pp. 119–27 (p. 123).
236 Ibid, p. 122.
237 Warnock, *A Question of Life*, p. 96. Emphasis added.

5

'A service to the community as a whole': the emergence of bioethics in British universities

Bioethics made inroads into British universities during the 1980s, thanks largely to those individuals, groups and political changes that we have already encountered. During the late 1970s and early 1980s members of medical groups and public figures such as Ian Kennedy called for greater emphasis on medical ethics in student training. They also stressed the benefits of 'non-medical' input, claiming that it relieved clinicians from teaching responsibilities and would help students become 'better doctors' in future.[1] This ensured that many prominent doctors supported new interdisciplinary ethics courses, which were aimed mainly at healthcare professionals.

Many of the academics who taught on these courses were increasingly located in dedicated bioethics centres from the late 1980s onwards. The establishment of these centres reflected a commitment to interdisciplinary teaching on the part of certain doctors, philosophers, lawyers and others. It also owed a great deal to cuts in government funding for universities, which encouraged academics in the humanities and social sciences to work on more 'applied' subjects such as bioethics. This combination of factors shaped the Centre for Social Ethics and Policy (CSEP) at the University of Manchester, which was established in 1986 by the philosopher John Harris, the lawyer Margaret Brazier, the theologian Anthony Dyson and the student health physician Mary Lobjoit. CSEP's establishment reflected Harris's interest in bioethics, Brazier's work on tort law and medical negligence, Dyson's belief that theology should engage with practical issues, and Lobjoit's conviction that student doctors and nurses needed formal ethics training.[2] It also reflected changing priorities in higher education, with CSEP's founding quartet stressing the applied nature of their work and that the new centre benefited doctors, patients and 'the community as a whole'.[3]

Involving non-doctors in medical ethics teaching

During the 1960s and 1970s, as Edward Shotter notes, 'there was no teaching in ethics in British medical education' and leading doctors believed that ethical questions were best 'discussed by consultants, with consultants and *in camera*'.[4] While groups such as the LMG provided a forum where students could discuss ethical issues, Shotter maintained that they 'never claimed to teach medical ethics'.[5] This stance stemmed from uncertainty over whether medical ethics was based on 'the moral codes of religion' or secular frameworks such as utilitarianism. It was hard to teach students, Shotter concluded, 'given the lack of any accepted body of knowledge in medical ethics, and the problems of definition involved'.[6]

While their refusal to teach ethics helped the medical groups appear 'non-partisan and independent of all interest groups', it did not satisfy those who wanted the subject to be included in the medical curriculum.[7] In a 1967 *Lancet* article, for instance, one student noted how controversies over organ transplants and the definition of death ensured that medical ethics had become 'an important facet of the profession's public image', but complained that 'in the training of medical students, however, the subject of a doctor's ethical commitment is presented in a haphazard manner'.[8] The author argued that little mention was 'made of ethics in normal teaching', where most doctors 'were unwilling to make firm generalizations and instead fall back on the time-worn apprenticeship principle – "you'll learn by experience"'.[9] This 'absence of guidance', they concluded, 'leaves many students confused and faintly dissatisfied'.[10]

This view was notably shared by the GMC, which issued a report in 1967 recommending that medical ethics should be given greater priority in 'basic medical education'.[11] Although medical schools were not obliged to follow this advice thanks to GMC rules that gave them autonomy in delivering the curriculum, many introduced lectures on medical ethics during the 1970s.[12] This was clear in the responses to a GMC survey from 1975, where twenty-five out of thirty-four medical schools claimed they now provided lectures on medical ethics, while eleven claimed to have a staff member dedicated to the subject.[13]

But while an increasing number of medical schools taught medical ethics, most continued to believe it was a job for doctors.

In a 1972 editorial, the *Lancet* acknowledged that 'ethical training in medicine' was justifiably getting greater attention in the medical syllabus. Although it praised the LMG's 'commendable attempts to stimulate interest', it nevertheless maintained that teaching ethics 'must surely in the first place remain the responsibility of the medical school themselves'.[14] This attitude was also evident in a *Journal of Medical Ethics* article on ethics teaching at the University of Nottingham, which stated that 'intracurricular aspects of medical ethics' should be taught 'mainly by medical members of staff'. If students wanted to encounter 'non-medical opinion', it continued, then they could do so in 'extracurricular discussion groups' that were open to staff and students of all faculties.[15]

But this is not to say that all medical schools left doctors in charge of ethics teaching. In Scotland, and Edinburgh in particular, non-doctors were increasingly involved in ethics teaching during the 1970s. While he was university chaplain and associated with the EMG, the Royal College of Nursing asked Alastair Campbell to teach an ethics course for postgraduate nurses.[16] These lectures formed the basis for *Moral Dilemmas in Medicine*, which Campbell aimed at qualified healthcare professionals and students following advice from the book's publisher, the medical press Churchill Livingstone.[17] In his foreword to the book, A. S. Duncan, the dean of Edinburgh's faculty of medicine, supported outside involvement in ethics teaching. 'The view from the sidelines can be more objective', Duncan argued, since doctors and nurses were 'only able to give a superficial view' of issues that was 'often biased by personal experience'.[18]

Following the publication of Campbell's book, and with support from Duncan, academics involved with the EMG set up an interdisciplinary research group to investigate whether ethics could be further incorporated into 'ordinary academic and clinical teaching'.[19]After consulting with students and staff in the medical faculty, they established a series of 'experimental' undergraduate courses on 'Social and Moral Issues in Health Care' in 1976. Aimed at student doctors, nurses and psychiatrists, these dedicated ethics courses were the first of their kind in Britain. While the content varied according to the students' disciplinary background, all courses reflected the belief that no viewpoint should dominate ethical discussions and that there was no obviously 'correct' answer to many problems. The philosopher and group member Ian Thompson argued that the courses should adopt a more 'Aristotelian approach' that was 'as closely integrated into the ordinary theoretical and practical training of doctors as possible'.[20] As a result, the courses overlooked formal lectures for group-based discussions of particular cases.[21]

Although the courses were not repeated due to a lack of funding, the heads of seven medical departments asked group members to provide ethics teaching on several undergraduate courses the following year. In a 1978 article for the *Journal of Medical Ethics*, they claimed that the positive response to their 'experimental' sessions highlighted a 'widely felt need for study and discussion of medical ethics in the medical school context'.[22] They argued that this necessitated the creation of more opportunities, both in Edinburgh and elsewhere, for 'multi-disciplinary and inter-professional discussion' in the medical curriculum.[23]

By the late 1970s small numbers of non-doctors were beginning to contribute to other undergraduate and postgraduate courses. At the University of Glasgow, for example, the lawyer Sheila MacLean taught compulsory modules in forensic medicine, which was where Scottish students traditionally discussed ethical issues.[24] In a 1977 piece for the *Journal of Medical Ethics*, doctors at the University of Southampton detailed how lawyers and philosophers made 'a considerable contribution' to a fourth-year ethics course and claimed that students 'benefited greatly from their inclusion'.[25] And the following year, the IME helped the Society for Apothecaries to establish a graduate Diploma in Philosophy of Medicine.[26]

But these changes did not impress Ian Kennedy, who used the prestigious Astor Memorial Lecture in 1979 to complain that 'the formal teaching of medical ethics is a desultory exercise'.[27] Kennedy stressed that more ethics teaching was vital for 'ensuring that medical students are properly exposed to medical ethics and moral debate'.[28] In his fourth Reith Lecture the following year, he repeated that doctors 'should have some educational grounding in ethical analysis' in order to meet the changing expectations of patients and the public.[29] Kennedy believed that this could be achieved by making medical ethics a 'central course' that was taught by someone 'who is not deafened by the rhetoric of medicine'. Prioritising ethics and involving non-doctors in teaching was the only way, he argued, of dragging medical schools 'back into our world and out of their hermetically sealed cocoon'.[30] As before, Kennedy dwelt on the benefits of these new arrangements. He stressed that involving non-doctors in ethics teaching would not threaten traditional relations between doctors and patients, but would instead strengthen them by 'exploring the conflicts and tensions between different ethical principles and suggesting ways of resolving them'.[31]

Kennedy was clear, however, that he did not regard the medical groups as an adequate forum for discussing medical ethics. In a 1981 edition of the *Journal of Medical Ethics*, he claimed that they perpetuated the 'unsatisfactory state of affairs' that his Reith Lectures condemned.[32] Their existence, he argued, 'allows medical schools to avoid incorporating medical ethics into the curriculum in all but a perfunctory manner'. Kennedy also claimed that since medical group seminars were voluntary, they only attracted 'those who are already interested in ethics and want to learn more. The rest, probably a majority of students, do not attend. Arguably it is these who need the education, yet they can and do avoid it.'[33] This was no doubt true for some universities in the 1970s, where doctors believed that the medical groups provided sufficient coverage of ethical issues and were a voluntary 'soft option'. But we should also bear in mind that EMG members were instrumental in developing more formal ethics courses during the 1970s and that the IME, as we shall see, argued that 'non-medical' perspectives should be incorporated into the medical curriculum during the 1980s.

Kennedy's Reith Lectures prompted considerable debate about how medical ethics should be taught, and who should teach it. Most commentators agreed with his claim that medical ethics should be prioritised more in teaching. In an editorial for the *Journal of Medical Ethics*, Raanan Gillon argued there was a '*prima facie* case for the claim that some formal and probably compulsory teaching of analytical medical ethics should be provided within the medical curriculum'.[34] Echoing Kennedy's Reith lectures, Gillon argued that students should undertake '*critical study* of moral decisions, attitudes and actions' in order to prepare themselves for the fact that 'patients in our plural society have various norms and expectations and are often unwilling to accept the moral standards and decisions of their doctors, at least without adequate discussion'.[35] In the same issue, the student president of the LMG praised 'Kennedy's important contribution to the health debate' and agreed that 'the need for ethical and humanitarian education in medical schools is now more necessary than ever'.[36] He also endorsed outside involvement in ethics teaching, claiming that the traditional focus on medical etiquette 'must be replaced by a more genuinely philosophical approach' where 'philosophers, doctors and students pool their expertise and apply it to both old and new issues and problems'.[37]

Alan Johnson, professor of surgery at Sheffield University, was one of several doctors who endorsed Kennedy's view that medical ethics should be 'no optional extra'.[38] Johnson agreed that there was 'no need for the doctor to feel threatened by advice and analysis from the non-medical [*sic*]', but stressed that they should not completely pass responsibility for ethics teaching to 'a solitary, departmentless, "professional ethicist"'.[39] He believed doctors should instead 'get together with non-medical experts' to deliver a course that best captured the realities of clinical practice in a 'pluralistic country and a pluralistic profession'.[40] Like the members of the EMG, Johnson believed that the absence of obviously 'correct' answers meant that 'formal lectures alone are not the best way to teach medical ethics' and should be secondary to 'small-group teaching'.[41]

By 1983, when Johnson's article was published, growing numbers of doctors and 'non-medical experts' were beginning to collaborate in medical ethics teaching. Following a chance meeting with the head of the University of Southampton's medical school, Ian Kennedy had the chance to put his ideas into practice when he co-designed five sessions on medical ethics, timetabled as part of the third-year course in general practice.[42] During each session, students read a book such as Glover's *Causing Death and Saving Lives*, 'which discussed the nature of moral argument, showing how moral arguments may be rationally defended or questioned', and then related it to a specific case study. Although Kennedy and his colleagues noticed that students initially expressed some antagonism towards the new format, which 'may have come from the common reaction to any "outsider" trying to teach doctors their job', they claimed it soon gave way to 'lively discussion'.[43] This encouraging start led the medical school to plan a similar venture for the next academic year, with staff from the law faculty taking over Kennedy's role 'so that in future we shall be able to rely on local sources'.[44]

During the early 1980s Kennedy was also involved in designing ethics courses at the University of Cambridge and St Thomas's Hospital, London, and planned to 'set up teaching at postgraduate level, and indeed at undergraduate level, as soon as possible' through the Centre of Medical Law and Ethics, which he had established at King's College in 1978.[45] In letters to Alan Johnson and Mary Lobjoit, Kennedy proposed that the Centre of Medical Law and Ethics could sponsor a 'meeting of all those interested in the teaching of medical ethics so that experience can be shared and a common policy adopted for the future'.[46] This echoed earlier proposals by Raanan Gillon, who was an affiliate of Kennedy's Centre. In the *Journal of Medical Ethics*, Gillon called for a meeting where representatives from different professions could discuss 'whether or not matters are satisfactory as they are', and also supported the establishment of a 'working party' to suggest new directions in ethics teaching.[47]

Both these demands were met in 1984, when the GMC's Education Committee held a multidisciplinary conference on 'Teaching Medical Ethics' and the IME convened a working party to make recommendations for ethics teaching. All speakers at the GMC conference argued that medical ethics should be an important part of medical training, and most agreed, like one young doctor, that 'non-medical teachers have a role to play'.[48] This viewpoint was captured by Edward Shotter, who conceded that the success of the medical groups may have hindered the incorporation of medical ethics into the curriculum. Shotter now argued that medical ethics should be taught as a central subject, providing that doctors collaborated with 'staff from departments of law, philosophy, the social sciences and theology'. This was vital, he claimed, to ensuring that ethics teaching was not 'the sole responsibility of those who might feel morally bound to express their own views to the detriment of others', and to accurately preparing students for dealing with ethical issues 'in a society where values are in conflict'.[49] Like Shotter, other speakers agreed that since there are many different viewpoints on ethical issues and no obviously 'correct' answer, 'tutorial type teaching, with interplay between tutor and students [is] the most productive method' of teaching medical ethics.[50]

These opinions were shared by an IME working party on medical ethics teaching, which Shotter claimed had been established to make recommendations 'at a time when there is no overriding moral viewpoint, and in the face of diversity of opinion about how the subject might be taught'.[51] The working party was chaired by the psychologist Sir Desmond Pond, then Chief Scientist at the DHSS, and the eighteen other members included doctors, nurses, lawyers, philosophers and theologians.[52] After deciding that reliable information on ethics teaching was 'not readily available', the working party sent out questionnaires to the deans of thirty British medical schools and received twenty-six replies.[53] The responses indicated considerable variety in how medical ethics was taught. While all schools claimed to teach medical ethics, some taught it more than others; some focused on traditional questions of a doctor's rights and duties, while others looked in greater depth at the ethical problems raised by medicine; and some taught it in formal lectures, while others used 'problem-oriented' methods and group discussion.[54]

The working party also found little consistency in who taught medical ethics, especially when it came to the involvement of 'non-medical teachers'.[55] When they asked if non-medical staff were involved in ethics teaching, only two schools replied in the negative. But the replies from the others indicated clear differences. Students in six medical schools only encountered non-medical opinions via the voluntary medical group seminars, while six others used health visitors in formal teaching, seven used hospital chaplains, five used representatives from patient groups, three used theologians, eight used philosophers and five used lawyers, although in four of these cases it was 'the same lawyer' (presumably Ian Kennedy).[56]

When the working party published their recommendations in 1987, which became known as the 'Pond report', they 'decided against recommending a specific syllabus for medical ethics' on account of GMC rules that gave medical schools flexibility in how they delivered the curriculum.[57] But they nevertheless issued 'general recommendations' – with the first being that 'medical ethics teaching should occur at regular intervals throughout medical training'.[58] They also recommended that ethics sessions 'should involve a teacher or teachers with training in the analytic disciplines (moral philosophy, theology or law)' working alongside doctors and 'representatives of the professions associated with medicine (nursing, social work, chaplaincy and others)'.[59] Like the speakers at the GMC conference, the Pond report argued that disciplinary collaboration was essential in order to 'avoid leaving ethics teaching in the hands of a teacher whose tendency is to promote a single, political, religious or philosophical viewpoint'.[60] This was as true for non-medical teachers as doctors, it continued, since 'some lawyers or philosophers may not be as even-handed as their profession suggests'.[61]

This last point was elaborated in the report's appendix, where Jonathan Glover wrote a short paper arguing that non-doctors should simply help students 'think more clearly about issues and understand them better'.[62] To do otherwise, he claimed, would 'leave people more dogmatic or muddled than they were before'.[63] Glover proposed that ethics teaching should thus 'start with cases, not theory' and was better taught in group-based sessions than lectures, since this allowed teachers and students to 'state their own approaches and think out the implications and problems'. Although he reiterated the commonly held view that there were 'no objective true answers', Glover argued that group-based argument could achieve the 'more modest' aim of giving students a 'clear grasp of the complex issues involved'. This approach, he concluded, would make future doctors more tolerant of different views and might also allow them to identify inconsistencies in certain arguments, which might encourage patients or colleagues to 'abandon or modify' particularly dogmatic lines of thought.[64]

Like Kennedy's Reith Lectures, there was nothing particularly new about the Pond report. By using the dismissal of 'correct' answers to support interdisciplinary and group-based teaching, it echoed the individualistic view of ethics held by many British philosophers. And it embodied the core philosophy of the medical groups and the IME, which believed that interdisciplinary debates made students better doctors by reconciling them to moral pluralism. Like other advocates of interdisciplinary teaching, Pond's working party also reflected the British view that medical ethics and bioethics was a 'partnership' between non-doctors and the medical profession.[65] They argued that in contrast to the United States, where medical ethics was often taught by a 'solitary' philosopher, it should be a collaborative enterprise where the doctor remained central.[66]

These arguments were well received before the IME's report was published, as medical support for the EMG group's proposals and Kennedy's Reith Lectures indicates. But the Pond report was certainly influential, partly thanks to the illustrious nature of its working party, which included renowned doctors, scientists, philosophers and lawyers, and partly thanks to the way that it framed 'non-medical' teaching as beneficial to doctors. Its proposals were endorsed by the GMC's 1993 report, *Tomorrow's Doctors*, which stated that sessions on 'the ethical and legal issues relevant to the practice of medicine' should become part of the core curriculum.[67] Although *Tomorrow's Doctors* did not promote multidisciplinary teaching as explicitly as the Pond report, it nevertheless noted that doctors were increasingly likely to rely on the 'social sciences and philosophy' to help them handle 'ethical issues that will increasingly impinge on the problems of health'.[68]

In 2007 Kenneth Boyd, a founding member of the EMG, recalled that the Pond and GMC reports 'gave teeth' to existing support for multidisciplinary teaching.[69] During the late 1980s and early 1990s greater numbers of non-doctors became involved in teaching medical students, which went from being something they did rarely and on an *ad hoc* basis to a major activity. By 1996, when the University of Bristol appointed Alastair Campbell to one of the first British chairs in medical ethics, it was clear that in universities, just as in public and Parliament, medical ethics was no longer considered solely a matter for doctors.[70]

Creating academic centres for bioethics

The non-doctors who taught medical students were initially based in law, philosophy and social science departments. Medical ethics constituted a small part of their workload and was secondary to general teaching in their respective subjects.[71] However, by the publication of the Pond report in 1987, and certainly by the publication of *Tomorrow's Doctors* in 1993, they were often located in new and interdisciplinary centres dedicated to medical ethics or bioethics. Many of these centres were established by those advocates of 'multidisciplinary teaching' we have already encountered, who believed they provided an institutional base for lawyers, philosophers, doctors and others to collaborate on undergraduate and postgraduate courses. Ian Kennedy opened the first dedicated Centre of Medical Law and Ethics at King's College in 1978, to facilitate interdisciplinary teaching and research, and members of local medical groups helped establish similar centres at the universities of Swansea and Manchester during the mid 1980s.[72]

But the promotion of 'multidisciplinary teaching' was not the sole reason that these centres emerged; nor was it the sole reason that increasing numbers of academic lawyers, philosophers and others engaged with practical issues during the 1980s and 1990s. We cannot understand the growth of academic bioethics without also appreciating the impact of budget cuts on higher education in the 1980s, which forced staff in the humanities and social sciences to work in more 'applied' areas and stress the utility of their work. As we saw in chapter 3, Margaret Thatcher's Conservative Party sought to reform many professions and public services after the 1979 general election, and universities were no exception. Thatcher believed that the guaranteed distribution of government money through the University Grants Committee (UGC) had insulated universities against the current recession and, crucially, had distanced them from the commercial approaches that the Conservatives believed would transform Britain. She made her views clear in 1981, when she informed Geoffrey Warnock, then vice-chancellor at Oxford, that universities were arrogant, elitist and indifferent to economic needs: 'wasting time and public money on such subjects as history, philosophy and classics'.[73]

The same year, the government announced that it was taking steps to reform universities by making them more competitive and self-sufficient. Mark Carlisle, Secretary of State for Education and Science, claimed they wanted to encourage a 'leaner university system ... better oriented to national needs and operating within the context of what the nation can afford'.[74] Carlisle announced that the government was cutting UGC funding, and that reductions would be imposed selectively between institutions and subject areas. In July 1981 the UGC informed all vice-chancellors of the cuts that were being imposed on their universities, and offered guidance on how they felt the reduction should be spread across particular disciplines.[75]

Given the government's emphasis on meeting 'national needs' and enthusiasm for commercial approaches, academics rightly predicted that the UGC would prioritise disciplines that were seen to contribute to economic growth, while penalising those they viewed as unproductive.[76] As one pro-vice-chancellor at the University of Manchester noted, the UGC's guidance clearly 'favoured "big science", particularly engineering, physics, chemistry and computer science'.[77] As expected, the UGC also advised vice-chancellors to use budget cuts to 'downgrade the arts'.[78] In a newsletter, the Society for Applied Philosophy outlined how the UGC had warned that fields such as philosophy 'will be receiving low priority in its deliberations in the coming years', and reported that they had advised philosophers to set up a review to 'assist in the closing of some departments'.[79]

In 1985 philosophers formed a committee to defend their subject against UGC cuts. This National Committee for Philosophy (NCP) submitted a report to the UGC arguing that cuts to philosophy 'should be proportionate to the cuts suffered by other disciplines'.[80] To support their argument, the NCP framed philosophy as an increasingly practical discipline, with growing numbers of philosophers now 'applying their insights to other disciplines, and to the philosophical and ethical problems of everyday life'.[81] At the same time, A. J. Ayer and Mary Warnock publicly asserted that philosophy was vital to maintaining a society that valued reasoned debate, analytical rigour and intellectual originality, and protested that the government and the UGC's 'new vocationalism' represented an assault on the notion of learning for its own sake.[82]

In spite of these protests, faced with severe financial pressure and fearing the consequences of ignoring the UGC's guidance, most universities did protect the sciences and 'downgrade' their arts and humanities departments. Senior academics in these fields were encouraged to take early retirement and were not replaced, which made it easier for politicians and administrators to criticise shrinking departments as 'weak and ineffectual'.[83] By the end of the 1980s seven philosophy departments had closed and many others faced an uncertain future.[84]

The pressure on philosophy departments was compounded when the government replaced the UGC with a new Universities Funding Council (UFC) in 1988. The government's enthusiasm for external oversight was reflected in the UFC's composition, in which academics were outnumbered by outsiders, and particularly businessmen, who shared the Conservative belief that 'the purpose of higher education was to satisfy the needs of industry'.[85] The UFC distributed money on the basis of 'research assessment exercises' that judged the quality of a department's academic work. 'Quality' here involved acquiring large grants from research councils, collaborating with industry and publishing regular articles; and these criteria, as Mary Warnock noted, clearly favoured applied fields such as the sciences and engineering.[86] In order to meet the new standards and combat declining government funds, academics in fields such as philosophy were obliged to seek money from outside sources such as research councils and charities.[87] Many now believed that they stood a better chance of gaining funding and meeting demands that research had to confer 'social benefits' if they worked in areas with obvious practical relevance.[88]

Budget cuts and the new assessment criteria also encouraged the reconfiguration of traditional disciplinary structures. University managers and some academics believed it was 'possible to improve both performance and image by casting down old-fashioned departmental barriers and abandoning worn-out subject divisions'.[89] This belief led to the creation of interdisciplinary institutes, both in the sciences and the humanities, which 'brought together scholars with shared interests to consider common problems'.[90] As Robin Downie and Jane MacNaughton note, during the 1980s this combination of factors encouraged growing numbers of academics to 'become involved with bioethics'.[91] Bioethics appealed to staff in the humanities who sought funding for applied work, and its presentation as a 'partnership' made it an obvious subject for interdisciplinary collaboration.

Support for 'multidisciplinary' ethics teaching and the changing political climate were both evident in the 1986 formation of the University of Manchester's Centre for Social Ethics and Policy (CSEP). CSEP brought together academics from medicine, theology, law and philosophy, who all believed that medical students would benefit from interdisciplinary ethics teaching. Although CSEP's founders did not directly cite UGC cuts as an influence, financial constraints nevertheless ensured that they stressed the applied nature of their work when seeking money and publicising new courses. Changing priorities for higher education also ensured that CSEP was well received by senior university figures, who praised it as good evidence of the 'transdisciplinary co-operation' that was increasingly expected of academics.[92]

One of CSEP's co-founders was Mary Lobjoit, a student health physician who had organised the Manchester Medical Group (MMG) since its formation in 1975. Lobjoit had long believed that increased ethics teaching, with input from several professions, was vital to helping medical students become 'better doctors'.[93] She supported Ian Kennedy's claims that medical ethics should be a 'central course' and corresponded with him after his Reith Lectures to discuss 'ideas concerning a more formal approach to teaching medical ethics'.[94] Between 1984 and 1987 Lobjoit was also a member of Desmond Pond's working party on ethics education, whose final report reflected her existing support for interdisciplinary teaching.

As organiser of the MMG, Lobjoit invited several non-doctors to talk on ethical issues. One of her speakers was John Harris, who had been appointed to the University of Manchester as lecturer in the philosophy of education during the late 1970s.[95] Although he mainly taught philosophy of education, Harris pursued his interest in bioethics by talking to the medical groups, arranging a 1983 workshop on 'Philosophical and Ethical Issues in Medicine', writing on the ethics of IVF, and extending his arguments from here and the 'Survival Lottery' in a 1985 book on *The Value of Life*.[96] This work ensured that Harris was often asked to comment publicly on bioethical issues by the mid 1980s, with the Society for Applied Philosophy selecting him as a preferred speaker for their 1985 'conference on bio-ethics'.[97]

In 1984, during the train journey to a debate on surrogacy at the University of Aberdeen, Harris met Anthony Dyson, professor of social and pastoral theology at the University of Manchester. Like Ian Ramsey and Gordon Dunstan, Dyson believed that theology needed to engage with contemporary concerns to remain relevant. He also shared their enthusiasm for interdisciplinary collaboration, which he had had the chance to satisfy when he served on the Warnock inquiry between 1982 and 1984.[98] These convictions were clear in a talk to the Manchester Literary and Philosophical Society, when Dyson outlined how:

> many different kinds of human experience from many different directions are necessary contributions to the future shape and content of medical ethics, as we explore wider definitions of the healing community, and seek to discover the social and political arrangements which will help the health care professionals to identify more fully their own central, crucial and indispensable role in the evolution of a medically moral society.[99]

Dyson shared his enthusiasm for interdisciplinary collaboration with Mary Lobjoit, who he knew through the MMG. He also shared it with John Harris; but despite working at the same university, they had never met or heard of each other before their trip to Aberdeen.[100] After returning to Manchester, Harris and Dyson began discussions with Lobjoit about establishing an interdisciplinary MA degree in Healthcare Ethics, which would bring together local academics with an interest in medical ethics. During 1984 and 1985 they set about gaining support from senior figures in the medical school and the faculty of science – 'without which', Dyson wrote to one scientist, 'we cannot easily go ahead'.[101]

These letters framed the potential MA as a timely venture and a catalyst for greater collaboration. As Lobjoit informed Max Elstein, head of the department of obstetrics and gynaecology, it was 'a response to increasing interest, both in Manchester and nationally, and we hope that such a course will both stimulate and focus the interests of many of our colleagues'.[102] Lobjoit claimed that the MA was needed to keep pace with new degrees being offered in places such as King's College, where Ian Kennedy's Centre started a Diploma in Medical Law and Ethics in 1984, and to meet the growing support for 'multidisciplinary teaching' from bodies such as the IME and the GMC. As they did nationally, local doctors and scientists supported proposals for an interdisciplinary ethics degree. Elstein viewed it as 'a splendid idea' and told Lobjoit that ethics was 'something that the students feel they need to be taught in greater depth'.[103] After seeing a draft of plans circulated in the medical school, Mark Ferguson, professor of basic dental sciences, wrote to Dyson and pledged his 'wholehearted support' for the new MA.[104]

Buoyed by the 'outstanding measure of support' from scientists and doctors, Dyson, Harris and Lobjoit broadened their plans to include the establishment of a 'more general forum in which the crucial problems facing our society can be examined from a wide variety of perspectives'.[105] They now proposed the establishment of an interdisciplinary centre where they and others could pursue their mutual interest in 'medical ethics and applied ethics broadly conceived'.[106] Planning for this centre soon included the lawyer Margaret Brazier, who had written on medical negligence as part of broader work in tort law and also knew Lobjoit through the MMG.[107] The title this quartet chose for their new venture indicated that they wanted it to be a 'more broadly "bioethics" venture' than Kennedy's Centre of Medical Ethics and a recently opened Centre for the Study of Philosophy and Health Care at the University of Swansea.[108]

This broad remit was apparent in a planning document for the prospective 'Centre for Social Ethics and Policy', which stated that it would look at 'the ethical and policy implications of developments in all areas of society'. These included 'the necessity for education about AIDS and for an educational response to medical and biotechnological advance', as well as more general topics such as 'race and gender issues in education and the wider social role of education and educational institutions'.[109] These broad aims reflected

Harris's interest in the moral consequences of new biotechnologies and philosophy of education, Brazier's interests in patient rights and Dyson's interest in 'the contribution of feminist thought to medical ethics'.[110] They also indicate that while new ethics centres may have been broadly shaped by demands for multidisciplinary teaching and 'applied research' in the 1980s, their orientation and focus differed according to specific local conditions and the interests of the staff involved.

CSEP was established at a time when the university hierarchy was promoting interdisciplinary collaboration and asserting the value of applied research. The vice-chancellor Mark Richmond, a microbiologist by training, had been appointed in 1981 to help the university deal with the UGC's budget cuts. Richmond was a member of the government's advisory group on genetic manipulation, the successor to GMAG, and had strong connections with the UGC. He shared the UGC and the government's enthusiasm for practical research, both in the arts and humanities, and was pictured in *The Times* wearing a train-driver's hat emblazoned with the logo 'Universities Work!'[111] Richmond also believed that the university would be better equipped to deal with budget cuts and changing priorities by establishing 'research units' that brought academics from different fields together to work on common problems.[112] He encouraged a wholesale reform of biology in 1986, when eleven separate departments merged into a unified School of Biological Sciences that prioritised 'applied' work in molecular biology and built links with pharmaceutical firms.[113] CSEP was formally established in the same year and was administered as a centre within the department of education, with Lobjoit acting as administrative director, Dyson as academic director and Harris as research director.[114] Although CSEP was a much smaller enterprise than the School of Biological Sciences, it nevertheless attracted praise from Richmond, who told John Harris that he considered it 'a most interesting example of transdisciplinary co-operation'.[115]

By dwelling on the practical benefits of interdisciplinary collaboration, publicity for CSEP reflected changing priorities in higher education during the 1980s. A front-page article in the university magazine *This Week*, which detailed the opening of this 'new ethics centre', asserted that the 'crucial problems facing society today' were best 'examined from a variety of perspectives'.[116] A promotional brochure from 1987 also claimed that CSEP was 'a product of, and committed to, inter-departmental and inter-institutional collaboration'.[117] This collaboration was vital, it asserted, 'if we are to harness the benefits while protecting ourselves from the all too probable catastrophes of a technological age'.[118] The brochure emphasised the practical benefits of CSEP's MA in Healthcare Ethics and of research projects on autonomy and consent, theological issues in nuclear disarmament and the ethics of biotechnology, which it argued would help professions and the public develop 'the ability to resolve the dilemmas they pose'.[119]

There were clear motives behind this emphasis on the applied nature of CSEP's work. In line with government and UGC criteria, the university hierarchy expected its departments and 'research centres' to generate their own income by obtaining money from charities, research councils and commercial firms. This was made clear in CSEP's promotional literature, which stated that the university had provided 'very limited resources' towards 'the work of the Centre and its future development'.[120] In order to obtain money for a series of lectures on 'Experiments on Embryos', which doubled as CSEP's public launch event, Lobjoit requested money from a variety of commercial sources and Brazier wrote to several charities and legal firms. Their letters acknowledged that any donation 'would be a different form of investment from the usual clinical research', but stressed that 'we are convinced of equivalent importance in improving healthcare practice'.[121]

Many firms turned these requests down, claiming that 'research departments in the pharmaceutical industry find themselves under great restraints these days for any sponsorship other than clinical research projects'.[122] But some were more forthcoming, and ICI, Hoechst, Boots and the charitable Hamlyn Trust all contributed to publicity for CSEP's embryo lectures and paid the costs of speakers, who were Robert Edwards, the neurologist John Marshall, the lawyer Douglas Cusine and the theologian Keith Ward.[123] Thanks largely to Edwards's attendance, these lectures attracted a large audience and, John Harris recalls, 'really put us on the map as dealing with something that was a public interest and general interest'.[124] Buoyed by this success, Mary Lobjoit told her contacts at ICI that CSEP intended 'to continue with this format for next year' and was planning a series of lectures on 'The Ethics of Experimentation on Children'.[125]

While they arranged the open lectures and lobbied potential

funders, CSEP's founders continued to promote the MA in Healthcare Ethics to the medical school and across the university. In letters to senior lecturers and the dean of the medical faculty, Lobjoit used her position on Desmond Pond's working party to claim that the MA would bring Manchester into line with forthcoming proposals that 'ethical teaching should take a more prominent place in medical education'.[126] Like Kennedy, she presented interdisciplinary teaching as a 'partnership' that would benefit doctors – since 'it is not either feasible or appropriate for your discipline to bear the whole responsibility for such enterprises if they are to be expanded'.[127] As with Kennedy's Reith Lectures, this tactic was successful. Several doctors agreed to teach on the MA and the Board of Studies for medicine decided that the medical school would be one point of entry for the new degree (the others were the departments of education and theology).[128]

CSEP's founders decided to aim the MA primarily at healthcare professionals and 'hoped the course will attract, *inter alia*, doctors, nurses, psychologists, health-care administrators, the whole range of social workers involved in care or therapy and all those, including teachers, involved with handicap'.[129] This was made clear in publicity, which portrayed the MA largely as a professional development course. A promotional brochure and press release claimed that it met 'the need for study training and research in healthcare ethics' and was 'centred firmly on healthcare practice'.[130] Perhaps unsurprisingly, when the course started in October 1987 its intake consisted mainly of 'nurses, including midwives, and a number of doctors from several branches of medicine'.[131] At the same time, publicity for CSEP was also aimed at non-medical professionals and humanities graduates, and claimed that the MA 'does full justice to the philosophical, legal, religious, historical and social dimensions to healthcare ethics'.[132] This ensured that the MA intake also included smaller numbers of 'solicitors, philosophy graduates and ministers of religion'.[133]

Launched the same year as an MA at Kennedy's Centre of Medical Law and Ethics, CSEP's MA course was one of the first dedicated medical ethics or bioethics degrees in Britain.[134] In a 1987 letter Kennedy told John Harris that it 'looks exciting' and hoped 'the course is a great success – in my view there can't be too many of them'.[135] As Kennedy and CSEP's founders had hoped, both degrees were successful from the outset. In a 1987 edition of the *Journal of*

Medical Ethics, Harris, Dyson, Lobjoit and Brazier claimed that the 'high rate of applications for the MA indicates a level of interest and enthusiasm which has astonished and pleased the degree's organisers'.[136] Indeed, the level of interest was so high that they discussed 'the question of a ceiling on student numbers'.[137] This ceiling was never implemented, however, and fees for the MA in Healthcare Ethics provided a regular source of income from 1987 onwards.

The degree's structure and focus, with input from many staff and departments across the university, reflected the British attitude that no one profession should dominate medical ethics or bioethics. Students took two compulsory modules in 'moral philosophy', taught by Harris and the philosopher Harry Lesser, and two modules in 'cases in healthcare practice', taught by Lobjoit and several medical staff. The handbook for Harris's module also reflected the general belief that 'the importance of medical ethics does not lie in its ability to provide any answers in advance to the difficult problems faced by healthcare professionals and others'. Harris claimed that it lay instead 'in its ability, first, to widen awareness of the issues involved and sensitivity to them; secondly, to clarify one's thinking about these issues'.[138] Like other British philosophers, Harris was sceptical of the principles-based approach that dominated American bioethics and taught it more as a 'problem-based subject'.[139] In addition to these courses and a 20,000 word dissertation, students also had to pick two options from courses in 'medico-legal problems', taught by Brazier, 'religious issues in medical ethics', taught by Dyson, 'medical ethics in historical context', taught by historians of medicine, and 'medicine in modern society', taught by staff in a department for science and technology policy.[140]

Following the publication of the Pond report, CSEP staff also became more involved in undergraduate medical ethics teaching. As Harris and Max Elstein outlined in 1990, they scheduled interdisciplinary ethics sessions during the third-year obstetrics and gynaecology course, since 'many of the problems which cause concern to students are within the field of human reproduction'.[141] Harris and Elstein acknowledged that this format was something of a compromise though, as students had 'requested a specialised course in medical ethics' but senior staff rejected the 'idea of "giving up valuable teaching time" within the medical curriculum'.[142] They nevertheless believed that wherever it was scheduled, 'an essential component of this teaching is its interdisciplinary nature'.[143]

In order to give students experience 'of moral argument and of the disciplines that make moral argument possible', the sessions were taught by a combination of 'theologians, social workers, senior nurses, philosophers and lawyers'.[144] They took up a single morning, with three lectures devoted to specific issues. Harris gave an introductory lecture that outlined the moral problems raised by an example such as the *Arthur* case, Brazier outlined the associated legal issues, and students were then split into small discussion groups led by 'a clinician and an ethicist from a non-clinical background'.[145] Harris and Elstein echoed the Pond report and previous discussion of ethics teaching when they argued that this format was essential to enabling students 'to think their way through a problem and come to an appropriate solution, or equally important, to the realisation this may not be possible'.[146]

In addition to these undergraduate sessions, the MA attracted increasing numbers of students each year. While this boosted CSEP's profile in and beyond the university, it also presented problems. In 1991 Mary Lobjoit wrote to a contact at Boots claiming that it was 'gratifying to find that we are getting known, that our students are regarded with respect and people actually enjoy getting involved in this particular form of study'. But she noted that this success meant that CSEP had ceased to be 'a part time enterprise'.[147] Teaching on the MA and undergraduate courses now constituted a significant workload for Harris, Dyson and Brazier, who Lobjoit claimed were already 'snowed under' with work for their own departments.[148]

Although she had taken early retirement in 1990, following a reform of the student health service, Lobjoit still taught her MA module on a voluntary basis. As her letters demonstrate, she also continued to seek outside funding for CSEP. In order to ease her colleagues' workloads and 'get the centre on a more secure footing', Lobjoit now wrote to Boots exploring the possibility 'of a number of interested parties donating a fixed sum, per year, rather like a covenant', which would pay for an administrator and additional staff.[149]

While Lobjoit's contact at Boots agreed to donate £500 'toward your costs', he was unable to sanction more permanent support and doubted whether any other firm 'would wish to sign a covenant guaranteeing support for any length of time'.[150] But CSEP did gain a more 'secure footing' in 1992 when it secured a major European Commission grant for a three-year project titled 'AIDS: Ethics,

Justice and European Policy'. The project was co-ordinated by Harris and brought together participants from fourteen European countries to look at issues associated with AIDS, such as the ethics of compulsory screening, confidentiality and euthanasia.[151] Crucially, the award of over £300,000 lightened the workload for CSEP's founders by funding the appointment of an administrator and research staff, who taught on the MA course and worked on the AIDS project. It also consolidated CSEP's growing reputation within the university, drawing praise from Martin Harris, who replaced Mark Richmond as vice-chancellor in 1992 and shared his predecessor's belief that academics needed to be 'entrepreneurial' in order to gain funding.[152] In a 1992 letter the new vice-chancellor congratulated John Harris on the AIDS grant and claimed that 'in the present financial climate, it is crucial that the University is able to attract significant recognition and funding of this kind if we are to maintain our momentum as a leading research institution'.[153]

By this point Margaret Brazier and John Harris had been promoted to chairs at the university. The fact that CSEP was now a recognised centre was reflected in the title of Harris's chair, which was not a professorship of applied ethics or philosophy of education, but of 'bioethics'.[154] CSEP gained further recognition and funding in 1995 with another European Commission grant for a project on 'Communicable Diseases, Lifestyles and Personal Responsibility: Ethics and Rights'. By the late 1990s this research income and an annual intake of around forty MA students ensured that CSEP was at the forefront of a growing network of centres for bioethics and medical ethics, which brought together staff from different fields in King's College, Cardiff, Liverpool, Bristol, Glasgow, Keele, Newcastle, Edinburgh, Nottingham, Swansea and Oxford.

Although they all broadly looked at ethical issues associated with medicine or biological science, institutional factors and the interests of particular staff ensured that these centres were often located in different departments and prioritised varying approaches to bioethics. Some were based in medical or veterinary schools, while others were based in law or philosophy departments. Many, such as CSEP, adopted a case-driven approach to bioethics, while a minority, such as the University of Nottingham's Centre for Applied Bioethics, organised their work around principles-based methods.[155] While these centres increasingly provided an institutional home for bioethics, this did not mean that it became a narrow academic activity.

In addition to teaching and carrying out research, their staff spoke publicly on ethical issues, established links with politicians and policymakers, and served on regulatory bodies such as the HFEA.[156]

Many academics in the humanities and social sciences believe that the focus on 'applied' topics such as bioethics ensured the future of their subjects, by providing funding and jobs in an increasingly competitive and austere climate. Writing on the emergence of academic bioethics, Downie and Macnaughton claim that it is hard to disagree with Stephen Toulmin's contention that 'medicine saved the life of ethics'.[157] Changing priorities in higher education certainly provided an opportunity for like-minded academics to engage with practical issues and collaborate across disciplinary lines. By establishing ethics courses and centres such as CSEP, which secured outside money and postgraduate fees, these academics won the approval of senior managers and ensured that academic bioethics became a 'growth industry' into the 1990s. In addition to proving their utility by helping 'doctors become better doctors', these courses also sustained the growth of bioethics by acting as an entry point to the field.[158] Several students on CSEP's MA in Healthcare Ethics, such as Søren Holm, have gone on to have successful careers as bioethicists, and the same is true of graduates from other centres.[159]

But not everyone took a positive view of these new centres and courses. Mary Warnock, for example, did not believe that philosophers should concentrate solely on medical issues or work 'in a special medical ethics department'.[160] In a 1990 lecture she claimed that if philosophers, lawyers and theologians were 'fed nothing but a diet of medical ethics, or, even worse, if they have taught nothing but this subject and have conducted all their research in it, they are likely to become as tunnel-visioned as the doctors and scientists themselves'.[161] Warnock argued that while they should be 'acquainted with some of the issues they are likely to encounter as members of ethical committees ... a moral philosopher who deserves the name must concern himself with the nature of morality in general, and must be prepared to consider examples from all kinds of areas, public and private'.[162] She also stressed that moral philosophy 'should not be studied separately from all the rest of philosophy, epistemology, for example, or the philosophy of mind', and defended these subjects against 'spiteful and short-sighted' budget cuts.[163]

Yet while Warnock viewed bioethics as an important component of philosophy, some regarded it as inferior to more theoretical approaches and often resented the prestige and money it attained.[164] This was not necessarily a problem for ethicists who worked in interdisciplinary centres, but it often left those in more traditional departments 'out in the cold'.[165] And while their interest in bioethics may have proved fruitful for staff in new centres, it was often not a central concern for staff or undergraduates in the schools and faculties in which these centres were based, such as law, philosophy or medicine. This ensured that in order to justify their existence, staff in bioethics centres often had to work harder to maintain a strong postgraduate intake and generate research income.[166]

Michael Whong-Barr has also argued that the emphasis on clinical matters in ethics teaching came at a price, marginalising those broader issues that were previously discussed in the medical group seminars and running the risk of 'complacently supporting social structures and assumptions'.[167] This criticism no doubt stems from the fact that the emergence of bioethics centres and courses contributed to the demise of the medical groups. As ethics teaching became a full-time occupation for many academics, their enthusiasm for organising medical groups diminished. And once ethics was increasingly taught on formal undergraduate and postgraduate courses, universities stopped subsidising the medical groups and student demand tailed off significantly. The LMG disbanded in 1989, after Edward Shotter was appointed Dean of Rochester Cathedral, and the regional medical groups followed suit during the 1990s.[168] Recalling the end of the Newcastle medical group in the early 1990s, the Revd Bryan Vernon states that 'as medical ethics became something that was taught more, so it became less something that was done outside [in medical groups]'.[169]

Whong-Barr is certainly right to claim that undergraduate courses discussed a narrower set of issues than the medical groups and 'lacked analysis of the social processes that … help generate moral dilemmas in the first place'.[170] As Harris and Elstein acknowledged, timetabling constraints ensured that they only had time to look at one or two issues that students were likely to encounter in clinical practice.[171] Nevertheless, we should bear in mind that some of the postgraduate courses that emerged in the 1980s looked at a broad range of issues, including the social and historical aspects of medicine and science. And we should also bear in mind that in contrast to the medical group seminars, compulsory undergraduate sessions guaranteed that all medical students encountered some interdisciplinary perspectives, which had been the original aim of the LMG's founders in the 1960s.

The realisation of these ambitions in the 1980s stemmed from the efforts of the individuals encountered in this chapter and reflects the importance, once again, of the political changes that followed the 1979 election. In stressing how bioethics benefited doctors and 'the community as a whole', figures such as Lobjoit, Harris, Brazier and Dyson were well placed to benefit from the increasing emphasis on 'applied' work and ensured that new ethics courses and centres were increasingly prized by students, doctors and university managers alike.

Conclusion

As elsewhere in Britain, the interplay between professional concerns and political changes underpinned the emergence of 'multidisciplinary' courses and bioethics centres in universities during the 1980s and 1990s. Support for interdisciplinary teaching first emerged in the 1970s, when medical group members argued that it would give students a greater awareness of moral issues, including the fact that they often lacked clear answers. This argument was reiterated in Ian Kennedy's 1980 Reith Lectures, and it also underpinned the 1987 Pond report. By presenting 'non-medical' input as beneficial to both practising and future doctors, these arguments secured the support of influential bodies such as the GMC and ensured that medical schools increasingly involved philosophers, lawyers and others in ethics teaching during the 1980s.[172]

Support for these multidisciplinary approaches was consolidated by budget cuts for universities, which favoured the growth of 'applied' subjects such as bioethics. This emphasis boosted those supporters of formal ethical training and applied ethics, making it easier for Lobjoit, Harris, Dyson and Brazier to establish and promote CSEP. And the success that CSEP and similar centres enjoyed in attracting students and research funding sustained the growth of bioethics, encouraging greater numbers of academics to engage with 'applied' issues and leading vice-chancellors to praise it as an important field 'in the present financial climate'.[173]

But we should nevertheless refrain from making broad assumptions about how and why bioethics emerged at particular universities. While the broad factors noted above were undoubtedly influential, local factors such as individual personalities and institutional politics also played a decisive role. This is illustrated by the fact that new bioethics centres emerged in different faculties, prioritised varying approaches and often had contrasting relationships with their local medical groups. While Mary Lobjoit's role as organiser of the MMG ensured that it had strong links to CSEP, for instance, Ian Kennedy's scepticism towards the medical groups ensured that the Centre of Medical Law and Ethics had little connection to the LMG.[174]

The interplay between national and local factors was also evident in the way that Mary Lobjoit helped shape broad policies as part of Desmond Pond's working party, and then used her position on this group to promote CSEP's courses to doctors at the University of Manchester. This demonstrates how academics in particular institutions both generated and utilised the growing enthusiasm for interdisciplinary teaching. They did so in order to advance their own agendas: to demonstrate the utility of particular approaches, to formalise ethical training for doctors and to work with like-minded colleagues in other disciplines. With this in mind, we should therefore see national factors as *enabling* local changes rather than simply directing them.

Notes

1 Margaret Brazier, Raanan Gillon and John Harris, 'Helping Doctors Become Better Doctors: Mary Lobjoit – An Unsung Heroine of Medical Ethics in the UK', *Journal of Medical Ethics*, Vol. 38 (2012) pp. 383–5.

2 Ibid.

3 Anon, 'Draft Begging Letter', March 1991. Archives of the Centre for Social Ethics and Policy, the University of Manchester (henceforth CSEP archives).

4 Edward Shotter, cited in Reynolds and Tansey (eds), *Medical Ethics Education in Britain*, p. 8; Edward Shotter, 'Director's Report: Can Medical Ethics be Taught?' (London: London Medical Group, 1974) pp. 1–4 (p. 2). Wellcome archives: GC/253/A/31/8.

5 Shotter, 'Director's Report: Can Medical Ethics be Taught?', p. 1.

6 Ibid.

7 Ibid.

8 J. B. MacDonald, 'Medical Ethics Today: A Student's Uncertainties', *Lancet*, Vol. 289 (1967) p. 563.

9 Ibid.
10 Ibid.
11 General Medical Council, *Recommendation as to Basic Medical Education* (London: General Medical Council, 1967) p. 18.
12 This GMC adopted a flexible approach to education in 1957. For more detail, see A. H. Crisp, 'The General Medical Council and Medical Ethics', *Journal of Medical Ethics*, Vol. 11 (1985) pp. 6–7.
13 General Medical Council, *Survey of Basic Medical Education in the British Isles* (London: Nuffield Provincial Hospitals Trust, 1977).
14 Anon, 'Ethics at Medical School', *Lancet*, Vol. 299 (1972) p. 1406.
15 J. S. P. Jones and D. H. H. Metcalfe, 'The Teaching of Medical Ethics', *Journal of Medical Ethics*, Vol. 2 (1976) pp. 83–6 (p. 83). These 'extracurricular' lectures were not part of the medical group network. Indeed, by the 1980s Nottingham remained one of the few medical schools not to have its own medical group seminars.
16 Campbell, interview with the author (2009).
17 Ibid.
18 A. S. Duncan, 'Foreword', in Campbell, *Moral Dilemmas in Medicine*, pp. v–vi (p. v).
19 Kenneth Boyd, Colin Currie, Ian Thompson and Alison J. Tierney, 'Teaching Medical Ethics: University of Edinburgh', *Journal of Medical Ethics*, Vol. 4 (1978) pp. 141–5 (p. 141).
20 Ian Thompson, 'The Implications of Medical Ethics', *Journal of Medical Ethics*, Vol. 2 (1976) pp. 74–82 (p. 76).
21 Boyd et al., 'Teaching Medical Ethics'. See also Edward Shotter, 'A Retrospective Study and Personal Reflection on the Influence of Medical Groups', in Reynolds and Tansey (eds), *Medical Ethics Education in Britain*, pp. 71–118 (p. 107).
22 Boyd et al., 'Teaching Medical Ethics', p. 145.
23 Ibid, p. 141.
24 MacLean, interview with the author (2009).
25 K. J. Dennis and M. R. P. Hall, 'The Teaching of Medical Ethics', *Journal of Medical Ethics*, Vol. 3 (1977) pp. 183–5 (p. 185).
26 Edward Shotter, *Twenty-Five Years of Medical Ethics: A Report on the Development of the Institute for Medical Ethics* (London: The Institute of Medical Ethics, 1988) p. 7. Wellcome archives: GC/253/A/31/8.
27 Kennedy, 'What is a Medical Decision?', p. 31.
28 Ibid, p. 30.
29 Kennedy, 'Medical Ethics are not Separate from but Part of Other Ethics', *Listener*, 27 November 1980, pp. 713–15 (p. 715).
30 Ibid.
31 Kennedy, *The Unmasking of Medicine*, p. 117.

32 Kennedy, 'Response to the Critics', p. 208.
33 Ibid.
34 Raanan Gillon, 'Medical Ethics and Medical Education', *Journal of Medical Ethics*, Vol. 7 (1981) p. 171–2 (p. 172). Emphasis in original. Gillon replaced Alistair Campbell as editor of the *Journal of Medical Ethics* in 1980.
35 Gillon, 'Medical Ethics and Medical Education', p. 172. Emphasis in original.
36 Alan Schamroth, 'A Medical Student's Response', *Journal of Medical Ethics*, Vol. 7 (1981) pp. 191–3 (pp. 193, 192).
37 Ibid, p. 193.
38 Alan G. Johnson, 'Teaching Medical Ethics as a Practical Subject: Observations from Experience', *Journal of Medical Ethics*, Vol. 9 (1983) pp. 5–7 (pp. 5–6).
39 Ibid, p. 5.
40 Ibid, pp. 5–6.
41 Ibid.
42 M. D. Jewell, 'Teaching Medical Ethics', *British Medical Journal*, Vol. 289 (1984) pp. 364–5 (p. 364).
43 Ibid, p. 364.
44 Ibid, p. 365.
45 Ian Kennedy to Mary Lobjoit, 14 March 1983. CSEP archives. For more on the Centre of Medical Law and Ethics, see Reynolds and Tansey (eds), *Medical Ethics Education in Britain*, pp. 46–7.
46 Kennedy to Lobjoit, 14 March 1983. CSEP archives.
47 Gillon, 'Medical Ethics and Ethics Education', p. 172.
48 S. A. T. Law, 'The View of a Young Doctor' (1984). Paper given at GMC 'Teaching Medical Ethics Conference'. Conference proceedings are held at the National Archives: FD7/3628.
49 Edward Shotter, 'The View of the Society for the Study of Medical Ethics' (1984). Paper given at the GMC 'Teaching Medical Ethics Conference'. National Archives: FD7/3628. This talk was reprinted as Edward Shotter, 'Self Help in Medical Ethics', *Journal of Medical Ethics*, Vol. 11 (1985) pp. 32–4.
50 Law, 'The View of a Young Doctor'.
51 Shotter, 'The View from the Society for the Study of Medical Ethics'.
52 For more on the working party's background, see Reynolds and Tansey (eds), *Medical Ethics Education in Britain*, pp. 58–9; Desmond Pond (chair), *Report of a Working Party on the Teaching of Medical Ethics* (London: Institute of Medical Ethics, 1987) p. vii. Pond was listed as a member of the FMG in Daniel Jenkins's *The Doctor's Profession*. He served as Chief Scientist at the DHSS between 1982 and 1985, and died shortly before his working party's report was

published in 1986. See Anon, 'Obituary: Sir Desmond Pond', *Journal of Medical Ethics*, Vol. 12 (1986) p. 130.

53 The deans were asked seven questions concerning their school's policy on medical ethics; timetabled periods; encouragement of informal discussion; involvement of non-medical teachers; assessment of student's familiarity with ethical issues; extra-curricular activities; and the respondent's views on ethics teaching. See Pond (chair), *Teaching of Medical Ethics*, pp. 18–33.

54 Ibid, pp. 19–20.

55 Ibid, p. 21.

56 Ibid, p. 22.

57 Ibid, p. 35.

58 Ibid.

59 Ibid, p. 37.

60 Ibid, p. 38.

61 Ibid.

62 Jonathan Glover, 'Appendix Two: Philosophy', in Pond (chair), *Teaching of Medical Ethics*, pp. 51–5 (p. 51).

63 Ibid, p. 51.

64 Ibid, p. 52.

65 Kennedy, *The Unmasking of Medicine*, p. 124.

66 See, for example, Johnson, 'Teaching Medical Ethics', p. 5.

67 General Medical Council, *Tomorrow's Doctors: Recommendations for Undergraduate Medical Education* (London: General Medical Council, 1993) p. 14. For more on the report's influence, see Kenneth Calman, *Medical Education: Past, Present and Future* (London: Churchill Livingstone, 2006). See also Brazier et al., 'Helping Doctors Become Better Doctors'; Reynolds and Tansey (eds), *Medical Ethics Education in Britain*, pp. 41–3, 59–60.

68 General Medical Council, *Tomorrow's Doctors*, p. 10.

69 Kenneth Boyd, quoted in Reynolds and Tansey (eds), *Medical Ethics Education in Britain*, p. 60.

70 Campbell, interview with the author (2009).

71 Ibid; MacLean, interview with the author (2009); O'Neill, interview with the author (2009).

72 Kennedy, interview with the author (2010).

73 Warnock, *People and Places*, p. 174.

74 Mark Carlisle, quoted in Brian Pullan and Michelle Abendstern, *A History of the University of Manchester, 1973–1990* (Manchester: Manchester University Press, 2004) p. 131.

75 The total sum distributed by the UGC was cut by a total of 17 per cent, although this varied considerably between universities. For example, the University of Salford, a former technical college, suffered

a 44 per cent reduction in its UGC grant, while prestigious universities such as Cambridge only experienced a 6 per cent reduction. For more background, see Peter Scott, *The Crisis of the University* (London: Croom Helm, 1984).

76 For example, see Dennis Austin, 'A Memoir', *Government and Opposition*, Vol. 17 (1982) pp. 469–82.

77 Ibid, p. 484. See also Pullan and Abendstern, *A History of the University of Manchester*, p. 121.

78 Austin, 'A Memoir', p. 477.

79 Anon, 'Philosophy Postponed?', Society for Applied Philosophy Newsletter (Spring 1987) p. 2. Society for Applied Philosophy archives.

80 NCP Submission to the UGC Philosophy Working Party (1988). Information on the formation of the NCP can also be found online at www.heacademy.ac.uk/resources/detail/subjects/prs/The_National_Committee_For_Philosophy-_Its_Origins_And_Activities (accessed 21 February 2014).

81 NCP Submission to the UGC.

82 Warnock, *People and Places*, pp. 185–8; Mary Midgley, *The Owl of Minerva: A Memoir* (London and New York: Routledge, 2005) p. 208. On criticism of the 'new vocationalism' from other fields in the arts and humanities, see Pullan and Abendstern, *A History of the University of Manchester*, pp. 159–60.

83 Midgley, *The Owl of Minerva*, p. 205.

84 The philosophy departments that closed were Surrey, Bangor, Exeter, Leicester, Newcastle, Aberystwyth and City. For a first-hand account on the closure of philosophy at Newcastle, see Midgley *The Owl of Minerva*, pp. 205–9. By 1988 other small departments also appeared threatened, as the NCP noted that 'Many UGC subject reviews have recommended departmental closures on the grounds that all departments must be at or above a certain minimum size.' NCP submission to the UGC (1988).

85 Warnock, *People and Places*, p. 185.

86 Ibid, pp. 185–8.

87 O'Neill, interview with the author (2009); Warnock, interview with the author (2009).

88 Pullan and Abendstern, *A History of the University of Manchester*, p. 249.

89 Ibid, p. 243.

90 Ibid. On how these new priorities helped change disciplinary structures in other fields, see Wilson and Lancelot, 'Making Way for Molecular Biology'.

91 Robin Downie and Jane MacNaughton, *Bioethics and the Humanities:*

Attitudes and Perspectives (Abingdon: Routledge-Cavendish, 2007) p. 32

92 Mark Richmond to John Harris, 3 March 1987. CSEP archives.

93 Brazier et al., 'Helping Doctors Become Better Doctors'.

94 Ian Kennedy to Mary Lobjoit, 14 March 1983. CSEP archives.

95 Harris, interview with the author (2011).

96 Harris, *The Value of Life*; John Harris, 'In Vitro Fertilization: The Ethical Issues', *The Philosophical Quarterly*, Vol. 3 (1983) pp. 202–27; Anon, 'Preliminary Information: Workshop on Philosophical and Ethical Issues in Medicine, 12 November 1983', 11 August 1983. Society for Applied Philosophy archives.

97 Minutes of the Fifth Meeting of the Executive Committee of the Society for Applied Philosophy, 25 May 1984. Society for Applied Philosophy archives.

98 Elaine Graham, 'Obituary: The Rev Professor Anthony Dyson', *Independent*, 17 October 1998, p. 23.

99 Anthony O. Dyson, 'Why Has Medical Ethics Become So Prominent?' *Manchester Memoirs*, Vol. 128 (1988–89) pp. 73–81 (p. 81).

100 Brazier et al., 'Helping Doctors Become Better Doctors'.

101 Anthony Dyson to Professor Mark Ferguson, School of Biological Sciences, 22 May 1985. CSEP archives.

102 Mary Lobjoit to Professor Max Elstein, 6 June 1984. CSEP archives.

103 Max Elstein to Mary Lobjoit, 14 June 1984. CSEP archives.

104 Professor Mark Ferguson to Anthony Dyson, 29 May 1985. CSEP archives

105 Anon, 'Centre for Social Ethics and Policy, University of Manchester' (1987). CSEP archives.

106 Ibid.

107 Brazier et al., 'Helping Doctors Become Better Doctors'. For more background on Brazier's career, see Lawrence O. Gostin, 'Foreword in Honour of a Pioneer of Medical Law: Professor Margaret Brazier OBE QC FMEDSCI', *Medical Law Review*, Vol. 20 (2012) pp. 1–5.

108 On the Swansea centre, see Donald Evans, 'Health Care Ethics: A Pattern for Learning', *Journal of Medical Ethics*, Vol. 13 (1987) pp. 127–31.

109 Centre for Social Ethics and Policy: Five Year Plan (1986). CSEP archives.

110 Harris, interview with the author (2011); Dyson, 'Why Has Medical Ethics Become So Prominent?', p. 78.

111 Duncan Wilson, *Reconfiguring Biological Sciences in the Late Twentieth Century: A Study of the University of Manchester* (Manchester: University of Manchester Faculty of Life Sciences, 2008) pp. 22–4.

112 Ibid, p. 25.
113 Ibid.
114 John D. Turner to John Harris, 'Memorandum: The Centre for Social Ethics and Policy', 3 October 1986. CSEP archives.
115 Mark Richmond to John Harris, 3 March 1987. CSEP archives.
116 Anon, 'New Ethics Centre', *This Week*, Vol. 12, no. 14 (2 February 1987) p. 1. CSEP archives.
117 *The Centre for Social Ethics and Policy (CSEP), University of Manchester* (1987). CSEP archives.
118 Ibid.
119 Ibid.
120 *The Centre for Social Ethics and Policy.*
121 Mary Lobjoit, draft letter to potential sources of funding, May 1987. CSEP archives.
122 Peter D. Stenier to Mary Lobjoit, 30 July 1986. CSEP archives.
123 Centre for Social Ethics and Policy, Annual Report 1986/87. Marshall was a former member of the Warnock inquiry and opposed any research on human embryos. See 'Expression of Dissent: B. The Use of Human Embryos in Research', in Warnock, *A Question of Life*, pp. 90–3. Ward often publicly commented on the ethics of IVF and embryo research, and was one of the contributors to the *Credo* programme discussed in the previous chapter.
124 Harris, interview with the author (2011). For a brief account of the lectures, see Margaret Brazier, Anthony Dyson, John Harris and Mary Lobjoit, 'Medical Ethics in Manchester', *Journal of Medical Ethics*, Vol. 13 (1987) pp. 150–2 (p. 151). The lectures were published as Anthony Dyson and John Harris (eds), *Experiments on Embryos* (London: Routledge, 1990).
125 Mary Lobjoit to Dr D. B. Newbould, 2 July 1987. CSEP archives.
126 Mary Lobjoit to Dr P. J. Haynes, 22 October 1986; Mary Lobjoit to Professor L. A. Turnberg, 28 October 1986. CSEP archives.
127 Lobjoit to Haynes, 22 October 1986. CSEP archives.
128 MA in Healthcare Ethics, preliminary meeting, 31 October 1986. CSEP archives.
129 MA Degree in Medical Ethics, June 1985. CSEP archives.
130 Anon, 'Press Release: New MA Degree in Healthcare Ethics' (1987); *MA in Healthcare Ethics, University of Manchester* (1987). CSEP archives.
131 Brazier et al., 'Medical Ethics in Manchester', p. 151. Centre for Social Ethics and Policy, Annual Report 1986/87; Harris, interview with the author (2011).
132 Anon, 'Press Release: New MA Degree in Healthcare Ethics' (1987). CSEP archives

133 Brazier et al., 'Medical Ethics in Manchester', p. 151.
134 The University of Swansea's Centre for the Study of Philosophy and Health Care launched its own MA in 1985. See Reynolds and Tansey (eds), *Medical Ethics Education in Britain*, p. 44.
135 Ian Kennedy to John Harris, 19 February 1987. CSEP archives.
136 Brazier et al., 'Medical Ethics in Manchester', p. 151.
137 MA in Healthcare Ethics/Centre for Social Ethics and Policy. Meeting minutes, 25 March 1987.
138 John Harris and Harry Lesser, 'Moral Philosophy' (1987). CSEP archives.
139 Harris, interview with the author (2011).
140 MA in Healthcare Ethics. Although it was listed in planning documents, the historical course on medical ethics does not appear to have ever run.
141 Max Elstein and John Harris, 'Teaching of Medical Ethics', *Medical Education*, Vol. 24 (1990) pp. 531–4 (p. 531).
142 Ibid, p. 531.
143 Ibid, p. 532.
144 Ibid.
145 Ibid.
146 Ibid.
147 Mary Lobjoit to Mervyn Busson, 31 January 1991. CSEP archives.
148 Ibid.
149 Ibid.
150 Mervyn Busson to Mary Lobjoit, 13 February 1991. CSEP archives.
151 These issues were considered in several publications and the project's final report. See Rebecca Bennett, Charles Erin and John Harris (eds), *AIDS: Ethics, Justice and European Policy. Final Report of a European Research Project* (Luxembourg: Office for Official Publications of the European Communities, 1998).
152 Wilson, *Reconfiguring Biological Sciences*, p. 76; Pullan and Abendstern, *A History of the University of Manchester*.
153 Martin Harris to John Harris, 16 December 1992. CSEP archives.
154 Profile of John Harris, in Bennett et al. (eds), *AIDS: Ethics, Justice and European Policy*, p. xxi. In line with the trend for securing outside funding, the businessman Sir David Alliance funded Harris's chair from 1997 onwards.
155 For an overview, see David Hunter, 'Bioethics – A Discipline without a Natural Home?', 18 April 2012, *Journal of Medical Ethics* blog. Available online at http://blogs.bmj.com/medical-ethics (accessed 14 February 2014).
156 Harris, interview with the author (2011); Campbell, interview with the author (2009); MacLean, interview with the author (2009).

157 Downie and Macnaughton, *Bioethics and the Humanities*, p. 32. See also Toulmin, 'How Medicine Saved the Life of Ethics'.
158 Brazier et al., 'Helping Doctors Become Better Doctors'.
159 Margaret Brazier, 'The Story of the Manchester Centre and UK Bioethics', paper given at GLEUBE Conference on 'The History of Bioethics', University of Manchester, January 2011. Available online at http://www.youtube.com/watch?v=mkPOSxdMD2M (accessed 14 February 2014).
160 Warnock, 'Embryo Therapy', p. 30.
161 Ibid.
162 Ibid, p. 31.
163 Ibid; Warnock, *People and Places*, p. 191.
164 O'Neill, interview with the author (2009).
165 Hunter, 'Bioethics – A Discipline without a Natural Home?'
166 Ibid; Brazier et al., 'Helping Doctors Become Better Doctors'.
167 Whong-Barr, 'Clinical Ethics Teaching in Britain', p. 81.
168 Ibid, p. 79. See also Brazier et al., 'Helping Doctors Become Better Doctors'.
169 The Revd Bryan Vernon, cited in Reynolds and Tansey (eds), *Medical Ethics Education in Britain*, p. 67.
170 Whong-Barr, 'Clinical Ethics Teaching in Britain', p. 81.
171 Elstein and Harris, 'Teaching of Medical Ethics'.
172 Pond (chair), *Teaching of Medical Ethics*, p. 21.
173 Martin Harris to John Harris, 16 December 1992. CSEP archives.
174 Kennedy, interview with the author (2010).

6

Consolidating the 'ethics industry': a national ethics committee and bioethics during the 1990s

During the 1980s many of the individuals who were pivotal to the making of British bioethics sought to establish what the *British Medical Journal* identified as a 'national bioethics committee'.[1] Ian Kennedy, for one, regularly called for a politically funded committee based on the American President's Commission, and his proposals were often endorsed by newspapers and other bioethicists. They were also endorsed by senior figures at the BMA, who believed a national bioethics committee would standardise decisions between different RECs and 'reassure the public'.[2] But plans for a national bioethics committee ultimately stalled in the late 1980s, after some politicians and doctors claimed that it would obstruct research and politicise bioethics.

The failure of these proposals led some bioethicists and senior scientists to argue that a national committee should have no links to government. After a series of conferences in 1990, the charitable Nuffield Foundation agreed to create an independent bioethics committee that included representatives from several professions. The resulting Nuffield Council on Bioethics embodied the belief that external oversight was vital to maintaining public confidence in biomedical research. Its establishment bolstered media support for outside involvement with medicine and science, leading the *Guardian* to claim that there was 'something of an ethics industry springing up'.[3] But while council members believed that their independence from government secured public trust and prevented political interference, it also ensured that their advice carried little influence.

External oversight of medicine and science increased under the 'New Labour' government that was elected in 1997 and shared the neo-liberal enthusiasm for 'empowered consumers'.[4] This arose largely in response to a public inquiry into paediatric heart surgery at Bristol Royal Infirmary. The inquiry's chair, Ian Kennedy, proposed and then chaired a Commission for Healthcare Audit and Inspection (CHAI), which he argued would empower patients by monitoring the performance of hospitals and healthcare trusts. In line with the continuing neo-liberal climate, he argued that the CHAI was needed because patients were 'not passive receivers of goods, but consumers with choices'.[5] But criticism of the CHAI reflected a growing backlash against bioethics and the 'audit society' at the beginning of the twenty-first century, with doctors, politicians and even some bioethicists now claiming that external oversight actually damaged public trust.

Essential or obstacle? Discussing a 'national bioethics committee'[6]

Ian Kennedy was the strongest advocate of bioethics in Britain during the late 1970s and early 1980s, so it is no surprise to find that he was also the first individual to endorse a national bioethics committee. Kennedy's enthusiasm for a permanent bioethics committee arose during his spell in the United States, after he became acquainted with the Yale lawyer Alexander Capron, who was a member and later executive director of the President's Commission for the Study of Ethical Problems in Medicine.[7] Kennedy and Capron believed that the President's Commission offered the 'perfect vehicle' for bioethics.[8] The majority of its members were non-doctors or scientists, it had the power to suggest legal changes, while members encouraged 'extensive media coverage' of their deliberations and ensured that all meetings were recorded and open to the public.[9]

Kennedy called for the establishment of a similar body in his final Reith Lecture, where he argued that a permanent 'board or committee' should be established to act 'as champion of the consumer's cause' in Britain.[10] He suggested that this committee should issue professional guidelines and enforce sanctions when these were breached, and should also have significant lay membership to enforce 'outside scrutiny, a key principle of consumerism'.[11] Kennedy expanded on these plans in the book *Unmasking Medicine* and a 1983 lecture on 'Emerging Problems in Science, Technology and Medicine', where he admitted that they were modelled on the President's Commission.[12] He outlined how the British committee, like its American counterpart, 'would issue opinions, publish working papers, seek opinions, and then propose law if necessary'.[13]

In a now familiar tactic, Kennedy promised doctors that a national bioethics committee would 'benefit the practice of medicine' by maintaining public confidence and forestalling legal challenges. He also argued that it would benefit politicians, by sparing them the time-consuming task of having to 'pass legislation on the consumer's behalf'.[14] In 1981 Kennedy was uncertain 'whether such a body should be created under the auspices of a government department or should be independently constituted', but claimed that this was a 'matter of detail' that could be settled once his proposal was accepted.[15] By 1983, however, he believed that any 'standing committee should be set up by Parliament'.[16]

Not content with endorsing a national committee in radio lectures and books, Kennedy also discussed his plans with politicians and the chairman of the Law Commission.[17] But he was not the only advocate of a government-sponsored ethics committee by the mid 1980s. In a 1984 editorial on IVF and embryo research, the *Mail on Sunday* also urged the government to establish 'a constant watchdog to involve ordinary people in the crucial decisions being made about our lives by men in white coats'.[18] Like Kennedy, it argued that a 'standing committee on the practice of medicine' was vital to ensuring that 'we, the public, know our interests are being considered'.[19]

At the same time, senior figures within the medical profession also endorsed a national ethics committee, albeit for different reasons. In 1984 members of the CEC and the BMA's general secretary, John D. Havard, began to promote a national ethics committee to the MRC and DHSS. While Kennedy and the *Mail on Sunday* were motivated by a desire for greater public influence over science and medicine, Havard and the CEC portrayed a national committee as the solution to professional concerns. They argued that it would help prevent the uneven decisions that sometimes arose in multi-centre trials, where 'one local committee may approve a study while another may reject the project'.[20]

In a meeting with the immunologist Sir James Gowans, secretary of the MRC, Havard endorsed a 'form of "line" relationship between the national and local ethics committees'.[21] He claimed that a national committee would take charge of all multi-centre proposals and issue broad 'guidelines for use by RECs on various areas of research, e.g., *in vitro* fertilization'.[22] The CEC expanded on this 'line relationship' in a document circulated to the DHSS and MRC, which claimed that the national committee would issue guidance for 'multi-centre or nationally based' projects, but stressed that it would 'not undermine the expertise of RECs'.[23] The CEC endorsed a more mutual relationship, where the national committee built a 'library of good practice filtering up from the RECs' and 'disseminated information on request' to local committees and researchers.[24]

Havard sought to improve the public image of medicine during his spell as the BMA's general secretary, and both he and the CEC believed that a national ethics committee would achieve more than simply standardising local decisions.[25] Their proposals asserted that 'ensuring consistency in approach … will reassure the public', with a strong network of ethical committees helping 'protect the good name of the medical profession'.[26] This desire to maintain public confidence led them to propose that the committee should be established and funded by the government rather than any medical organisation, since 'it is very important for the national ethical research committee to be seen to be independent'.[27]

Several groups and individuals welcomed Havard's and the CEC's proposals. Writing to the MRC, the chair of one REC claimed that a national committee would achieve a substantial 'saving in man-hours' by ensuring that different local committees did not have to consider 'experiments which will be carried out on a nation-wide basis'.[28] In reply, the MRC admitted that 'there may be a need for a national ethics committee to take broad decisions of principle as to whether or not research in a particular area (e.g., research on human embryos) is desirable and permissible', although it emphasised 'the importance of local ethics committees in representing local considerations and interests'.[29] James Gowans also expressed cautious support for a national committee in meetings with Havard, suggesting that it 'might take the form of a standing royal commission', but he suggested that the BMA needed to learn more about other European committees before it drew up firm plans for Britain.[30]

Perhaps most significantly, Sir Desmond Pond, then Chief Scientist at the DHSS, also expressed support for Havard and the CEC's proposal. In March 1985 Pond sent a letter to the MRC, the

BMA, the GMC and prominent doctors such as Sir Douglas Black, in which he claimed that the often 'perfunctory arrangements' for ethical oversight 'point to the need for a national committee'.[31] Pond's view of a national committee was similar to Ian Kennedy's. He argued that it could act 'firstly to set up a proper system of ethical supervision of research using human subjects (and possibly a subgroup embracing animal research); secondly to have a monitoring role to ensure that what is proposed is actually happening'. He also suggested that it might emulate the President's Commission by proposing new legislation and 'considering particular ethical issues on a case law basis'.[32] Although Pond ventured no firm opinion on who should establish and fund a national committee, he proposed that the DHSS might 'take the initiative' by convening a meeting of politicians, medical figures and 'interested laymen' in order to 'get agreement on the need for a national body and general guidelines on its membership and remit'.[33]

Yet while they may have agreed on the need for a national committee, these groups and individuals had differing views on its composition. Havard and the CEC believed that public trust in a national committee could be ensured by staffing it mainly with senior doctors and scientists who were 'people of distinction in their field'.[34] They proposed that only two out of eleven possible members should be laymen, nominated by the DHSS, while the other nine should be professionals nominated by medical societies and the pharmaceutical industry.

Desmond Pond, on the other hand, believed that the national committee should be more 'inter-disciplinary ... with a 'lay (i.e., non science) chairman'.[35] James Gowans also claimed that public trust could only be ensured through 'strong lay representation'.[36] In endorsing a 'lay-dominated' committee, Gowans drew on information that the MRC had obtained from France and Sweden.[37] In a letter to Donald Acheson, the government's CMO, Gowans claimed that documents on French and Swedish ethics committees showed that they included significant numbers of philosophers, lawyers, theologians, journalists and patient representatives, and acted more as vehicles for confronting 'public misgivings' than for standardising professional behaviour.[38] These European documents also supported Havard's and the CEC's belief that a British committee should be linked to the government. They detailed how the Swedish and French committees were both organised and funded by their respective governments, with the Secretary of State for Social Affairs selecting members of the Swedish committee, and President François Mitterand and the Secretaries of State for Health and Research selecting members of the French committee.[39]

But despite broad support from senior officials at the DHSS, MRC and BMA, as well as from some newspapers and public figures, plans for a British ethics committee never progressed beyond letter writing and 'low level informal meetings'.[40] An MRC report from October 1985 noted that Pond's calls for a meeting had received a lukewarm response and claimed that the DHSS was 'trying to hold things at arm's length'.[41] This stemmed partly from scepticism towards political involvement with a national ethics committee. As the *Lancet* noted, some politicians rejected the idea of a government-sponsored committee on the grounds that it would turn ethical issues 'into party political questions, which they are not'.[42]

These political misgivings were compounded by growing medical opposition. In a letter to Desmond Pond, one pharmacologist admitted that there 'may be a problem' with decision-making between different RECs, but claimed that he was 'not persuaded that a DHSS sponsored national committee is the right answer'.[43] He instead argued that existing bodies such as the Royal College of Physicians could effectively take on the role of 'a consultative body for district ethical committees'.[44] 'I am not confident', he continued, 'that a national committee containing DHSS and lay (political?) members would be more effective than the best of the existing committees.'[45] He closed the letter by predicting that since the lay members would 'face a difficult task and have themselves to acquire considerable expertise before they can give useful opinions', a national committee would achieve little more than 'slowing down the progress of clinical research'.[46]

In March 1986 the *British Medical Journal* reported that the BMA's Central Committee for Hospital Medical Services (CCHMS) had also 'objected to a national committee'.[47] The CCHMS questioned who a national committee would be accountable to and claimed that the current system of ethical review could be improved by simply staffing RECs with more members who were 'in day to day contact with patients'.[48] At a BMA meeting the following month, the CCHMS argued that proposals for a national ethics committee 'did not have the support of many people actually involved in research' and criticised the CEC for not consulting widely enough.[49] Crucially, the CCHMS persuaded the BMA's governing council to withhold its support after they claimed 'we can see no effective role for a national committee that did not disenfranchise local groups'.[50]

Plans for a national ethics committee had completely stalled in 1986, with the BMA withholding support, Desmond Pond retiring as Chief Scientist and the DHSS now exhibiting what the MRC called 'masterly inactivity'.[51] This clearly annoyed Ian Kennedy, who complained that calls for a national committee were now 'greeted with a deafening silence'.[52] His frustration was made clear in a 1987 letter to *The Times*, concerning cases in which judges had permitted abortions to be performed on two mentally handicapped adults who were unable to give or withhold consent. Kennedy argued that judges were 'being asked to decide questions of fundamental importance without guidance', and claimed that it was 'doubtful whether the complex arguments on both sides of such a moral dilemma can be marshalled and explored in depth' without expert advice.[53] He maintained it was 'increasingly hard to justify the failure to establish an appropriate body to investigate these dilemmas in a considered and detached fashion', and 'once again' urged the government to establish 'a national commission on medical law and ethics ... so that our elected politicians would have an informed basis on which to provide authoritative guidance on these fundamental questions'.[54]

But Kennedy's frustration was to prove short-lived. Calls for a national ethics committee re-emerged and gained momentum in 1988, thanks to growing debates on gene therapy and the transplantation of foetal brain tissue into adults with Parkinson's Disease. Although the government convened *ad hoc* inquiries into each of these issues, politicians, public figures and senior doctors now criticised this as a 'piecemeal' response and argued that Britain urgently needed 'a single body to which bioethical problems can be referred for assessment – and which can anticipate new ones'.[55]

While foetal tissue transplants had received REC approval, MPs claimed that they should not have been undertaken without a broader debate and called for 'national control and additional guidance'.[56] Several politicians endorsed a national committee in Commons debates, and set out their views on its possible links to government. Gerard Vaughan, a former doctor and Minister of Health, believed that it should function as 'statutory body' that was 'responsible to Parliament', while the former neurosurgeon Sam Galbraith argued that it would function better as 'a quasi non-governmental organization'.[57] No politician, tellingly, spoke in favour of the current ethical and regulatory framework.

Supporters of a national committee found a high-profile ally in Mary Warnock, who claimed that public interest in 'a growing number of topics' such as gene therapy and embryo research justified the formation of 'a permanent royal commission with a rolling membership'.[58] Writing in the *British Medical Journal*, Warnock endorsed a national committee that resembled the 'monitoring body' her committee had proposed for IVF and embryo experiments. She argued that it would scrutinise professional actions in order to meet the 'growing need for public candour', with a broad membership and a 'lay chairman' ensuring it was 'sufficiently detached' from the medical profession.[59] Warnock stated that a national committee should develop guidelines and publish an annual report in which it justified its decisions to Parliament and the public – helping create 'an ethical framework widely seen to be secure and sensible'.[60]

As before, and like Ian Kennedy, Warnock stressed that this 'ethical framework' would benefit science and medicine. She warned that biomedical research would suffer in the continued absence of a national bioethics committee, with the public relying on 'often partial and scaremongering items in the press to form their opinions', and Parliament likely to be 'rushed into wholly restrictive legislation'.[61] Warnock asserted that a national bioethics committee would ease ongoing tensions by issuing guidance to researchers while also proving that 'research can be regulated without being banned, that knowledge can be pursued without being put to morally intolerable uses'. This, she concluded, was vital to ensuring 'that we continue, as we must, to push back the frontiers of science'.[62]

Warnock's plans were endorsed in several letters to the *British Medical Journal*, with one doctor claiming that a national committee would 'add reassurance to the public about what was going on in research ... and would be of enormous benefit in dealing with the approval of numerous multi-centre projects, where conflicting advice is sometimes given by different local ethics committees'.[63] Stephen Lock, the journal's editor, also supported a permanent committee that would 'have a strategic and advisory role, dealing with broad bioethical issues as they arise and assessing their impact on our lives'.[64] Lock reported how the BMA's governing council now endorsed a national committee in their 1988–89 report, where they claimed that it was needed to 'develop guidelines on complex new areas of biomedical research' and to overcome 'widespread disagreements among local committees on multi-centre trial proposals'.[65] While the BMA council stated that their support arose from recent debates on foetal tissue transplants, they had also come under pressure from some regional divisions who 'deplored the delay' in approving Havard's and the CEC's proposals.[66]

Yet despite this groundswell of support, the government again rejected claims that it should establish a national bioethics committee. This was made clear during parliamentary questions in October 1988, when the Conservative MP Sir David Price asked Margaret Thatcher if she would make it government policy to set up a permanent House of Commons Select Committee 'to study ethical problems arising out of new developments in the practice of science, technology and medicine, and to make recommendations'.[67] The Prime Minister dismissed this suggestion by claiming that consideration of ethical issues was 'already within the terms of reference of the appropriate select committees'. She warned MPs that establishing a national bioethics committee would therefore entail expensive, time-consuming and unnecessary 'changes in the existing select committee structure'.[68]

The government's stance is hardly surprising given the difficulties it faced in legislative debates over IVF and embryo research. Politicians were still discussing embryo experiments and dealing with various interest groups over four years after the Warnock report had been published, with supporters and opponents of research both criticising them for failing to arrive at a satisfactory solution.[69] The government was therefore reluctant to become too involved with emerging problems such as foetal tissue transplants and gene therapy, and were content to pass responsibility to *ad hoc* inquiries or, preferably, to a non-governmental ethics committee.[70]

The government's stance was again made clear in 1989, when the *British Medical Journal* reported that civil servants at the Cabinet Office had discussed establishing a national bioethics committee, only for their plans to be 'squashed' by senior Conservatives.[71] It was also made clear to attendees at a CIBA Foundation meeting held in September 1989, where several speakers, including Ian Kennedy, spoke in support of a 'national review body to identify current and future issues'.[72] In his closing remarks, Sir Patrick Nairne, Chancellor of Essex University and a former Chief Secretary to the DHSS, told attendees there was 'no perception in Government of the need for a national body such as a national standing commission'.[73] Nairne concluded that 'it may be best for those who feel the need for such a body to set a lead' and look elsewhere for support.[74] In following this advice, one speaker at this meeting succeeded where Kennedy, Havard and others had failed, and ensured that Britain gained its 'urgently needed' bioethics committee in 1991.[75]

Establishing the Nuffield Council on Bioethics

The first speaker at the CIBA meeting was Sir David Weatherhall, a clinical geneticist at the University of Oxford, who had pioneered genetic testing for thalassaemic blood disorders and took a keen interest in the ethical issues associated with gene therapy techniques. In his paper, Weatherhall bemoaned the continued absence of a national forum that would stimulate public discussion and offer advice to researchers.[76] After the CIBA meeting, he took Patrick Nairne's advice and approached the charitable Nuffield Foundation to see if it would establish an advisory body on bioethics.[77]

Weatherhall's lobbying led members of the Nuffield Foundation to undertake informal soundings on the need for a national bioethics committee.[78] After encouraging feedback, in April 1990 the Foundation held a two-day 'Conference on Bioethics' at Cumberland Lodge, Windsor, where thirty participants debated if 'new machinery' was needed to handle ethical issues arising from scientific and medical research.[79] Attendees were drawn from a variety of disciplines and included Margaret Brazier, Cecil Clothier, Gordon Dunstan, Anne McLaren, Stephen Lock, Patrick Nairne, David Weatherhall and Mary Warnock.[80] A conference summary detailed how they 'generally agreed that a new national bioethics body, probably in the form of a national bioethics committee or council, was needed … to anticipate, or at least respond with speed to new bioethical problems'.[81] Attendees also agreed that the 'national bioethics body' should have two main roles: providing guidance 'to those engaged in biological and clinical research', and 'placing bioethical issues higher on the public agenda, promoting fuller public understanding and confidence'.[82]

As in earlier plans for a national committee, attendees believed that public confidence in the committee could be established by ensuring that 'a majority (however bare) of the members should be lay in the sense of being neither professional scientists nor clinically qualified' and that 'the chair should be lay in the same sense'.[83] They also argued, notably, that since the new committee was unlikely to have government support and 'would have no formal powers', its authority and influence would derive largely from the 'standing and quality of the individual members and the Chair'.[84]

The Nuffield Foundation issued a summary of the meeting's main conclusions as a consultation document in July 1990, which it sent to organisations such as the BMA, the GMC and the MRC, to patient groups and pharmaceutical firms, as well as to sixty academics working in the biomedical sciences and bioethics.[85] The consultation noted that following 'reservations' from the GMC and the BMA, the proposed committee would not consider the ethics of clinical practice and would focus instead on 'the ethical issues posed by emerging *research*'.[86] As the IME's *Bulletin of Medical Ethics* noted, the consultation was greeted with 'widespread support'.[87] At the end of the consultation period, in December 1990, Patrick Nairne led a small steering group that assessed responses and sent the Nuffield trustees proposed terms of reference for the new bioethics council. The terms of reference were, broadly, that the council would identify issues raised by new technologies 'in order to anticipate public concern'; would examine these questions and 'promote public understanding'; and would set up specialist working parties to scrutinise particular issues and provide guidance for government and regulatory bodies.[88]

In line with the BMA's and GMC's misgivings, the steering group maintained that the council would 'take into account the responsibilities for oversight and regulation that fall to professional and other relevant bodies' and would not investigate clinical treatment.[89] After Foundation trustees approved the steering group's terms of reference, the Nuffield Council on Bioethics was officially established in May 1991. The Nuffield Foundation was the sole source of income for an initial three-year period, estimated at £150,000 per annum, which paid for quarterly meetings and a permanent secretariat (the Wellcome Trust and the MRC joined the Foundation as co-sponsors from 1994 onwards).[90]

The council's founding membership bore the hallmarks of

proposals made in the April 1990 meeting and the subsequent consultation paper. First, it was weighted in favour of 'lay' expertise. Of the fifteen original members, seven were doctors and scientists, while the majority included two civil servants, two lawyers, a philosopher, a theologian, an educational consultant and a journalist. Secondly, it reflected the belief that independence from government meant that the council's authority needed to stem from the 'public standing' of its individual members.[91] The majority of the founding members were drawn from the 'Great and Good', and currently or had previously served on government inquiries into science and medicine. The inaugural chair, for example, was Patrick Nairne, who had been a renowned 'cardinal of bureaucracy' during his time as a civil servant.[92] Scientific members included Anne McLaren, who had been 'indispensable' to Warnock's committee and now sat on the HFEA, and David Weatherhall, who had recently been appointed to a public inquiry into the ethics of gene therapy, which was chaired by Cecil Clothier.[93] Lay members included the lawyer Sir David Williams and Gordon Dunstan, who had both served on Home Office committees on animal experiments, as well as Ian Kennedy, who had just served on a government review of guidelines for research on foetuses and foetal tissues.[94]

The Nuffield Council on Bioethics received a warm welcome from newspapers and politicians, including the new Conservative Prime Minister, John Major, who urged it to address issues raised by agriculture and environmental biotechnology in addition to biomedical science.[95] A long article in the *Independent* greeted the Nuffield Council as 'a brave attempt ... to get away from the piecemeal approach that has characterised previous British attempts to come to terms with the public policy implications of scientific advance, such as the Warnock committee on embryology and *in vitro* fertilization'.[96] The *Independent* also stated that the council would look at issues that 'concern us all and are far too important to be left to the practitioners', and welcomed the fact that its membership was 'not dominated by scientists and doctors'.[97] The *New Scientist* similarly dwelt on the council's membership and predicted that their 'eminent' backgrounds would carry 'enough authority to influence the government and Parliament'.[98]

But although newspapers portrayed the Nuffield Council as Britain's first national bioethics committee, in reality it was one of many bodies dedicated to medical and scientific policymaking.

This was clear from the council's first annual report, where a 'brief survey' of similar groups ran to over four pages.[99] Many of those listed were medical organisations, such as the GMC, the Royal Colleges and RECs; but others, such as the HFEA, shared the Nuffield's template of a 'lay'-dominated membership and looked at ethical issues.[100] The council's members and secretariat thus had to ensure that their work did not duplicate that of these existing bodies. This led them to select genetic screening as the council's first topic in 1991, closely followed by a review of the ethics of research on human tissues in 1992.[101] The review of genetic screening was prompted by growing concerns over the possible misuse of genetic information by employers, insurers and government, while the human tissue review was prompted by medical concerns over the ownership of tissues and cells, after an American patient had argued that he was the rightful owner of a cell line that UCLA researchers had derived from his spleen.[102]

These choices reflected the Nuffield Council's remit of anticipating public concerns and issuing advice for researchers in areas that lacked firm guidelines. Yet the response to both reports raised questions about its impact on policymaking. For example, the Nuffield Council's 1993 report on genetic screening argued that the government should establish an advisory committee to monitor the implementation of genetic testing.[103] But in 1995 the council's secretary, David Shapiro, noted that the government had 'not acted' on this and other recommendations.[104] The government only elected to form an advisory committee later in 1995, following advice from a House of Commons Select Committee on 'human genetics', which had notably been convened less than eighteen months after the Nuffield Council's report was published. This advice led to the establishment of several bodies, including an Advisory Committee on Genetic Testing (ACGT) and a Human Genetics Advisory Commission (HGAC), which both had a similar remit to the Nuffield Council but retained closer links to government.[105]

A similar fate befell the council's 1995 report on human tissue, which Onora O'Neill claimed 'wasn't as influential as it should have been'.[106] The report argued that the legal status of tissue used in research was unclear and called on the government to review or update the law, 'as uncertainty may impede legitimate teaching, treatment, study or research'.[107] While the Department of Health responded by claiming that it would review the law in

1996, ministers did not consider the issue until 2000, following a public scandal that arose when newspapers reported that hospitals had retained organs from infant cadavers without parental permission.[108] When the government looked to review the law, however, it overlooked the Nuffield Council's report and convened its own inquiry, led by the lawyer Michael Redfearn, and then established a 'Retained Organs Commission', led by Margaret Brazier. Like the select committee on human genetics, the Retained Organs Commission also proposed the creation of another advisory body, the Human Tissue Authority (HTA), which had a similar remit to the Nuffield Council but again enjoyed closer links to government.

The Nuffield Council's limited influence was further evident in debates on human–animal 'xenotransplants', which emerged following news that scientists had genetically modified pig organs to reduce the chance of rejection in humans.[109] Council members responded to concerns surrounding animal welfare and the possible transmission of diseases by establishing a working party in January 1995. As before, they instructed the working party to formulate proposals that Patrick Nairne hoped would 'be fully considered by the Government'.[110] But the government nevertheless convened its own inquiry into xenotransplants in September 1995 and appointed Ian Kennedy, who remained a member of the Nuffield Council, as its chairman.[111] As before, the *Bulletin of Medical Ethics* noted that the Nuffield Council's report was 'ignored by the government', which based its policies solely on the recommendations of Ian Kennedy's inquiry.[112]

Members of the Nuffield Council claimed that independence from government protected it from the political interference and budget constraints that affected other national committees. But it also ensured, as Onora O'Neill remarked, 'that you can't always achieve influence in government departments'.[113] While the guidelines produced by national committees in countries such as the United States and France formed the basis for professional guidelines and sometimes led to new laws, the Nuffield Council's reports influenced politicians indirectly at best and were often ignored.[114] This ensured that while the Nuffield Council on Bioethics became known for raising public awareness of certain issues, it was criticised by those who believed that a national committee should have political influence.[115] A 1997 editorial in the *Bulletin of Medical Ethics* claimed, for instance, that it was not comparable to national committees elsewhere because 'there remains the problem of whether the government takes notice of its reports'.[116]

Bioethics under 'New Labour'

The Nuffield Council's limited impact on policymaking ensured that bioethicists continued to have greater authority as members of *ad hoc* public inquiries. This was certainly the case with Ian Kennedy, who had more policy influence as chair of the government's xenotransplants inquiry than as a member or chairman of the Nuffield Council. While Kennedy was frustrated at the Nuffield Council's lack of influence by the mid 1990s, he was also dismayed by the fact that the government only convened inquiries into new procedures such as IVF, gene therapy and xenotransplants, and left the governance of clinical treatment to doctors. This reflected the way in which the Conservative challenge to medical paternalism had 'evaporated' by the late 1980s, following claims that giving patients and outsiders a greater say in medical treatment would 'destabilise' the NHS.[117] Medicine continued to be largely self-regulating and the far-reaching 'inspectorate' that Kennedy had championed in his Reith lectures remained conspicuous by its absence.

This, however, looked set to change following the May 1997 election of Tony Blair's 'New Labour' Party. The architects of 'New Labour' based their policies on a strategy known as the 'Third Way', which they used to differentiate themselves from the Conservatives and what Blair called 'the fundamentalist Left'.[118] This involved rejecting the leftist assumption that a strong state was a vital component of civil society and the Thatcherite belief that freedom could only be achieved by 'rolling back the state'. Proponents of the 'Third Way' instead argued that a fair and open society could be achieved through an 'enabling' state that incorporated market incentives and encouraged 'partnership' between public services and private enterprise.[119] This worldview led New Labour politicians to view some Conservative policies as 'necessary acts of modernization' – including the neo-liberal belief that external oversight was vital to constructing 'responsive public services to meet the needs of citizens, not the convenience of service providers'.[120]

This strategy was perhaps most obvious in medicine, where the new government promised to use what it called 'clinical governance' to construct a 'healthcare service built around the patient'.[121] Its enthusiasm for external oversight and 'empowered consumers' was given impetus by the disclosure of malpractice at Bristol Royal Infirmary in October 1997, which Rudolf Klein claims 'transformed the policy landscape as far as relations between the State and medicine were concerned'.[122] The case centred on the deaths of twenty-nine babies and young children between 1984 and 1995, either during or shortly after surgery, which were brought to public attention after a whistleblower contacted *The Times*. A GMC hearing in May 1998 concluded that two surgeons were guilty of operating on children when they knew that death rates were unacceptably high, and also found a hospital manager guilty of failing to act after colleagues had raised concerns.[123] These decisions and intense media criticism led the *British Medical Journal* to conclude that the case had irreparably damaged 'the trust that patients place in their doctors'. 'British medicine', it argued, was likely to 'be transformed by the Bristol case.'[124]

During May and June 1998 MPs called for a broad public inquiry that should examine not only the events at Bristol, but 'the appropriateness of professional self-regulation'.[125] Following these demands, the *British Medical Journal* predicted that New Labour would use the Bristol case to justify ending 'self regulation for doctors'.[126] Its suspicions were compounded when Frank Dobson, Secretary of State for Health, chose Ian Kennedy as chairman of a public inquiry.[127] Medical journals reprinted sections of *Unmasking Medicine* and identified Kennedy as a 'critic of vested medical interests', who endorsed oversight as 'an important check on medical standards'.[128]

In many respects, Kennedy was a logical choice to chair the Bristol inquiry. He was a trusted member of the 'Great and Good' by the time New Labour won the general election, having served on several government inquiries into science and medicine. He was well acquainted with Conservative and Labour politicians, and had urged members of the new government to establish a statutory body to represent patients' interests shortly after they came to power.[129] But while some medical journals dwelt on his support for oversight, others noted that Kennedy had also been 'a good ally' to the medical profession during his time on the GMC, where he had helped doctors with 'difficult decisions and ethical dilemmas'.[130] Despite his presentation as a 'critic of vested interests', it is more likely that the government chose Kennedy because he provided an intermediary between doctors and politicians.

In addition to Kennedy, the Bristol inquiry included Mavis MacLean, another academic lawyer, as well as the doctor Sir Brian Jarman and Rebecca Howard, director of nursing for Manchester hospitals.[131] Once the inquiry began formal proceedings in March 1999, it was clear that it would be a detailed and wide-ranging exercise, which the *Lancet* described as 'the largest ever independent public investigation into clinical practice'.[132] It took place in a specially constructed hearing chamber, had a budget of over £14 million, was scheduled to hear from over 500 witnesses and was due to assess over 600,000 pages of documents.[133] Its terms of reference, set out by Frank Dobson, were to investigate the care of children undergoing heart surgery at Bristol and then make broad 'recommendations which could help secure high quality care across the NHS'.[134] Ian Kennedy encapsulated this ambitious remit when he told the *British Medical Journal* that 'we are not seeking to focus on individuals but rather we are looking at the whole system'.[135]

Kennedy later admitted that the 'lacerating' evidence he heard during the Bristol inquiry strengthened his belief that medicine should be 'carefully monitored' by outsiders.[136] The case for external scrutiny of doctors was strengthened further by events that occurred while the Bristol inquiry was underway, including the 'retained organs scandal', reports that premature babies had been used in clinical trials without parental consent, and the trial and prosecution of the serial-killing doctor Harold Shipman.[137] Perhaps unsurprisingly, then, demands for external oversight permeated the Bristol inquiry's report when it was published in July 2001. The report, which ran to over 500 pages, began by stating that that while staff at Bristol were generally 'dedicated and well motivated', they were nevertheless representative of a paternalistic 'club culture' that fostered an 'imbalance of power' between doctors and patients.[138] Most of its 198 recommendations were designed to establish 'a new culture for the NHS' in which doctors worked to 'agreed standards, compliance with which is regularly monitored'.[139] In passages redolent of *Unmasking Medicine*, the report claimed that this could be achieved by implementing 'a system of external surveillance to review patterns of performance over time and to identify good and failing performance'.[140]

The report argued that this surveillance should be performed by two new 'overarching bodies', which would 'bring together the various bodies that regulate healthcare' in order to create a 'patient-centred' NHS.[141] The first was a 'Council for the Quality of Healthcare' that would incorporate several of the regulatory bodies that the government had established in its first years of office, including the National Institute for Clinical Excellence (NICE) and a Commission for Health Improvement (CHI).[142] The second was a 'Council for the Regulation of Healthcare Professionals', which would absorb bodies such as the GMC and the Nursing and Midwifery Council.

The report argued that these new organisations would 'ensure that there is an integrated and co-ordinated approach to setting standards, monitoring performance, and inspection and validation'.[143] Tellingly, these plans were similar to the politically funded 'inspectorate' that Kennedy had advocated throughout the 1980s. The Bristol report claimed that the government should 'establish and fund the Councils' and, crucially, that both 'must involve and reflect the interests of patients, the public and healthcare professionals'.[144]

The Bristol report also dovetailed with New Labour's own view of a 'patient-centred' NHS. Later in 2001 ministers sought to implement its main proposal when they circulated plans for a new 'Council for the Regulation of Healthcare Professionals'.[145] These plans claimed that the main priority for the new council was to 'explicitly put patients' interests first'.[146] It would be 'open and transparent and allow for robust public scrutiny', and would ensure that professional bodies 'conform to principles of good regulation'.[147] While the Bristol report had proposed that the GMC should be absorbed into the new council, it was saved by reforms that introduced a scheme to regularly check doctors' fitness to practice and increased the proportion of lay members to 40 per cent.[148] These changes indicated that politicians did not share Kennedy's enthusiasm for disbanding the GMC, but they nevertheless argued that it should 'be accountable to a new Council for the Regulation of Healthcare Professionals, and through the Council to Parliament'.[149] The government closed its plans by stressing that the council should 'have a broadly based membership to ensure key stakeholder interests are represented'. This would ensure, it continued, that it 'will be representative of the regulatory bodies, health service and public'.[150]

New Labour's plans for 'modernised regulation' here drew wholeheartedly on the Bristol report which, in turn, echoed much of Kennedy's work from the 1980s. As Rudolf Klein notes, once the government outlined plans for a new regulatory council in its 2003 Health and Social Care Act, it appeared that Ian Kennedy's 'vision for regulation was on the way to being achieved'.[151] This body, now known as the Commission for Healthcare Audit and Inspection (CHAI), was charged with 'encouraging improvement in the provision of care by and for NHS bodies'.[152] Its main task was to undertake annual reviews of the care provided by each NHS trust, using standards devised by the Department of Health to publish reports and alert ministers to failing trusts.

As part of the government's efforts to reduce the number of regulatory bodies, known as 'rationalization through amalgamation', the CHAI was scheduled to replace or assume the roles of several bodies when it formally became operational in April 2004.[153] It completely replaced the CHI, 'swallowed up' the National Care Standards Commission, which had been set up to regulate the independent healthcare sector, and took on the Audit Commission's task of assessing value-for-money in the NHS.[154] Given his role as chair of the Bristol inquiry, and his status as what the *Guardian* called 'the expert outsider', few people were surprised when Frank Dobson announced that the CHAI's first chairman was to be Sir Ian Kennedy – recently knighted for his 'services to bioethics'.[155]

Following Kennedy's appointment, the *Guardian* presented him as a 'man on a mission' who would use the CHAI to 'champion patients and protect the rights of the vulnerable'.[156] Here, as before, Kennedy promised that the CHAI would 'look at the NHS from the patients' perspective'.[157] He maintained that outside scrutiny of medicine was the best way of ensuring 'social justice and fair and equal treatment for the vulnerable'.[158] One major difference between 2004 and the 1980s, however, was that Kennedy now held a position with considerable influence. This was made clear when the *Guardian* quoted Labour's recently departed Health Secretary, Alan Milburn, as saying: 'Ian Kennedy? He runs the NHS, doesn't he?'[159] Although the *Guardian* noted that this was a private joke, it claimed that others in the NHS and government wondered whether Kennedy was now 'on course to do exactly that'.[160]

Kennedy used a guest editorial in the *British Medical Journal* to outline his vision for the CHAI, which showed how his concern with patient rights dovetailed with New Labour's neo-liberal enthusiasm for 'active citizens'.[161] Kennedy claimed that 'a better educated population, exposed to the idea of choice and impressed by the language of rights', had ensured that patients 'were no longer passive receivers of goods, but consumers with choices who were entitled to expect good quality and to complain if they were not satisfied'.[162] He argued that this shift ensured doctors alone were no longer 'the best judge of a patient's interests' – with good medical care now incorporating the views of patients, their families and professionals from other fields.[163] As he did throughout the 1980s, Kennedy stressed that involving others in setting standards was not designed to 'criticise or blame professionals', but would 'help them through the barriers that prevent them seeing patients as interactive partners'.[164] His goal, as set out in the Reith Lectures, was to facilitate 'a subtle negotiation between professional and patient as to what each wants and can deliver'.[165]

Kennedy's proposals were endorsed in the same issue by New Labour's CMO, Liam Donaldson, who claimed that the 'expert patient' had become central to the NHS.[166] Like Kennedy and Labour politicians, Donaldson argued that doctors benefited from treating their patients as consumers or 'experts', since it encouraged them to take responsibility for their own health rather than 'leaving it all to the doctor'.[167] Donaldson reiterated that this neo-liberal view of the 'empowered patient' would foster 'a new era of optimism and opportunity' by improving public trust in doctors and creating 'a new generation of patients who are empowered to take action to improve their health in an unprecedented way'.[168]

'A Question of Trust' and the bioethics backlash

But despite the synergy between Kennedy's and the government's views of empowered patients, the CHAI was short-lived and contested. Its brief existence coincided with a backlash against external oversight, as doctors and public figures increasingly turned on the 'audit society'. Senior doctors such as Bruce Charlton, a psychiatrist at the University of Newcastle, began to challenge the assumption that 'increased accountability is self-evidently a desirable goal'.[169] Charlton claimed, by contrast, that 'the meaning behind the accountability mantra is the opposite to that implied by its democratic, egalitarian, radical and "empowering" rhetoric'. He argued that the growing emphasis on accountability and oversight had simply provided a Trojan horse for new professional elites, such as lawyers, philosophers and healthcare managers, to exercise 'hierarchical domination' of doctors.[170]

The 'ideology of accountability' also came under fire, notably from the philosopher and bioethicist Onora O'Neill. Like Kennedy, O'Neill had encountered bioethics while working in the United States during the 1970s, and was also a founding member of the Nuffield Council on Bioethics. But she held markedly differing views on the merits and consequences of external oversight. This was clear from her 2002 Reith Lectures, entitled *A Question of Trust*, which offered a telling contrast to *Unmasking Medicine*. In her third lecture, O'Neill questioned the widespread belief that public trust could be improved through external forms of audit and regulation. She argued that externally imposed standards were 'surrogates' for professional actions, and prevented members of a profession from 'pursuing the intrinsic requirements for being good nurses and teachers, good doctors and police officers, good lecturers and social workers'.[171] The pursuit of 'ever more perfect accountability', O'Neill concluded, had damaged rather than repaired public trust and created a 'culture of suspicion and low morale'.[172]

While her Reith Lectures did not focus on specific examples, O'Neill targeted bioethics in a series of lectures at the University of Edinburgh, which were published in 2002 as *Autonomy and Trust in Bioethics*. She claimed here that:

> Although the decades since the beginning of contemporary bioethics have seen a lot of effort to improve the trustworthiness of public institutions and of experts, culminating in the UK in the additional demands for accountability, audit and openness of the 1990s, this is quite compatible with a decline in public trust, and specifically with a decline of public trust in medicine, science and biotechnology.[173]

O'Neill argued that bioethicists' efforts 'to improve trustworthiness' by calling for increased oversight of animal research, gene therapy and reproductive medicine had consistently failed to work.[174] She outlined how newspapers continued to voice unease at embryo research and gene therapy and, more seriously, how opponents of animal experiments continued to engage in 'intimidation, criminal trespass, vandalism and even terrorism'.[175]

O'Neill believed that the authority of bioethicists was undermined by the fact that professional or social status no longer guaranteed public trust.[176] She argued that like their colleagues in medicine, science and politics, bioethicists could no longer rely on a place among the 'Great and Good' for their arguments to be accepted. O'Neill even claimed that bioethicists had probably damaged trust themselves, by presenting scientists and doctors as untrustworthy figures 'who pursue their own interests rather than those of patients and the public'.[177] She argued that their demands for increased regulation and public accountability led 'not to a restoration of trust but to claims of escalating mistrust', and caused suspicion to be 'directed inaccurately in trustworthy persons and institutions'.[178]

O'Neill specifically claimed that the bioethical emphasis on 'empowered consumers' played a major role in escalating mistrust, by encouraging 'ethically questionable forms of individualism' and marginalising the other principles and duties that were vital to restoring confidence in professions.[179] 'We need', she stated, 'to identify more convincing patterns of ethical reasoning, and more convincing ways of choosing policies and action for medical practice and dealing with advances in the life science and biotechnology.'[180] O'Neill believed that the remedy to mistrust lay in focusing more on responsibilities instead of rights, and replacing an individualistic worldview with 'one that takes *relationships* as central'.[181] She argued that this shift to a more 'principled autonomy' would provide stronger 'reasons for seeking to establish, maintain and respect trustworthy institutions and relationships'.[182] This led O'Neill to conclude that encouraging doctors to be more open, asserting the obligations and duties involved in medical care or research, was a far better guarantee of public trust than simply imposing 'an audit trail'.[183]

O'Neill's position differed markedly from that of Ian Kennedy, who steadfastly believed in the value of oversight and viewed 'the language of rights' as fundamental to bioethics.[184] But despite Kennedy's faith in 'external scrutiny', O'Neill's claims resonated with now regular criticism of the CHAI.[185] In a long open letter to Kennedy, Richard Horton, editor of the *Lancet*, argued that the CHAI's labelling of certain hospitals as 'poor' and 'underperforming' was 'likely to undermine public confidence in a health system that enjoys an unparalleled commitment from its doctors, nurses and allied health workers'.[186] Horton echoed O'Neill's work when he stated that the CHAI was damaging patient care by introducing 'an environment of prejudice, anxiety and resignation into the workplace'.[187] He claimed that this could only be rectified by giving the CHAI's ratings 'scientific rigour' and increasing the number of commissioners with 'daily responsibilities for front-line patient care'.[188]

Kennedy responded that the 'current system of ratings performance has undoubtedly produced beneficial effects for patients' and reminded Horton that performance indicators were set by the government, not the CHAI.[189] He also dismissed Horton's belief that 'if there is a regulatory mechanism established by Parliament to monitor the performance of the NHS only healthcare professionals can operate it'. 'For real accountability', Kennedy countered, 'Horton must recognise that the voices of others, not least patients and the public, must be heard.'[190]

But despite Kennedy's regular protests, doctors continued to argue that the CHAI was demoralising NHS workers and damaging patient care. Perhaps more significantly, it also came under fire from politicians and provided a scapegoat when poor medical conduct was exposed. Following an outbreak of *Clostridium difficile* at the Maidstone and Tunbridge Wells NHS trust in 2007, which caused the death of ninety patients, the Health Secretary Alan Johnson criticised the CHAI for being 'slow to act' in identifying potential problems.[191] Labour politicians remained eager to reduce the number of regulatory bodies, which was apparent when they merged the ACGT and the HGAC into a new Human Genetics Commission (HGC), and claimed that the CHAI's failings were symptomatic of a 'burdensome' regulatory sector.[192] When the government announced plans to replace the CHAI and other health inspectorates with a single Care Quality Commission in 2008, Kennedy bemoaned the increasing tendency to see external oversight 'as part of the problem rather than part of the solution'.[193]

This complaint could just as well have been aimed at a new generation of bioethicists, who argued that the field was undergoing 'something of a mid-life crisis'.[194] Angus Dawson, for one, claimed that bioethics had 'become stale and tedious' thanks to a preoccupation with 'a consumerist model of the professional–client relationship'.[195] Like O'Neill, Dawson argued that the 'unthinking consensus view that autonomy is the dominant value' may 'impact in a negative way' on healthcare, by ignoring evidence that suggested that patients did not necessarily want to be seen as empowered consumers but rather wanted 'to be able to trust their doctor, seeking not just *information* but *help* with sometimes complex decisions'.[196]

Dawson claimed that bioethics could be revitalised by attending to the 'social reality of the doctor–patient relationship', which would encourage bioethicists to acknowledge the importance of values such as reciprocity and community, and appreciate that 'some forms of paternalism may be justified'.[197] These new arguments suggest that bioethics is already changing in a new public and political climate, and force us to reassess its history. We may well come to see the 1980s and 1990s not simply as the beginnings of bioethics in Britain, but as the high-water mark of an early incarnation: when the emphasis on oversight, public accountability and rights complemented widespread demands for audit and consumer choice.

Conclusion

Just as in the United States, where various presidents have established or closed national commissions, and staffed them with politically sympathetic figures, the debates surrounding a British ethics committee again show how bioethics is shaped by social and political factors.[198] The failure to establish a national bioethics committee during the 1980s stemmed partly from medical resistance but also, critically, from the government's belief that it would 'politicise' bioethics. But the continued absence of a politically sponsored national committee did not limit the opportunities for bioethicists to assist in policymaking. The government's continued preference for *ad hoc* committees on issues such as IVF, gene therapy and xenotransplants, and the subsequent creation of advisory bodies such as the HFEA, provided greater scope for official bioethics than would have arisen through a single national council. If anything, then, we might conclude that the continued growth of the 'ethics industry' during the 1980s and 1990s, which comprised several advisory bodies and the independent Nuffield Council, was due in no small part to the *absence* of a national ethics committee.

In not looking at clinical treatment, the Nuffield Council highlights how bioethics in Britain tended to examine questions raised by new biomedical technologies and neglected those issues that impact more on the day-to-day lives of patients. Yet while some criticised the Nuffield Council for not scrutinising medical care, this position was, in fact, a precondition of its establishment. As David Shapiro noted in a 1990 letter to the medical lawyer Jonathan Montgomery, transgressing on the work of bodies such as the GMC and BMA would undoubtedly have led them to oppose the council's formation. 'Our nascent body may seem a small infant', Shapiro wrote, '[and] we have to avoid at least one attempt at abortion.'[199]

When bodies such as the CHAI did begin to monitor clinical treatment following the Bristol inquiry, sustained medical resistance and a political desire to reduce the number of 'arm's-length' bodies ensured that they were contested and ultimately short-lived. The fact that growing opposition to the 'ideology of accountability' also came from Onora O'Neill reaffirms that bioethics is not a unified field that stands apart from medicine or politics. It is, rather, a diverse set of participants and ideas whose scope and influence are constituted, and renegotiated, through relations with other disciplines and their broader sociopolitical climate.

Notes

1 Stephen Lock, 'Towards a National Bioethics Committee. Wanted: A New Strategic Body to Deal with Broad Issues', *British Medical Journal*, Vol. 300 (1990) pp. 1149–50.
2 British Medical Association Central Ethical Committee, 'Local Research Ethics Committees', 1 April 1984. National Archives: FD 7/3273.
3 Nigel Williams, 'To the Heart of a Clinical Matter', *Guardian*, 17 April 1991, p. 21.
4 Brian Salter, *The New Politics of Medicine* (Basingstoke: Palgrave Macmillan, 2004).
5 Ian Kennedy, 'Patients are Experts in their Own Field', *British Medical Journal*, Vol. 326 (2003) p. 1276. Emphasis added.
6 Lock, 'Towards a National Bioethics Committee'.
7 Kennedy, interview with the author (2010). See also Capron, 'Looking Back at the President's Commission'.
8 Kennedy, interview with the author (2010).

9 On the public nature of the Commission's work, see Capron, 'Looking Back at the President's Commission', p. 8.
10 Kennedy, 'Consumerism in the Doctor–Patient Relationship', *Listener*, 11 December 1980, pp. 777–80 (pp. 780, 777).
11 Ibid, p. 777.
12 Ian Kennedy, 'Emerging Problems in Science, Technology and Medicine', in Kennedy, *Treat Me Right*, pp. 17–18. See also Kennedy, *Unmasking Medicine*, pp. 122–3.
13 Kennedy, 'Emerging Problems', p. 18.
14 Kennedy, *Unmasking Medicine*, pp. 123, 125.
15 Ibid, p. 139.
16 Kennedy, 'Emerging Problems', p. 17.
17 Kennedy, interview with the author (2010); Kennedy, 'Emerging Problems', p. 18.
18 Anon, 'Why We Must ALL Have a Say on Test-Tube Babies', *Mail on Sunday*, 20 May 1984, p. 16.
19 Ibid.
20 John D. Havard to Mr B. Taylor, secretary, Committee of Vice-Chancellors and Principals, 14 February 1985. National Archives: FD7/3273.
21 Malcolm Godfrey, 'Note for File', 9 October 1985. National Archives: FD 7/3273.
22 Havard to Taylor, 14 February 1985. National Archives: FD7/3273.
23 British Medical Association Central Ethics Committee, 'Improving the Network of Local Research Ethics Committees and the Establishment of a National Ethical Research Committee', 8 January 1986. National Archives: FD 7/3273.
24 Ibid.
25 On Havard's efforts to raise the profile of the BMA and improve medicine's public image, see Caroline Richmond, 'Obituary: John D. Havard', *British Medical Journal*, Vol. 340 (2010) p. 1361.
26 British Medical Association Central Ethics Committee, 'Local Research Committees', 1 April 1984. National Archives: FD 7/3273.
27 British Medical Association Central Ethics Committee, 'Improving the Network of Local Research Ethics Committees'.
28 J. A. D. Cropp to Sir James Gowans, 4 November 1985. National Archives: FD 7/3273.
29 Malcolm Godfrey to J. A. D. Cropp, 26 November 1985. National Archives: FD 7/3273.
30 Godfrey, 'Note for File'.
31 Sir Desmond Pond, 'Ethical Aspects of Research', 28 March 1985. National Archives: FD 7/3273.

32 Ibid.
33 Ibid.
34 British Medical Association Central Ethics Committee, 'Improving the Network of Local Research Ethics Committees.
35 Pond, 'Ethical Aspects of Research'.
36 Godfrey, 'Note for File'.
37 Malcolm Godfrey to Sir James Gowans, 'Meeting with Dr Havard and Dr Dawson (British Medical Association) about BMA Proposals for a National Ethical Committee', 1 October 1985. National Archives: FD 7/3273.
38 Sir James Gowans to Donald Acheson, 21 February 1986. National Archives: FD 7/2209.
39 Swedish Department of Social Affairs, 'Appointment to the Board on Medical/Ethical Questions', 14 March 1985. National Archives: FD 7/2209. Documents sent to the MRC by the French Institute for Medical Research (INSERM) detailed how its national committee's 'recommendations have no binding [*sic*] but are made known through the Secretaries of State for Research and for Health and widely circulated to the public'. Anon, 'National Advisory Ethics Committee for the Life Sciences and Health' (1985). National Archives: FD 7/2209. See also Claire Ambroselli, 'France: A National Committee Debates the Issues', *Hastings Center Report*, Vol. 14 (1984) pp. 20–1.
40 Malcolm Godfrey to Sir James Gowans, 'British Medical Association and Ethical Committees', 7 March 1985. National Archives: FD 7/3273.
41 Godfrey, 'Meeting with Dr Havard and Dr Dawson'.
42 John Lewis, 'A Commons Subcommittee on Medical Ethics?', *Lancet*, Vol. 331 (1988) p. 1005. The case for linking any committee to government was further hampered by the fate of the President's Commission, which had been disbanded in 1983 following cuts in federal funding. Senators Edward Kennedy and Al Gore eventually persuaded Congress to create a new bioethics commission in 1986, despite an attempted veto by President Reagan. See Anon, 'Congress Overrides Veto to Create New Bioethics Commission', *The Hastings Center Report*, Vol. 16 (1986) p. 2. This resulted in the formation in 1988 of a Biomedical Ethical Advisory Committee, which did not issue any reports and was disbanded by President George Bush Sr in 1990.
43 Colin Dollery to Sir Desmond Pond, 4 April 1985. National Archives: FD 7/3273.
44 Ibid.
45 Ibid.
46 Ibid. Emphasis added.

47 Anon, 'Approval of Ethical Research Committees Delayed', *British Medical Journal*, Vol. 292 (1986) p. 779.
48 Ibid.
49 Anon, 'More Talks on Research Committee', *BMA News Review*, April 1986, p. 3.
50 Ibid.
51 Godfrey to Cropp.
52 Kennedy, 'Emerging Problems', p. 18.
53 Ian Kennedy, 'Call for Guidance on Medical Ethics', *The Times*, 13 June 1987, p. 15.
54 Ibid.
55 Lock, 'Towards a National Bioethics Committee', p. 1149
56 Lewis, 'A Commons Subcommittee on Medical Ethics?'
57 Ibid.
58 Mary Warnock, 'A National Ethics Committee', *British Medical Journal*, Vol. 297 (1988) pp. 1626–7.
59 Ibid.
60 Ibid, p. 1627.
61 Ibid, pp. 1626, 1627.
62 Ibid, p. 1627.
63 Gordon W. Taylor, 'Letters to the Editor', *British Medical Journal*, Vol. 300 (1990) p. 395.
64 Lock, 'Towards a National Bioethics Committee', p. 1149.
65 British Medical Association, *Annual Report of Council, 1988–89* (London: British Medical Association, 1989) p. 30. See also Stephen Lock, 'Monitoring Research Ethical Committees', *British Medical Journal*, Vol. 300 (1990) pp. 61–2.
66 Taylor, 'Letters to the Editor', p. 395.
67 Rt Hon. Sir David Price, quoted in *Parliamentary Debates: House of Commons*, Vol. 140 (10 November 1988) col. 254.
68 Rt Hon. Margaret Thatcher, quoted in *Parliamentary Debates: House of Commons*, Vol. 140 (10 November 1988) col. 255.
69 Mulkay, *The Embryo Research Debate*.
70 Lewis, 'A Commons Subcommittee on Medical Ethics?'
71 Lock, 'Monitoring Research Ethical Committees', p. 62. See also Anon, 'National Ethics Committee – A Mirage?', *Bulletin of Medical Ethics*, no. 47 (February 1989) p. 9.
72 Ian Kennedy, 'The Role and Relevance of Legislation', in CIBA Foundation, *Medical Scientific Advance: Its Challenge to Society* (London: CIBA Foundation, 1990) pp. 18–23 (p. 21).
73 'Discussion', in CIBA Foundation, *Medical Scientific Advance*, p. 36.
74 Ibid.
75 Lock, 'Towards a National Bioethics Committee', p. 1149.

76 Sir David Weatherhall, 'The Problems as Perceived by Medical Researchers', in CIBA Foundation, *Medical Scientific Advance*, pp. 13–18 (p. 17).
77 Kennedy, interview with the author (2010). Weatherhall's role was acknowledged in a brief history of the Nuffield Council on Bioethics. See Ian Kennedy, 'Foreword', in *Nuffield Council on Bioethics: 1992–99* (London: Nuffield Council on Bioethics, 2000) p. 3.
78 David Shapiro, 'Nuffield Council on Bioethics', *Politics and the Life Sciences*, Vol. 14, no. 2 (1995) pp. 263–6; Jasanoff, *Designs on Nature*, pp. 185–6.
79 Shapiro, 'Nuffield Council on Bioethics', p. 263.
80 Nuffield Foundation Conference on Bio-Ethics, 20–22 April 1990. List of Guests, Trustees and Staff Attending.
81 Anon, 'Nuffield Foundation Conference on Bioethics. The Need for a New National Bioethics Body: Consultation Document', July 1990, p. 1.
82 Ibid, pp. 1, 3. See also Anon, 'National Ethics Committee Comes a Step Closer', *Bulletin of Medical Ethics*, no. 57 (April 1990) pp. 3–4.
83 Anon, 'Nuffield Foundation Conference on Bioethics. The Need for a New Bioethics Body', p. 3.
84 Ibid. See also Anon, 'National Ethics Committee Comes a Step Closer', p. 4.
85 O'Neill, interview with the author (2009); Shapiro, 'Nuffield Council on Bioethics', p. 263.
86 Anon, 'The Need for a New Bioethics Body', p. 3. Emphasis in original. O'Neill, interview with the author (2009); Anon, 'Editorial', *Bulletin of Medical Ethics*, no. 61 (August 1990) p. 1.
87 Anon, 'Nuffield Council on Bioethics', *Bulletin of Medical Ethics*, no. 85 (May 1991) p. 6.
88 *Nuffield Council on Bioethics, 1992–99*, p. 5; Shapiro, 'Nuffield Council on Bioethics', pp. 263–4.
89 Anon, 'Nuffield Council Starts Work', *Bulletin of Medical Ethics*, no. 73 (August 1991) pp. 3–4 (p. 4).
90 Shapiro, 'Nuffield Council on Bioethics', p. 264.
91 David Shapiro, permanent secretary to the Nuffield Council on Bioethics, quoted in Gail Vines, 'Research Ethics Face National Scrutiny', *New Scientist*, 2 March 1991, p. 14. As Sheila Jasanoff notes, the Nuffield Foundation believed that the council would in time rely less on the reputation of its members, with authority stemming more from the quality of its reports and public consultation exercises. See Jasanoff, *Designs on Nature*, pp. 186–7.
92 On Nairne's career in the civil service, see Hennessy, *Whitehall*, p. 253.

93 Anon, 'Nuffield Council Starts Work'. On the gene therapy inquiry, see Cecil Clothier (chair), *Report of the Committee on the Ethics of Gene Therapy* (London: HMSO, 1992).

94 Anon, 'Nuffield Council Starts Work'. On the foetal tissue inquiry, see John C. Polkinghorne (chair), *Review of the Guidance on the Research Use of Foetuses and Foetal Material* (London: HMSO, 1989).

95 *Nuffield Council on Bioethics, 1992–99*, p. 5.

96 Tom Wilkie, 'Whose Genes Are They Anyway?', *Independent*, 6 May 1991, p. 19.

97 Wilkie, 'Whose Genes Are They Anyway?'

98 Vines, 'Research Ethics Face National Scrutiny'.

99 Shapiro, 'Nuffield Council on Bioethics', p. 264.

100 Ibid.

101 Ibid; Jasanoff, *Designs on Nature*, p. 187.

102 On concerns over gene therapy, see Wilkie, 'Whose Genes Are They Anyway?' On concerns over tissue research, see Diana Brahams, 'Ownership of a Spleen', *Lancet*, Vol. 336 (1990) p. 239.

103 Nuffield Council on Bioethics, *Genetic Screening: Ethical Issues* (London: Nuffield Council on Bioethics, 1993).

104 Shapiro, 'Nuffield Council on Bioethics', p. 265.

105 Information on these and other bodies was given in the Labour government's report on *The Advisory and Regulatory Framework for Biotechnology* (London: Cabinet Office and Office of Science and Technology, 1999).

106 O'Neill, interview with the author (2009). In addition to being a council member, O'Neill served as a member of the working party that wrote the *Human Tissue* report. See Onora O'Neill, 'Medical and Scientific Uses of Human Tissue', *Journal of Medical Ethics*, Vol. 22 (1996) pp. 2–7.

107 Nuffield Council on Bioethics, *Human Tissue: Ethical and Legal Issues* (London: Nuffield Council on Bioethics, 1995) p. 124. For more on the background and reception to this report, see Richard Tutton, 'Person, Property and Gift: Exploring the Languages of Tissue Donation', in Richard Tutton and Oonagh Corrigan (eds), *Genetic Databases: Socio-ethical Issues in the Collection and Use of DNA* (London: Routledge, 2004) pp. 19–39.

108 *Nuffield Council on Bioethics, 1992–99*, p. 12. The DHSS was split into separate Departments of Health and Social Security in 1988.

109 Xenotransplantation was by no means a new procedure. It was the subject of professional and popular interest in the early twentieth century, thanks to tissue and gland grafting, and prompted considerable debates in the 1960s following unsuccessful attempts to

transplant chimpanzee hearts into humans. For more on these earlier debates, see Nathoo, *Hearts Exposed*; Susan Lederer, *Flesh and Blood: Organ Transplantation and Blood Transfusion in Twentieth-Century America* (Oxford: Oxford University Press, 2008).

110 Sir Patrick Nairne, 'Foreword', in Nuffield Council on Bioethics, *Animal-to-Human Transplants: The Ethics of Xenotransplantation* (London: Nuffield Council on Bioethics, 1996) pp. ii–iii (p. iii).

111 Anon, 'Reports on Xenografting', *Bulletin of Medical Ethics*, no. 114 (December 1995) p. 3.

112 Anon, 'Editorial', *Bulletin of Medical Ethics*, no. 125 (1997) p. 1. See also Advisory Group on the Ethics of Xenotransplantation, *Animal Tissue into Humans* (London: HMSO, 1997).

113 O'Neill, interview with the author (2009).

114 Advisory Group on the Ethics of Xenotransplantation, *Animal Tissue into Humans*, p. 8. On how the President's Commission influenced policymaking, see Tamar W. Carroll and Myron P. Guttmann, 'The Limits of Autonomy: The Belmont Report and the History of Childhood', *Journal of the History of Medicine and Allied Sciences*, Vol. 66 (2011) pp. 82–115; Evans, *The History and Future of Bioethics*, pp. 53–5. On how French committees influenced the development of 'bioethics laws' and vetoed some government policies, see Donna Dickenson, 'A National Bioethics Committee: Lessons from France', *BioNews*, 6 December 2004. Available online at http://www.bionews.org.uk/page_37776.asp (accessed 14 February 2014).

115 Kennedy, interview with the author (2010); MacLean, interview with the author (2009).

116 Anon, 'Editorial', p. 1.

117 Salter, *The New Politics of Medicine*, p. 58.

118 Tony Blair, quoted in Klein, *The New Politics of the NHS*, p. 188.

119 Ibid, pp. 189–90.

120 Tony Blair, quoted in Ibid, p. 190. See also Michael Marinetto, 'Who Wants to be an Active Citizen? The Politics and Practice of Community Involvement', *Sociology*, Vol. 37 (2003) pp. 103–20.

121 Salter, *The New Politics of Medicine*, p. 59; see also Klein, *The New Politics of the NHS*, pp. 188–205.

122 Klein, *The New Politics of the NHS*, p. 198.

123 Salter, *The New Politics of Medicine*, pp. 123–4.

124 Richard Smith, 'All Changed, Changed Utterly', *British Medical Journal*, Vol. 316 (1998) pp. 1917–18 (p. 1917).

125 Ibid, p. 1917.

126 Ibid, p. 1918; Klein, *The New Politics of the NHS*, p. 230.

127 John Warden, 'High Powered Inquiry into Bristol Deaths', *British Medical Journal*, Vol. 316 (1998) p. 1925.

128 Anon, 'Long-Time Critic of Vested Medical Interests', *Health Service Journal*, 8 July 1999. Accessed online at http://www.hsj.co.uk/long-time-critic-of-vested-medical-interests/30416.article (accessed 14 February 2014).
129 Kennedy, interview with the author (2010).
130 Anon, 'Long-Time Critic of Vested Medical Interests'. Kennedy served as a member of the GMC between 1984 and 1993.
131 Annabel Ferriman, 'Bristol Inquiry Appoints Doctor to its Panel', *British Medical Journal*, Vol. 318 (1999) p. 283.
132 Sarah Ramsey, 'UK "Bristol Case" Inquiry Formally Opened', *Lancet*, Vol. 353 (1999) p. 987.
133 Ibid.
134 John Warden, 'Cardiac Surgery Inquiry Given Wide Remit', *British Medical Journal*, Vol. 317 (1998) p. 498.
135 Ian Kennedy, quoted in Ferriman, 'Bristol Inquiry Appoints Doctor', p. 283.
136 Clare Dyer, 'Ian Kennedy Refused to Read the General Medical Council's Reports', *British Medical Journal*, Vol. 323 (2001) p. 183.
137 For more detail on the Shipman case, see Salter, *The New Politics of Medicine*, pp. 124–9. The 'retained organs scandal' was prompted by a witness statement at the Bristol inquiry, when a doctor recounted how many British hospitals retained organs and tissues without the consent of patients or families. For more detail, see Wilson, *Tissue Culture in Science and Society*, pp. 112–15.
138 Ian Kennedy (chair), *The Report of the Public Inquiry into Children's Heart Surgery at Bristol Royal Infirmary, 1984–1995: Learning from Bristol* (London: HMSO, 2001) pp. 1–2.
139 Ibid, p. 435.
140 Ibid, p. 3.
141 Ibid, pp. 442, 452.
142 Ibid. For more on the establishment of NICE, the CHI and other regulatory bodies under New Labour, see Klein, *The New Politics of the NHS*, pp. 199–200, 231–2.
143 Kennedy (chair), *Learning from Bristol*, p. 442.
144 Ibid, p. 443.
145 Department of Health, *Modernising Regulation in the Healthcare Professions: Consultation Document* (London: Department of Health, 2001) p. 4.
146 Ibid.
147 Ibid.
148 Mark Hunter, 'GMC Agrees New Structure', *British Medical Journal*, Vol. 323 (2001) p. 250
149 Department of Health, *Modernising Regulation*, p. 4.

150 Ibid, p. 9. Emphasis in original.
151 Klein, *The New Politics of the NHS*, p. 231. Indeed, the consultation document only differed from the Bristol plans in one or two areas. Ministers did not believe that the GMC and other bodies should be absorbed into the new council; and they also believed that standards for doctors should be set by the Department of Health, not the new council, whose job would be to ensure that these standards were adhered to. Thanks to New Labour's drive to shrink the regulatory sector in its second term, ministers were also prepared to create only one of the two 'overarching bodies' outlined in the Bristol report.
152 *Health and Social Care (Community Standards) Act, 2003: Chapter 43* (London: HMSO, 2003) p. 19.
153 Klein, *The New Politics of the NHS*, p. 231.
154 Ibid. See also Anon, 'In Brief', *British Medical Journal*, 326 (2003) p. 464.
155 Sarah Boseley, 'Man on a Mission', *Guardian*, 25 June 2003, p. C2; Laura Donnelly, 'Kennedy Takes the Helm at CHAI', *Health Service Journal*, Vol. II2, no. 5813 (2002) pp. 4–5.
156 Sarah Boseley, 'NHS Patients' Champion Puts Focus on Rights not Needs', *Guardian*, 25 June 2003, p. 9.
157 Boseley, 'Man on a Mission'.
158 Ibid.
159 Ibid.
160 Ibid.
161 Marinetto, 'Who Wants to be an Active Citizen?'
162 Kennedy, 'Patients are Experts'.
163 Ibid.
164 Ibid.
165 Ibid.
166 Liam Donaldson, 'Expert Patients Usher in a New Era of Opportunity for the NHS', *British Medical Journal*, Vol. 326 (2003) p. 1279.
167 Ibid.
168 Ibid. See also Kennedy, 'Patients are Experts'; Kennedy, 'The Patient on the Clapham Omnibus', pp. 446–7.
169 Bruce Charlton, 'The Ideology of Accountability', *Journal of the Royal College of Physicians of London*, Vol. 33 (1999) pp. 33–5 (p. 33).
170 Charlton, 'The Ideology of Accountability', p. 33.
171 Onora O'Neill, *A Question of Trust* (Cambridge: Cambridge University Press, 2002) p. 56.
172 O'Neill, *A Question of Trust*, p. 57.
173 O'Neill, *Autonomy and Trust in Bioethics*, p. 137.
174 Ibid, p. 139.

175 Ibid, p. 138.
176 Ibid, pp. 139–40.
177 Ibid, p. 3.
178 Ibid, pp. 144, 141.
179 Ibid, p. 73.
180 Ibid.
181 O'Neill, *A Question of Trust*, p. 83. Emphasis in original.
182 Although individual autonomy remained a component of 'principled autonomy', O'Neill argued that it should only be 'one minor aspect'. See Ibid, p. 97.
183 Ibid, p. 158.
184 See, for example, Kennedy, 'Preface', p. vii. See also Ian Kennedy, 'Doctors, Patients and Human Rights', in Kennedy, *Treat Me Right*, pp. 385–415.
185 Kennedy, 'Consumerism in the Doctor–Patient Relationship', p. 777.
186 Richard Horton, 'Why Is Ian Kennedy's Healthcare Commission Damaging NHS Care?', *Lancet*, Vol. 364 (2004) pp. 401–2 (p. 401).
187 Ibid, p. 402.
188 Ibid.
189 Ian Kennedy, 'Setting of Clinical Standards', *Lancet*, Vol. 364 (2004) p. 1399.
190 Ibid.
191 Charlotte Santry, 'Healthcare Regulator Longed for Government's Embrace', *Health Service Journal*, 12 November 2008, available online at www.hsj.co.uk (accessed 6 February 2014).
192 Charlotte Santry, 'Sir Ian Kennedy Champions "Fearless" NHS Regulator', *Health Service Journal*, 12 November 2009. Available online at www.hsj.co.uk (accessed 14 February 2014).
193 Santry, 'Healthcare Regulator Longed for Government's Embrace'. The Care Quality Commission was established by the government's 2008 Health and Social Care Act, fulfilling New Labour's goal of a single, integrated regulator for medicine and social care. In addition to the CHAI, it replaced the Commission for Social Care Inspection and the Mental Health Act Commission. See Klein, *The New Politics of the NHS*, pp. 270–1.
194 Richard Ashcroft, 'Futures for Bioethics?' *Bioethics*, Vol. 24, no. 5 (2010) p. ii.
195 Angus Dawson, 'Futures for Bioethics: Three Dogmas and a Cup of Hemlock', *Bioethics*, Vol. 25, no. 4 (2010) pp. 218–25 (pp. 218, 221).
196 Ibid, pp. 221, 222. Emphasis in original.
197 Ibid, p. 224.
198 On how various presidents have staffed national commissions with

politically sympathetic figures, see Jonathan D. Moreno, *The Body Politic: The Battle Over Science in America* (New York: Bellevue Literary Press, 2011).

199 David Shapiro to Jonathan Montgomery, 10 September 1990. Private papers of Jonathan Montgomery.

Conclusion

While she became associated with British bioethics following her engagement with IVF and embryo research in the 1980s, Mary Warnock is better known today for her views on euthanasia.[1] Warnock first engaged with this issue in 1993, when she was appointed to a House of Lords Select Committee that investigated whether there were circumstances in which 'assisted dying' might be permissible, when a doctor would not be prosecuted for ending a patient's life or helping them end their own lives. After deliberating for a year, Warnock and her fellow committee members agreed that the law surrounding euthanasia should remain the same. They argued that doctors who actively killed a patient should continue to be charged with murder, while those who helped a patient end their lives should continue to be prosecuted for aiding or abetting suicide. In line with the legal distinction between killing and letting die, which underpinned the verdict in the *Arthur* case, the committee also agreed that doctors should not be prosecuted for withdrawing or withholding treatment, or for administering a drug to relieve pain knowing it would shorten life.[2]

Despite the Select Committee's conclusions, euthanasia remained a contentious and high-profile issue. This was due in no small part to the campaigns of terminally ill patients such as Diane Pretty, who suffered from Motor Neurone Disease and sought legal assurances that her husband would not be prosecuted for helping her commit suicide. Although Pretty lost her 2002 case at the Court of Appeal and the European Court of Human Rights, continuing support for the 'right to die' led Lord Joffe, a human rights lawyer, to try and get the House of Lords to approve several Private Member's Bills that permitted assisted dying under strictly defined circumstances.[3]

Assisted dying also remained 'in the public eye' thanks to the regular and very public arguments of Mary Warnock.[4] In 1998 Warnock claimed that bioethics had a vital role to play in debates on assisted dying, by 'helping lift some of the burdens of value judgements from the shoulders of individual practitioners' and determining whether the law was 'too much out of line with ethical beliefs'.[5] As she increasingly contributed to public debates, it soon became clear that the law was now out of line with Warnock's own ethical beliefs. Although she and her fellow committee members had rejected calls for assisted dying in the early 1990s, she changed her views after watching Geoffrey Warnock suffer and eventually die from the lung disease fibrosing alveolitis in 1995.[6]

In a 2003 piece for the *Sunday Times*, Warnock voiced regret at having supported a 'bad law' and threw her weight behind Lord Joffe's and Diane Pretty's campaigns for legal change.[7] In newspapers, television programmes and books such as *Easeful Death*, she now claimed that it was 'inhumane' to deny people the right to die.[8] She argued that this included not only terminally ill patients, but also individuals who felt they were a burden on their families due to disability or old age.[9] When it came to assisted dying, Warnock argued, doctors had a pressing duty, 'unless their religion forbids it', to respect the autonomy of dying, elderly and disabled patients.[10] In a 2008 column for the *Observer*, she stressed that 'we have a moral obligation to take other people's seriously reached decisions with regard to their lives equally seriously, not putting our judgement of the value of their lives above theirs'.[11]

Warnock's support for assisted dying is significant in a number of respects. It shows, first, how an individual's ethical views are not fixed and can change according to what the *Observer* called 'the lessons of life'.[12] Secondly, and more importantly, it shows just how much authority bioethicists are thought to wield over public affairs. Whether they were for or against the 'right to die', journalists and politicians all agreed that Warnock's involvement was hugely important because she had been 'Britain's chief moral referee for the past thirty years'.[13] As the *Daily Telegraph* noted, opponents of assisted dying feared that her arguments 'may find wider support because of her influence on ethical matters'.[14] Members of a 'Right to Life' campaign group, for example, believed that Warnock's influence 'would sway people' towards her views. And the Conservative MP Nadine Dorries similarly worried that 'because of her previous experiences and well-known standing on contentious moral issues,

Baroness Warnock automatically gives moral authority to what are entirely immoral viewpoints'.[15]

Warnock's views were seen as highly significant not for their rightness or wrongness, then, but more for the weight they carried thanks to her 'moral authority'. We must not presume that this authority derived solely from her status as a member of the 'Great and Good' either, for other bioethicists are also regarded as high-profile and authoritative figures. In 1982 the BBC chose Jonathan Glover to present a *Horizon* programme on genetic engineering and enhancement, entitled 'Brave New Babies', in which he discussed ethical issues with scientists, students, members of the public and his own children. The fact that a philosopher fronted an episode of the BBC's flagship science series again shows how bioethicists emerged as a 'new epistemic power' in Britain from the 1980s onwards.[16] Some years later, in 2006, the *Independent* newspaper included John Harris in its 'Good List' of the 'fifty men and women who help make the world a better place'. Like Glover, Harris was a high-profile figure and the *Independent* claimed that his inclusion on the 'Good List' resulted from his status as 'a key player in the shaping of the moral debates around human fertility and bioethics'.[17]

The profile and authority that Warnock, Glover, Harris and others have attained demonstrates that bioethicists now play an equal and sometimes greater role than doctors and scientists in publicly discussing the ethics of issues such as assisted dying, embryo research and genetic engineering. Although the notion of moral expertise remains contested and many bioethicists refuse to acknowledge it, they are often portrayed as what the *Guardian* called 'ethics experts'.[18] Thanks to escalating mistrust of club regulation, both in public and, crucially, in government, they derive their authority from being 'expert outsiders' who are independent from the profession or procedure under scrutiny.[19]

Their portrayal as 'ethics experts' confirms that bioethicists have indeed contributed to a shift in the location and exercise of biopower in Britain. The days of 'club regulation' are a thing of the past and we no longer believe that expertise in medical and scientific ethics is inscribed solely within doctors or scientists, imbued through professional training and the acquisition of specialist knowledge.[20] As previous chapters have shown, and as Nikolas Rose outlines, 'multiple forces now encircle' the work of doctors and scientists, and bioethicists are widely expected to help 'shape the paths taken, or not taken' by research.[21] The public demand for bioethics also substantiates Brian Salter's claim that the late twentieth century has witnessed the growth of a 'cultural biopolitics', where bioethicists do not directly control bodies or populations *per se*, but help clarify the ethical values that determine the fate of technologies and procedures which can subsequently impact on individual and collective health.[22]

But historical research can do more than simply corroborate these broad claims. Rabinow and Rose outline how contemporary biopower, in which we can include 'cultural biopolitics', combines three overlapping 'dimensions or planes'. These comprise a form of public discourse about living beings and the array of authorities that are considered competent to speak the truth; modes of subjectification in the name of individual and collective health; and strategies for intervention upon patients and populations. They argue that these planes are 'continually recombining and transforming one another' and claim that we therefore need to work on a 'small scale' in order to detail how and why configurations of biopower differ markedly over time and in different locations.[23] When we consider bioethics, history is vital for showing how bioethicists have had a different impact on these various aspects of biopower in specific locations, thanks to the outlook of individuals and professions, and their interplay with social and political factors.

If we take each of these 'dimensions or planes' in turn, it is clear that British bioethicists exerted their greatest influence in public debates. From the 1980s onwards they came to shape the discussion of issues that were once left to doctors and scientists. They were clearly deemed more than competent authorities in these debates, as their portrayals as influential 'moral referees' and the reward of knighthoods for 'services to bioethics' demonstrates.[24] Bioethicists also played a decisive role in the second domain of biopower and helped create new subjectivities when they led these public debates. In their discussion of assisted dying, IVF and other issues, they regularly framed patients and the public as empowered stakeholders who were 'entitled to know, and to control' biomedical practices.[25] These portrayals carried weight thanks to the way they reflected, and reinforced, the neo-liberal conception of individuals as 'active citizens' who were entitled to a greater say in the running of professions and public services.[26]

Charting these links is central to understanding why bioethics emerged as a recognised term and approach in Britain. While calls for external involvement were by no means new, they gained traction in the 1980s because they dovetailed with the Conservative government's enthusiasm for oversight, transparency and public accountability. Yet bioethics was not simply the top-down result of political pressure, and owes as much to the agency of specific individuals and groups as it does to changing sociopolitical contexts. Figures such as Ian Kennedy and Mary Warnock endorsed outside involvement for specific reasons, such as empowering patients, introducing American forms of oversight and applying philosophy to practical affairs. Their public rhetoric was not simply a reaction to growing calls for external involvement but was fundamentally constitutive of it, which shows how these public figures generated and helped perpetuate the demand for bioethics, and played a major role in their own transformation into 'ethics experts'. At the same time, this changing context also led prominent doctors and journals to accept calls for external oversight, in the belief that it would maintain public and political trust in research.

The political demand for 'non-expert' involvement also allowed lawyers, philosophers and others to play a vital role in regulatory inquiries that had been traditionally dominated by doctors and scientists. This ensured that they had some influence in the third domain of biopower: helping to define 'strategies for intervention upon collective existence in the name of life and health' by developing guidelines for procedures such as IVF, gene therapy and xenotransplants.[27] These appointments gave bioethicists a major say in shaping legal guidelines for new techniques and allowed them to help determine whether, and when, entities such as *in vitro* embryos were entitled to legal protection.

But in contrast to their American counterparts, British bioethicists had far less say over medical treatment and research, which remains the major source of strategies for determining individual and collective health. This difference was not lost on participants at a 1987 'Anglo-American Conference on Biomedical Ethics', which was held in New York and co-sponsored by the New York Academy of Sciences, the Hastings Center and the Royal Society of Medicine.[28] During a planning meeting, members of the British organising committee, which included Gordon Dunstan and Sir Douglas Black, had suggested that an 'interesting and fruitful' approach would be to look at 'topics that reveal differences between the UK and USA'.[29] Staff at the Hastings Center claimed that discussing the 'marked differences' between Britain and the United States would 'foster an understanding and mutual appreciation for the work that goes on in each country and, moreover, should enable participants to return to their countries with new insights for dealing with the issues confronting them'.[30] Both organising committees initially proposed that papers could look at contrasting approaches to embryo research, which 'continues in the UK, but has been almost completely stopped in the USA'. But they eventually agreed that the most fruitful discussion would result from concentrating on the role that bioethicists played in determining 'the rights and duties of physicians and patients' in either country, since it was here that the greatest differences were apparent.[31]

The legal philosopher Gerald Dworkin, then working at Queen Mary University in London, highlighted the major differences in his paper on the 'delicate balance' between ethics, law and medicine in Britain and the United States. Dworkin claimed that British bioethicists exerted less influence over medical practices thanks partly to the ongoing lack of a 'permanent review body' such as the President's Commission, which drew up guidelines for new procedures and also issued guidelines for medical treatment in the United States.[32] Shortly before it was disbanded in the early 1980s, the President's Commission had recommended the establishment of hospital ethics committees which included bioethicists, doctors and community representatives. Although this recommendation was not legally binding, many hospitals established their own ethics committees to conduct reviews whenever a family and their doctor disagreed about life-sustaining treatments.[33] Dworkin noted that in Britain, by contrast, where there was a 'long overdue' need for a national committee, bioethicists only influenced practices as members of occasional *ad hoc* committees into new technologies, remained in the minority of professional bodies such as the BMA and were completely absent from NHS hospitals.[34]

Dworkin and other speakers also outlined how the growth of hospital ethics committees, and the influence bioethicists had on clinical care, stemmed from an increasingly litigious culture in the United States, where judges often ruled against doctors and 'the traditional paternalistic role of the adviser is being more and more questioned'.[35] With American doctors 'exposed to increased liability to litigation', they viewed ethics committees and bioethical advice as a vital way of resolving conflict and forestalling damaging legal cases.[36] Dworkin and other speakers noted that the British courts, by contrast, continued to endorse the *Bolam* ruling and 'prefer to accept rather than condemn the practices and ethical standards of the medical profession'.[37] As Ian Kennedy acknowledged in *Unmasking Medicine*, and as the endocrinologist Sir Raymond Hoffenberg detailed in his paper, the continued 'hands-off' stance in Britain was widely seen as a precondition of the welfare state, where 'it is accepted that limitation of choice, indeed of the range of treatment, is part of the price we pay for open and free access to our health service'.[38]

Faced with little threat of legal interference, British doctors continued to assert that involving outsiders in clinical treatment would harm patient care and damage the morale of doctors.[39] These arguments ensured that politicians did not take 'an obtrusive role' when it came to medical practices, despite their enthusiasm for oversight, while they were also supported by figures such as Ian Kennedy, who promised that they were not looking to judge medical care on a 'case-by-case basis'.[40] As Gordon Dunstan outlined in his conference paper, this ensured that while philosophers and lawyers played a central role in determining clinical practices in the United States, 'in the United Kingdom we locate those functions firmly within the profession of medicine: the doctors themselves are the moral agents we hold accountable for their decisions'.[41]

The papers from this 'Anglo-American Conference' illustrate that while bioethics has certainly played a role in reshaping our contemporary 'politics of life', its impact has varied according to social, political and professional factors in different locations. The influence that bioethicists exert over the various settings where biomedical knowledge is generated and deployed, such as the clinic, the courtroom, regulatory commissions and the public sphere, differs considerably within and between countries. If we are to draw any broad conclusion from this evidence, then, it is that British bioethicists enjoyed considerable influence in public debates and *ad hoc* committees that drew up guidelines for new procedures, but had less influence in governing medical treatment than their colleagues elsewhere. Yet this also comes with an important caveat, for by showing how the influence and contours of bioethics are 'fluid and changing with context', history reminds us that its status and authority are likely to change in future, both in Britain and elsewhere.[42]

Remaking bioethics?

Indeed, there are signs that such a change is underway if we look at British bioethics today. While the broader climate has been conducive in the past, changing political and financial priorities now threaten bioethics with 'retrenchment and decline'.[43] Criticism of the 'ideology of accountability' has grown since the CHAI was shelved and continues to prompt questions about who should judge medical and scientific practices. This was clear in March 2010, when Labour's Health Secretary, Andy Burnham, responded to claims that regulation was 'stifling' innovation by inviting the Academy of Medical Sciences to review the governance of biomedical research.[44] The Academy's working party was chaired by the pharmacologist Sir Michael Rawlins and was composed predominantly of senior doctors and scientists, with only three of its nineteen members drawn from outside the NHS or the biomedical research sector.[45]

Their report, published in January 2011, argued that medical progress was being 'seriously undermined by an overly complex regulatory and governance environment'.[46] The committee echoed Onora O'Neill when they claimed that there was no evidence to suggest that increased oversight had 'enhanced the safety and well-being of either patients or the public'.[47] They argued instead that oversight had 'the potential to undermine public health by delaying important medicines being investigated in clinical trials'.[48] Their solution to this 'cumbersome' environment notably lay in 'streamlining regulation'.[49] This entailed the formation of an overarching Health Research Agency (HRA) that would 'bring together the regulatory functions that are currently fragmented across multiple bodies'.[50] While the committee's report claimed that the HRA should contain some patient representatives and a mixture of 'medical, scientific and ethical' professionals, it stressed that oversight worked best when conducted by 'an informed committee with relevant expertise'.[51]

Crucially, the proposals of this 'Rawlins report' dovetailed with the policies of the Conservative–Liberal Democrat coalition that had been formed after no party gained an outright majority in the 2010 general election. Coalition politicians quickly agreed that a sizeable national deficit should be reduced by cutting government spending and rationalising public services. Partly thanks to this austerity programme, and partly thanks to an ideological rejection of 'big government', they pledged 'a radical simplification of the regulatory landscape for medical research' after the election.[52] Writing in the *Observer* shortly after this announcement, the Prime Minister David Cameron notably promised 'no interference – just real power for professionals'.[53]

In October 2010 coalition politicians announced the planned abolition of 192 advisory bodies, including the HFEA, the HTA and the HGC.[54] They proposed that the HFEA's regulation of fertility treatments should move to the Care Quality Commission, while its research licensing work and the functions of the HTA were to be absorbed by 'a new super-regulator'.[55] After the coalition approved the Rawlins report in its March 2011 budget, it announced that this super-regulator would be a new Health Research Authority, which was established in December 2011 and began work on creating 'a unified approval process' for biomedical research (after concerted appeals, the HFEA was ultimately saved from this 'bonfire of the quangos' in 2013).[56] The government's 'simplification' of the regulatory landscape also underpinned the reconstitution of the HGC as an Emerging Bioethics and Advisory Committee (ESBAC) in 2011, which was tasked with advising ministers and relevant stakeholders on 'emerging healthcare scientific developments and their ethical, legal, social and economic implications'.[57]

While ESBAC had a wider remit than the HGC, committee members and the Department of Health were clear it should not be regarded as a national bioethics committee.[58] The government's austerity programme, meanwhile, ensured that ESBAC had fewer financial resources than its predecessor and was expected to undertake 'adept networking and collaborations with external bodies in order to pool resources'.[59] Unusually for a body with the word 'bioethics' in its title, ESBAC was chaired by the clinical pharmacologist Sir Alasdair Beckenbridge, and only seven of its seventeen members were non-doctors or scientists.[60]

At the same time, David Cameron's 'Big Society' initiative, which encouraged charities and private firms to assume roles previously fulfilled by the state, meant that politicians were less likely to establish public inquiries into science and medicine and appeared to limit the scope for 'official bioethics'.[61] This was made clear in a recent Commission on Assisted Dying, which by was established not by the government but by the entrepreneur Jonathan Lewis and the author Terry Pratchett. To retain their policy influence, bioethicists will have to accommodate these changing structures. And if future commissions or inquiries are no longer state-supported, bioethicists will have to ask questions about their legitimacy and possible bias – especially following criticism of industry-funded bioethics in the United States, where critics argued that bioethicists were unwilling to 'bite the hand that feeds them'.[62]

The future of many academic bioethics centres is also threatened by the coalition government's decision to cut university funding and increase student fees.[63] In an obituary of Mary Lobjoit, who died in 2011, Margaret Brazier, Raanan Gillon and John Harris warned that 'cash strapped' institutions were likely to attach 'less value' to postgraduate degrees in bioethics, which attract fewer students than courses in medicine, science or engineering.[64] At the time of writing, some of these centres, which act as entry points to the field, are scheduled to close and others face an uncertain future.[65]

Brazier, Gillon and Harris defended these academic centres by claiming that they provided 'a service to the community as a whole and not just an indulgence for our own academic passions'.[66] While some assert the continued utility of their work, others claim that bioethics must change if it is to survive. Richard Ashcroft, for example, criticised the 'formalism' and predictability of existing approaches and urged the adoption of new strategies 'to reinvigorate research and debate on our field'.[67] This included the incorporation of methods from fields such as sociology and anthropology, which Angus Dawson argued should encourage consideration of the commercial, political and cultural factors that 'go beyond the autonomous decisions of individuals' and would help shift the focus in bioethics to 'the complex web of social relations that binds us all'.[68]

These arguments found a receptive audience among social scientists, who were also encouraged to undertake practically oriented work thanks to the continued enthusiasm for 'impact' among universities and research councils. Sociologists and anthropologists, among others, now moved beyond previous critiques of bioethics and outlined how it might benefit from empirical methods. They argued that a more 'bottom-up' approach, based on dense knowledge of particular social settings, would help connect bioethics to the actual expectations of doctors and patients, who regularly displayed preferences, values and forms of reasoning different to those prioritised in bioethical texts.[69] As social scientists increasingly published in bioethics journals and sat on regulatory commissions, many talked of an 'empirical turn' in bioethics and described it as a 'dynamic, changing and multi-sited field'.[70]

But others contested this shift and claimed that the different approaches that now constituted bioethics were 'not all complementary'. The expanding number of methods and participants, they argued, 'begin from distinctive premises about human nature, justice and social organization and often proceed to different normative conclusions'.[71] To Ruth Macklin, 'increased subspecialization' and the potential for disagreements threatened nothing less than 'the death of bioethics (as we once knew it)'.[72] As these arguments show, there is still no consensus on what bioethics is or how it should function, and it continues to be remade in line with changing professional and political outlooks. What form and emphasis bioethics will take in future remains unclear, but it is likely to be different from the approach that took hold in the 1980s and 1990s. As Onora O'Neill remarked in 2002, it appears there is still 'no *complete* answer to the old question: "who will guard the guardians?"'[73]

Notes

1 Jasanoff, *Designs on Nature*, p. 152.
2 For more background, see Nick Kemp, *Merciful Release: The History of the British Euthanasia Movement* (Manchester: Manchester University Press, 2002).
3 Joffe's conditions were that two doctors must certify that a patient was mentally competent; that the patient must sign a document stating their wish to die; that they must have alternatives explained to them; and that they must be able to change their mind in the period between signing the form and death taking place. For more detail, see Mary Warnock, *Dishonest to God: On Keeping Religion Out of Politics* (London: Continuum, 2010) pp. 45–6.
4 Mary Warnock and Elisabeth MacDonald, *Easeful Death: Is There a Case for Assisted Dying?* (Oxford: Oxford University Press, 2009) p. vii.
5 Warnock, 'Commentary on "Suicide, Euthanasia and the Psychiatrist"', p. 129.
6 In a 2005 interview with the *Observer*, Warnock outlined how her views on euthanasia stemmed from Geoffrey Warnock's illness, and admitted that his death had been hastened after a medically qualified

friend gave him morphine. See Robin McKie, 'There's Something About Mary', *Observer*, 12 June 2005. Available online at www.guardian.co.uk/uk/2005/jun/12/schools.education/print (accessed 15 February 2014).

7 Mary Warnock, 'I Made a Bad Law – We Should Help the Ill to Die', *Sunday Times News Review*, 23 December 2003, p. 9. See also Jonathon Carr-Brown, 'Peers Recant on Mercy Killing', *Sunday Times*, 23 December 2003, p. 14.

8 Mary Warnock, quoted in McKie, 'There's Something About Mary'.

9 Mary Warnock, 'Legalise Assisted Suicide for Pity's Sake', *Observer*, 19 October 2009; Anon, 'Dementia Patients' "Right to Die"', BBC News, 19 September 2008. Available online at http://news.bbc.co.uk/1/hi/7625816.stm (accessed 21 February 2014).

10 Warnock, 'I Made a Bad Law'. For more on Warnock's views on assisted dying and autonomy, see Warnock, *Dishonest to God*, pp. 46–50.

11 Warnock, 'Legalise Assisted Suicide for Pity's Sake'.

12 McKie, 'There's Something About Mary'.

13 Ibid.

14 Martin Beckford, 'Baroness Warnock: Dementia Sufferers May Have a "Duty to Die"', *Daily Telegraph*, 18 September 2008, p. 11.

15 Nadine Dorries, quoted in Beckford, 'Dementia Sufferers May Have a "Duty to Die"'.

16 Salter and Salter, 'Bioethics and the Global Moral Economy', p. 564.

17 Anon, 'The Good List 2006: The Fifty Men and Women who Make Our World a Better Place', *Independent*, 1 September 2006. Available online at http://www.independent.co.uk/news/people/profiles/the-good-list-2006-414252.html (accessed 15 February 2014).

18 Claire Dyer, 'Ethics Expert Calls for Legal Euthanasia', *Guardian*, 26 April 1994, p. 3. This article refers to Ian Kennedy, who urged the House of Lords Select Committee to support a change in the law to allow assisted dying in specific circumstances.

19 Sarah Boseley, 'Man on a Mission', *Guardian*, 25 June 2003, p. C2. In this sense, we might conclude that bioethicists offer a prime example of the 'interactional experts' who Harry Collins and Steve Evans claim have emerged in recent decades. 'Interactional experts', in their view, complement or replace 'specialist experts' in public debates and decision-making processes. They have specialist expertise pertaining to a particular field, such as philosophy or law, and apply this expertise to fields in which they have no specialist expertise, such as science or medicine, to satisfy growing demands that policymaking should include 'the opinion of the consumer'. See Collins and Evans, *Rethinking Expertise*, p. 134. While the emergence of bioethics appears to support elements

of this work, more research is needed, thanks in part to ongoing questions over whether there is such a thing as 'bioethical expertise'. See, for example, Archard, 'Why Philosophers Are Not and Should Not be Moral Experts'; Warnock, *A Question of Life*.

20 Rose, *The Politics of Life Itself*, pp. 30–1.
21 Nikolas Rose, 'Democracy in the Contemporary Life Sciences', *Biosocieties*, Vol. 7 (2012) pp. 459–72 (p. 461).
22 Salter, 'Cultural Biopolitics, Bioethics and the Moral Economy'.
23 Rabinow and Rose, 'Biopower Today', pp. 203–5.
24 Ibid, p. 203.
25 Warnock, *A Question of Life*, p. xiii.
26 Reubi, 'The Human Capacity to Reflect and Decide', p. 355. See also Rose, *Powers of Freedom*, pp. 164–7.
27 Rabinow and Rose, 'Biopower Today', p. 203.
28 Papers related to the organisation of this conference are held at the National Archives: FD7/2677.
29 Gordon Dunstan (chair), 'Proposed Anglo-American Conference on Biomedical Ethics. A Planning Meeting held at the Royal Society of Medicine', 26 February 1986. National Archives: FD7/2677.
30 Daniel Callahan, 'The Hastings Center: Anglo-American Conference on Biomedical Ethics' (1986). National Archives: FD7/2677.
31 Dunstan (chair), 'Proposed Anglo-American Conference'; Callahan, 'Anglo-American Conference on Biomedical Ethics'.
32 Gerald Dworkin, 'The Delicate Balance: Ethics, Law and Medicine', *Annals of the New York Academy of Sciences*, Vol. 530 (1988) pp. 24–36 (p. 34).
33 On how the percentage of hospitals with ethics committees doubled between 1983 and 1985, see Rothman, *Strangers at the Bedside*, pp. 255–7.
34 Dworkin, 'The Delicate Balance', p. 35.
35 Ibid.
36 Ibid, p. 21. See also Stevens, *Bioethics in America*.
37 Dworkin, 'The Delicate Balance', p. 24.
38 Sir Raymond Hoffenberg, 'A Tale of Two Cultures; Do We Differ?' *Annals of the New York Academy of Sciences*, Vol. 530 (1988) pp. 16–23 (p. 19). See also Kennedy, 'Consumerism in the Doctor–Patient Relationship', *Listener*, 11 December 1980, pp. 777–80 (p. 778).
39 Salter, *The New Politics of Medicine*, p. 58.
40 Hoffenberg, 'A Tale of Two Cultures', p. 19; Kennedy, *The Unmasking of Medicine*, p. 119.
41 Gordon Dunstan, 'Two Branches from One Stem', *Annals of the New York Academy of Sciences*, Vol. 530 (1988) pp. 3–6 (p. 5).

42 De Vries et al., 'Social Science and Bioethics', p. 5.
43 Turner, 'Does Bioethics Exist?', p. 779; Montgomery, 'Lawyers and the Future of UK Bioethics'.
44 Academy of Medical Sciences, 'Media Statement', March 2010. Available online at www.acmedsci.ac.uk (accessed 15 February 2014). Calls for a review of research governance were originally made in the Academy of Medical Sciences document on *Reaping the Rewards: A Vision for UK Medical Science* (London: Academy of Medical Sciences, 2011). Available online at http://www.acmedsci.ac.uk.
45 The full list of members is given in Appendix A of the Academy of Medical Science's final report. See Professor Sir Michael Rawlins (chair), *A New Pathway for the Regulation and Governance of Health Research* (London: Academy of Medical Sciences, 2011) p. 114. The non-medical or scientific members were the bioethicist Mike Parker, chair of the Ethox Centre at the University of Oxford; Genevra Richardson, a lawyer from King's College, London; and Paddy Storie, a teacher from Harpenden.
46 Ibid, p. 2.
47 Ibid, p. 3.
48 Ibid, p. 84.
49 Ibid, pp. 30, 4.
50 Ibid, p. 4.
51 Ibid, pp. 85, 82.
52 Daniel Cressey, 'UK Embryo Agency Faces the Axe', *Nature*, Vol. 466 (2010) p. 674.
53 David Cameron, 'This is a Government that Will Give Power Back to the People', *Observer*, 11 September 2010, p. 18.
54 Cressey, 'UK Embryo Agency Faces Axe', p. 674. See also Amelia Gentleman, 'A Quiet Battle for Life: IVF Regulator Aims to Show its Worth', *Guardian*, 10 January 2011, p. 7.
55 Cressey, 'UK Embryo Agency Faces Axe'.
56 Geoff Watts, 'New Body Aims to Streamline Approval and Regulation of Research in NHS', *British Medical Journal*, Vol. 343 (2011) p. 7950; Ian Sample, 'Two Healthcare Regulators Spared Axe But Face Efficiency Review', *Guardian*, 25 January 2013, p. 16.
57 Anthony Blackburn-Starza, 'UK Government's New Bioscience Advisory Body Starts Recruiting', *BioNews*, 19 March 2012. Available online at www.bionews.org.uk (accessed 15 February 2014).
58 Meeting minutes and terms of reference for ESBAC can be found online at www.gov.uk/government/policy-advisory-groups/emerging-science-and-bioethics-advisory-committee (accessed 21 February 2014).
59 Stuart Hogarth and Brian Salter, 'Emerging Science and Established Problems – Governing Biomedical Innovation in the UK', *BioNews*,

11 April 2012. Available online at www.bionews.org.uk (accessed 15 February 2014).

60 Anthony Blackburn-Starza, 'Alasdair Beckenbridge Appointed Chair of Emerging Science and Bioethics Advisory Committee', *BioNews*, 21 May 2012. Available online at www.bionews.org.uk (accessed 15 February 2014).

61 Montgomery, 'Lawyers and the Future of UK Bioethics'.

62 Carl Elliot, 'Should Journals Publish Industry-Funded Bioethics Articles?', *Lancet*, Vol. 366 (2005) pp. 422–4.

63 The coalition imposed an upper limit of £9,000 per year on undergraduate fees, although postgraduate fees can be much higher. Since the majority of bioethics courses are offered at postgraduate level, applications are likely to be affected by this increase.

64 Brazier et al., 'Helping Doctors Become Better Doctors', pp. 2–3.

65 See, for example, John Morgan, 'Keele's Philosophers Stare into the Abyss', *Times Higher Education Supplement*, 18 March 2011. Available online at www.timeshighereducation.co.uk/story.asp?storycode=415539 (accessed 15 February 2014).

66 Brazier et al., 'Helping Doctors Become Better Doctors'.

67 Ashcroft, 'Futures for Bioethics', p. ii.

68 Dawson, 'Futures for Bioethics', p. 224.

69 See, for examples, Adam Hedgecoe, 'Critical Bioethics: Beyond the Social Science Critique of Applied Ethics', *Bioethics*, Vol. 18 (2004) pp. 120–43; Jose Lopez, 'How Sociology Can Save Bioethics ... Maybe', *Sociology of Health and Illness*, Vol. 26 (2004) pp. 875–96; Raymond de Vries, 'How Can We Help? From "Sociology in" to "Sociology of" Bioethics', *Journal of Medical Law and Ethics*, Vol. 23 (2003) pp. 279–92; Erica Haimes, 'What Can the Social Sciences Contribute to the Study of Ethics? Theoretical, Empirical and Substantive Considerations', *Bioethics*, Vol. 16 (2002) pp. 89–113; Arthur Kleinman, 'Moral Experience and Ethical Reflection: Can Ethnography Reconcile Them? A Quandary for "The New Bioethics"', *Daedalus*, Vol. 128, no. 4 (1999) pp. 69–99.

70 De Vries et al., 'Social Science and Bioethics', pp. 2–3. See also Pascal Borry, Paul Schotmans and Kris Dierickx, 'The Birth of the Empirical Turn in Bioethics', *Bioethics*, Vol. 19 (2005) pp. 49–71.

71 Turner, 'Does Bioethics Exist?', p. 778; see also Bracanovic, 'Against Culturally Sensitive Bioethics'.

72 Ruth Macklin, 'The Death of Bioethics (As We Once Knew It)', *Bioethics*, Vol. 24 (2010) pp. 211–17 (p. 211).

73 O'Neill, *Autonomy and Trust in Bioethics*, p. 6. Emphasis in original.

Bibliography

Archives

British Film Institute, London
Durham Cathedral Archive: Ian Ramsey papers (uncatalogued)
National Archives, London: Medical Research Council files
Society for Applied Philosophy papers (uncatalogued)
University of Manchester: Centre for Social Ethics and Policy papers (uncatalogued)
University of Manchester Library: Special Collections
University of Virginia Historical Collections: Joseph Fletcher papers
Wellcome Library, London, Archives and Manuscripts: London Medical Group papers; Maurice Henry Pappworth papers; Strangeways Research Laboratory papers

Interviews

Professor Alastair Campbell, telephone interview (May 2009)
Professor John Harris, University of Manchester (November 2011)
Sir Ian Kennedy, Portland House, London (July 2010)
Professor Sheila MacLean, University of Glasgow (July 2009)
Baroness Onora O'Neill, The Royal Society, London (June 2009)
Professor Jerome Ravetz, Oxford (August 2010)
Baroness Mary Warnock, House of Lords, London (March 2009)
Baroness Shirley Williams, Millbank House, London (September 2010)

Bulletins, newspapers and popular sources

Bulletin of Medical Ethics; *Bulletin of the Churches' Council of Healing*; *Daily Express*; *Daily Mirror*; *Daily Telegraph*; *Guardian*; *Independent*; *Listener*; *London Review of Books*; *Mail on Sunday*; *New Scientist*; *New Statesman*; *Observer*; *Progress in Medical Ethics*; *The Sunday Times*; *The Times*; *Times Literary Supplement*

Official reports and guidelines

Academy of Medical Sciences, *Reaping the Rewards: A Vision for UK Medical Science* (London: Academy of Medical Sciences, 2011).

Ad Hoc Committee of Harvard Medical School, 'A Definition of Irreversible Coma', in P. Singer and H. Kushe (eds), *Bioethics: An Anthology* (London: Blackwell, 1999) pp. 287–92.

Advisory Committee on Animal Experiments, *Report to the Secretary of State on the Framework of Legislation to Replace the Cruelty to Animals Act 1876* (London: Home Office, 1981).

Advisory Group on the Ethics of Xenotransplantation, *Animal Tissue into Humans* (London: HMSO, 1997).

Anon, 'Nuffield Foundation Conference on Bioethics. The Need for a New National Bioethics Body: Consultation Document' (July 1990).

British Medical Association, *Annual Report of Council, 1988–89* (London: British Medical Association, 1989).

Cabinet Office and Office of Science and Technology, *The Advisory and Regulatory Framework for Biotechnology: Report from the Government's View* (London, 1999).

Clothier, C. (chair), *Report of the Committee on the Ethics of Gene Therapy* (London: HMSO, 1992).

Cross, G. (chair), *Report on the LD50 Test* (London: Home Office, 1979).

Department of Health, *Modernising Regulation in the Healthcare Professions: Consultation Document* (London: Department of Health, 2001).

General Medical Council, *Recommendation as to Basic Medical Education* (London: General Medical Council, 1967).

General Medical Council, *Survey of Basic Medical Education in the British Isles* (London: Nuffield Provincial Hospitals Trust, 1977).

General Medical Council, *Tomorrow's Doctors: Recommendations for Undergraduate Medical Education* (London: General Medical Council, 1993).

Griffiths, R. (chair), *NHS Management Inquiry* (London: Department of Health and Social Security, 1983).

Health and Social Care (Community Standards) Act, 2003 (London: HMSO, 2003).

Kennedy, I. (chair), *The Report of the Public Inquiry into Children's Heart Surgery at Bristol Royal Infirmary, 1984–1995: Learning from Bristol* (London: HMSO, 2001).

Littlewood, S. (chair), *Report of the Departmental Committee on Experiments on Animals* (London: HMSO, 1965).

Jones, A., and Bodmer, W., *Our Future Inheritance: Choice or Chance?*

A Study by a British Association Working Party (Oxford: Oxford University Press, 1974).
Morley, D., *The Sensitive Scientist: Report of a British Association Study Group* (London: SCM Press, 1978).
Nuffield Council on Bioethics, *Genetic Screening: Ethical Issues* (London: Nuffield Council on Bioethics, 1993).
Nuffield Council on Bioethics, *Human Tissue: Ethical and Legal Issues* (London: Nuffield Council on Bioethics, 1995).
Nuffield Council on Bioethics, *Animal-to-Human Transplants: The Ethics of Xenotransplantation* (London: Nuffield Council on Bioethics, 1996).
Nuffield Council on Bioethics, *Nuffield Council of Bioethics, 1992–99* (London: Nuffield Council on Bioethics, 2000).
Polkinghorne, J. C. (chair), *Review of the Guidance on the Research Use of Foetuses and Foetal Material* (London: HMSO, 1989).
Pond, D. (chair), *Report of a Working Party on the Teaching of Medical Ethics* (London: Institute of Medical Ethics, 1987).
Rawlins, M. (chair), *A New Pathway for the Regulation and Governance of Health Research* (London: Academy of Medical Sciences, 2011).
Rosenheim, M. (chair), *Report of the Committee on the Supervision of the Ethics of Clinical Investigation in Institutions* (London: Royal College of Physicians, 1967).
Shotter, E., 'Director's Report: Can Medical Ethics be Taught?' (London: London Medical Group, 1974).
Shotter, E., *Twenty-Five Years of Medical Ethics: A Report on the Development of the Institute of Medical Ethics* (London: Institute of Medical Ethics, 1988).
Warnock, M., *A Question of Life: The Warnock Report on Human Fertilisation and Embryology* (London: Basil Blackwell, 1985).
Williams, R. (chair), *Report of the Working Party on the Practice of Genetic Manipulation* (London: HMSO, 1976).

Other works cited

Abbott, A., *The System of Professions: An Essay on the Division of Expert Labour* (Chicago and London: University of Chicago Press, 1988).
Ackroyd, E., 'Mr Kennedy and Consumerism', *Journal of Medical Ethics*, Vol. 7 (1981) pp. 180–1.
Allerton, G. T., 'Working Hours in the NHS', *British Medical Journal*, Vol. 1 (1947) p. 112.
Ambroselli, C., 'France: A National Committee Debates the Issues', *Hastings Center Report*, Vol. 14 (1984) pp. 20–1.
Anon, 'British Medical Association: Proceedings of Council', *British Medical Journal*, Vol. 1 (1947) pp. 103–14.

Anon, 'Controversy: Human Guinea Pigs', *World Medicine* (1967) pp. 86–7.
Anon, 'Responsibilities of Research', *Lancet*, Vol. 289 (1967) p. 1144.
Anon, 'What Comes After Fertilization?' *Nature*, Vol. 221 (1969) p. 613.
Anon, 'Ethics at Medical School', *Lancet*, Vol. 299 (1972) p. 1406.
Anon, 'Merrison Committee: Report of Inquiry', *British Medical Journal*, Vol. 2 (1975) pp. 183–8.
Anon, 'Complaints by Patients', *Lancet*, Vol. 310 (1977) p. 1238.
Anon, 'Medicine and the Media', *British Medical Journal*, Vol. 2, issue 6148 (1978) p. 1361.
Anon, 'Embryology Needs Rules, Not New Laws', *Nature*, Vol. 302 (1983) pp. 735–44.
Anon, 'Research Ethics Committees', *Lancet*, Vol. 321 (1983) p. 1026.
Anon, 'Confused Comment on Warnock', *Nature*, Vol. 312 (1984) p. 389.
Anon, 'A Welcome Report', *British Medical Journal*, Vol. 249 (1984) pp. 207–8.
Anon, 'Approval of Ethical Research Committees Delayed', *British Medical Journal*, Vol. 292 (1986) p. 779.
Anon, 'Congress Overrides Veto to Create New Bioethics Commission', *Hastings Center Report*, Vol. 16 (1986) p. 2.
Anon, 'Obituary: Sir Desmond Pond', *Journal of Medical Ethics*, Vol. 12 (1986) p. 130.
Anon, 'Who's for Bioethics Committees?', *Lancet*, Vol. 327 (1986) p. 1016.
Anon, 'In Brief', *British Medical Journal*, Vol. 326 (2003) p. 464.
Archard, D., 'Why Philosophers Are Not and Should Not Be Moral Experts', *Bioethics*, Vol. 25, no. 3 (2011) pp. 119–27.
Ashcroft, R., 'Emphasis Has Shifted from Medical Ethics to Bioethics', *British Medical Journal*, Vol. 332 (2001) p. 302.
Ashcroft, R., 'Bioethics and Conflicts of Interest', *Studies in the History and Philosophy of the Biological and Biomedical Sciences*, Vol. 35 (2004) pp. 155–65.
Ashcroft, R., 'Why the UK Doesn't Need a National Ethics Committee', *BioNews*, 29 November 2004. Available online at http://www.bionews.org.uk.
Ashcroft, R., 'Futures for Bioethics', *Bioethics*, Vol. 24, no. 5 (2010) p. ii.
Austin, D., 'A Memoir', *Government and Opposition*, Vol. 17 (1982) pp. 469–82.
Austoker, J., and Bryder L. (eds), *Historical Perspectives on the Role of the MRC* (Oxford: Oxford University Press, 1989).
Ayer, A. J., *Philosophical Essays* (London: Macmillan, 1965).
Ayer, A. J., *Language, Truth and Logic* (Harmondsworth: Penguin, 2001).
Baker, R. B., 'From Meta-Ethicist to Bioethicist', *Cambridge Quarterly of Healthcare Ethics*, Vol. 11 (2002) pp. 369–79.
Baker, R., Porter, D., and Porter, R., 'Introduction', in R. Baker, D. Porter

and R. Porter (eds), *The Codification of Medical Morality: Historical and Philosophical Studies of the Formalization of Western Medical Morality in the Eighteenth and Nineteenth Centuries* (Dordrecht: Kluwer Academic Publishers, 1993) pp. 1–14.

Bennet, R., Erin, C., and Harris, J. (eds), *AIDS: Ethics, Justice and European Policy. Final Report of a European Research Project* (Luxembourg: Office for Publications of the European Communities, 1998).

Bennett, D., Glasner, P., and Travis, D., *The Politics of Uncertainty: Regulating Recombinant DNA Research in Britain* (London: Routledge and Kegan Paul, 1986).

Berlin, I., *The Crooked Timber of Humanity: Chapters in the History of Ideas* (London: Pimlico, 2003).

Berlingeur, G., 'Bioethics, Health and Inequality', *Lancet*, Vol. 364 (2004) pp. 1086–91.

Black, D., 'Both Sides of a Public Face', *British Medical Journal*, Vol. 282 (1981) pp. 2044–5.

Blackburn-Starza, A., 'Alastair Beckenbridge Appointed Chair of Emerging Science and Bioethics Advisory Committee', *BioNews*, 21 May 2012. Available online at www.bionews.org.uk.

Blackburn-Starza, A., 'UK Government's New Bioscience Advisory Body Starts Recruiting', *BioNews*, 19 March 2012. Available online at www.bionews.org.uk.

Bonnar, A. B., *The Catholic Doctor* (London: Burns & Oates, 1948).

Borry, P., Schotmans, P., and Dierickx, K., 'The Birth of the Empirical Turn in Bioethics', *Bioethics*, Vol. 19 (2005) pp. 49–71.

Bosk, C. L., 'Professional Ethicist Available: Logical, Secular, Friendly', *Daedalus*, Vol. 128, no. 4 (1999) pp. 47–69.

Bowler, P. J., *Science for All: The Popularization of Science in Early Twentieth Century Britain* (Chicago: University of Chicago Press, 2009).

Boyd, K., 'The Discourses of Bioethics in the United Kingdom', in R. B. Baker and L. B. McCullough (eds), *The World History of Medical Ethics* (Cambridge: Cambridge University Press, 2009) pp. 486–90.

Boyd, K., Currie, C., Thompson, I., and Tierney, A. J., 'Teaching Medical Ethics: University of Edinburgh', *Journal of Medical Ethics*, Vol. 4 (1978) pp. 141–5.

Bracanovic, T., 'Against Culturally Sensitive Bioethics', *Medical Health Care and Philosophy* (advance access 2013) DOI 10.1007/s11019-013-9504-2.

Brahams, D., 'Ownership of a Spleen', *Lancet*, Vol. 336 (1990) p. 239.

Brandt, A. M., 'Racism and Research: The Case of the Tuskegee Syphilis Study', *The Hastings Center Report*, Vol. 8 no. 6 (1978) pp. 21–9.

Brazier, M., *Medicine, Patients and the Law* (Harmondsworth: Penguin, 3rd edn, 2003).

Brazier, M., 'The Story of the Manchester Centre and UK Bioethics', paper given at GLEUBE Conference on 'The History of Bioethics' (University of Manchester, January 2011). Available online at www.youtube.com/watch?v=mkPOSxdMD2M.

Brazier, M., Dyson, A., Harris, J., and Lobjoit, M., 'Medical Ethics in Manchester', *Journal of Medical Ethics*, Vol. 13 (1987) pp. 150–5.

Brazier, M., Gillon, R., and Harris, J., 'Helping Doctors Become Better Doctors: Mary Lobjoit – An Unsung Heroine of Medical Ethics in the UK', *Journal of Medical Ethics*, Vol. 38 (2012) pp. 383–5.

Bresalier, M., 'Uses of a Pandemic: Forging the Identities of Influenza and Virus Research in Interwar Britain', *Social History of Medicine*, Vol. 25 (2012) pp. 400–24.

Brooke, J. H., *Science and Religion: Some Historical Perspectives* (Cambridge: Cambridge University Press, 1991).

Brookes, B., *Abortion in England, 1900–1967* (London: Croom Helm, 1988).

Brown, A., *J. D. Bernal: The Sage of Science* (Oxford: Oxford University Press, 2005).

Brown, C. G., *The Death of Christian Britain: Understanding Secularisation, 1800–2000* (London and New York: Routledge, 2001).

Brown, M., *Performing Medicine: Medical Culture and Identity in Provincial England, c. 1760–1850* (Manchester: Manchester University Press, 2011).

Brownsword, R., 'Bioethics: Bridging from Morality to Law?', in M. Freeman (ed), *Law and Bioethics* (Oxford: Oxford University Press, 2008) pp. 12–30.

Bud, R., 'Penicillin and the New Elizabethans', *British Journal for the History of Science*, Vol. 31 (1998) pp. 305–33.

Burnham, J. C., 'How the Concept of a Profession Evolved in the Work of Historians of Medicine', *Bulletin of the History of Medicine*, Vol. 70 (1996) pp. 1–24.

Bynum, B., 'The McKeown Thesis', *Lancet*, Vol. 371 (2008) pp. 644–5.

Callahan, D., 'Bioethics as a Discipline', *Hastings Center Studies*, Vol. 1 (1973) pp. 66–73.

Callon, M., 'Some Elements of a Sociology of Translation: Domestication of the Scallops and the Fishermen of St. Brieuc Bay', in J. Law (ed), *Power, Action and Belief: A New Sociology of Knowledge?* (New York: Routledge, 1986) pp. 196–223.

Calman, K., *Medical Education: Past, Present and Future* (London: Churchill Livingstone, 2006).

Campbell, A., *Moral Dilemmas in Medicine* (Edinburgh and London: Churchill Livingstone, 1972).

Campbell, A., 'Philosophy and Medical Ethics', *Journal of Medical Ethics*, Vol. 2 (1976) pp. 1–2.
Campbell, J., *Margaret Thatcher, Volume One: The Grocer's Daughter* (London: Vintage, 2007).
Campbell-Smith, D., *Follow the Money: The Audit Commission, Public Money and the Management of Public Services* (London: Allen Lane, 2008).
Canguilhem, G., *The Normal and the Pathological* (New York: Zone Books, 1991).
Capron, A., 'Looking Back at the President's Commission', *The Hastings Center Report*, Vol. 13, no. 5 (1983) pp. 7–10.
Carroll, T. W., and Guttman, M. P., 'The Limits of Autonomy: The Belmont Report and the History of Childhood', *Journal of the History of Medicine and Allied Sciences*, Vol. 66 (2011) pp. 82–115.
Charlton, B., 'The Ideology of Accountability', *Journal of the Royal College of Physicians London*, Vol. 33 (1999) pp. 33–5.
Clements, K. (ed), *The Moot Papers: Faith, Freedom and Society* (London: T&T Press, 2010).
Collini, S., *Absent Minds: A History of Intellectuals in Britain* (Oxford: Oxford University Press, 2007).
Collins, H., and Evans, S., *Rethinking Expertise* (Chicago and London: University of Chicago Press, 2007).
Cooter, R., 'The Resistible Rise of Medical Ethics', *Social History of Medicine*, Vol. 8 (1995) pp. 257–70.
Cooter, R., 'The Ethical Body', in R. Cooter and J. V. Pickstone, (eds), *Medicine in the Twentieth Century* (Amsterdam: Harwood Academic Press, 2000) pp. 451–67.
Cooter, R., 'After Death/After "Life": The Social History of Medicine in Post-Postmodernity', *Social History of Medicine*, Vol. 20 (2007) pp. 441–64.
Cooter, R., 'Inside the Whale: Bioethics in History and Discourse', *Social History of Medicine*, Vol. 23 (2010) pp. 662–73.
Cooter, R., and Stein, C., 'Cracking Biopower', *History of the Human Sciences*, Vol. 23 (2010) pp. 109–28.
Cressey, D., 'UK Embryo Agency Faces Axe', *Nature*, Vol. 466 (2010) p. 674.
Crisp, A. H., 'The General Medical Council and Medical Ethics', *Journal of Medical Ethics*, Vol. 11 (1985) pp. 6–7.
Crossley, N., *Making Sense of Social Movements* (Buckingham: Open University Press, 2002).
Crossley, N., *Contesting Psychiatry: New Social Movements in Mental Health* (Oxford: Routledge, 2006).
Culliton, B., 'National Research Act: Restores Training, Bans Fetal Research', *Science*, Vol. 185 (1974) pp. 426–7.

Culliton, B., and Waterfall, W. K., 'The Flowering of American Bioethics', *British Medical Journal*, Vol. 2 (1978) pp. 1270–1.

Curran, C. R., 'The Catholic Moral Tradition in Bioethics', in J. K. Walter and E. P. Klein (eds), *The Story of Bioethics: From Seminal Works to Contemporary Explorations* (Washington, DC: Georgetown University Press, 2003) pp. 113–27.

Daston, L., and Galison, P., *Objectivity* (New York: Zone Books, 2007).

Daube, D., 'Transplantation: Acceptability of Procedures and the Required Legal Sanctions', in G. Wolstenholme (ed), *The Law and Ethics of Transplantation* (London: CIBA Foundation, 1968) pp. 188–201.

Davies, M., *Textbook on Medical Law* (London: Blackstone Press, 2nd edn, 1998).

Davis, G., 'The Medical Community and Abortion Law Reform: Scotland in National Context, c. 1960–1980', in I. Goold and C. Kelly (eds), *Lawyer's Medicine: The Legislature, the Courts, and Medical Practice, 1760–2000* (Oxford: Hart Publishing, 2009) pp. 143–65.

Davis, G., and Davison, R., '"A Fifth Freedom" or "Hideous Atheistic Expediency"? The Medical Community and Abortion Law Reform in Scotland *c.*1960–1970', *Medical History*, Vol. 50 (2006) pp. 29–48.

Dawson, A., 'Futures for Bioethics: Three Dogmas and a Cup of Hemlock', *Bioethics*, Vol. 25, no. 4 (2010) pp. 218–25.

De Styrap, J., *A Code of Medical Ethics: With General and Special Rules for the Guidance of the Faculty and the Public in the Complex Relations of Professional Life* (London: Lewis, 1890).

De Vries, R., 'How Can We Help? From "Sociology in" to "Sociology of" Bioethics', *Journal of Law and Medical Ethics*, Vol. 23 (2003) pp. 279–92.

De Vries, R., Turner, L. Orfali, K., and Bosk C. L., 'Social Science and Bioethics: The Way Forward', in R. De Vries, L. Turner, K. Orfali and C. L. Bosk (eds), *The View from Here: Bioethics and the Social Sciences* (Oxford: Blackwell, 2007) pp. 1–13.

Dean, M., *Governmentality: Power and Rule in Modern Society* (London: Sage, 2nd edn, 2010).

Dennis, K. J., and Hall, M. R. P., 'The Teaching of Medical Ethics', *Journal of Medical Ethics*, Vol. 3 (1977) pp. 183–5.

Dewey, J. T., *Ethics* (New York: Henry Holt, 1909).

Dewey, J. T., *Reconstruction in Philosophy* (London: University of London Press, 1920).

Diaz-Amado, E., 'Bioethicization and Justification of Medicine in Colombia in the Context of Healthcare Reform of 1993', PhD thesis, University of Durham, 2011.

Dickenson, D., 'A National Bioethics Committee: Lessons from France',

BioNews, 6 December 2004. Available online at http://www.bionews.org.uk/page_37776.

Dixon, B., *What is Science For?* (London: Collins, 1973).

Dixon-Woods, M., and Ashcroft, R. E., 'Regulation and the Social Licence for Medical Research', *Medical Health Care and Philosophy*, Vol. 11 (2008) pp. 381–91.

Donald, I., 'Introduction', in I. Donald (ed), *Test Tube Babies: A Christian View* (Oxford: Order of Christian Unity, 1984) pp. 1–15.

Donaldson, L., 'Expert Patients Usher in New Era of Opportunity for NHS', *British Medical Journal*, Vol. 326 (2003) p. 1279.

Donnelly, C., 'Inquiring Mind. The HSJ Interview: Sir Ian Kennedy', *Health Services Journal*, Vol. 113 (2003) pp. 22–3.

Downie, R. S., 'Ethics, Morals and Moral Philosophy', *Journal of Medical Ethics*, Vol. 6 (1980) pp. 33–4.

Downie, R. S., 'Cases and Casuistry', in S. MacLean (ed), *First Do No Harm: Law, Ethics and Healthcare* (Aldershot: Ashgate, 2006) pp. 17–29.

Downie, R. S., and MacNaughton, J., *Bioethics and the Humanities: Attitudes and Perspectives* (Abingdon: Routledge-Cavendish, 2007).

Dunstan, G., *The Artifice of Ethics* (London: SCM Press, 1974).

Dunstan, G., 'The Authority of a Moral Claim: Ian Ramsey and the Practice of Medicine', *Journal of Medical Ethics*, Vol. 13 (1987) pp. 189–94.

Dunstan, G., 'Two Branches from One Stem', *Annals of the New York Academy of Sciences*, Vol. 530 (1988) pp. 3–6.

Durbach, N., *Bodily Matters: The Anti-Vaccination Movement in England, 1853–1907* (Durham, NC: Duke University Press, 2005).

Dworkin, G., 'The Delicate Balance: Ethics, Law and Medicine', *Annals of the New York Academy of Sciences*, Vol. 530 (1988) pp. 24–36.

Dyer, C., 'Ian Kennedy Refused to Read the General Medical Council's Reports', *British Medical Journal*, Vol. 323 (2001) p. 183.

Dyson, A. O., 'Why Has Medical Ethics Become So Prominent?', *Manchester Memoirs*, Vol. 128 (1988–89) pp. 73–81.

Dyson, A. O., and Harris, J. (eds), *Experiments on Embryos* (London: Routledge, 1990).

Edgerton, D., and Pickstone, J. V., 'Science, Technology and Medicine in the United Kingdom, 1750–2000' (2009). Available online at https://workspace.imperial.ac.uk/humanities/Public/files/Edgerton%20Files/edgerton_science_technology_medicine.pdf.

Edwards, D. L., *Ian Ramsey, Bishop of Durham: A Memoir* (London: Oxford University Press, 1973).

Edwards, D. L., 'Ramsey, Ian Thomas (1915–1972)', *Oxford Dictionary of National Biography* (Oxford: Oxford University Press, 2004).

Edwards, R. G., 'Aspects of Human Reproduction', in W. Fuller (ed), *The*

Social Impact of Modern Biology (London: Routledge and Kegan Paul, 1971) pp. 108–22.

Edwards, R. G., 'Fertilization of Human Eggs *In Vitro*: Morals, Ethics and the Law', *Quarterly Review of Biology*, Vol. 49 (1974) pp. 3–26.

Edwards, R. G., and Steptoe, P. C., *A Matter of Life: The Story of a Medical Breakthrough* (London: Finestride and Crownchime, 2nd edn, 2011).

Edwards, R. G., Bavister, B. D., and Steptoe, P. C., 'Early Stages of Fertilization *in vitro* of Human Oocytes Matured *in vitro*', *Nature*, Vol. 221 (1969) pp. 632–5.

Elliot, C., 'Should Journals Publish Industry-Funded Bioethics Articles?', *Lancet*, Vol. 366 (2005) pp. 422–4.

Elstein, M., and Harris, J., 'Teaching of Medical Ethics', *Journal of Medical Ethics*, Vol. 24 (1990) pp. 531–4.

Elston, M. A., 'Women and Vivisection in Edwardian England', in N. Rupke (ed), *Vivisection in Historical Perspective* (London and New York: Routledge, 1990) pp. 259–95.

Emmerich, N., 'Literature, History and the Humanization of Bioethics', *Bioethics*, Vol. 25 (2011) pp. 112–18.

Engelhardt, T. E., Jr, *The Foundations of Bioethics* (New York and Oxford: Oxford University Press, 1986).

Epstein, M., 'How Will the Economic Downturn Affect Academic Bioethics?' *Bioethics*, Vol. 24, no. 5 (2010) pp. 226–33.

Evans, D., 'Health Care Ethics: A Pattern for Learning', *Journal of Medical Ethics*, Vol. 13 (1987) pp. 127–31.

Evans, E. J., *Thatcher and Thatcherism* (London: Routledge, 2004).

Evans, J. H., 'A Sociological Account of the Growth of Principalism', *Hastings Center Report*, Vol. 30, no. 5 (2000) pp. 31–8.

Evans, J. H., *Playing God: Human Genetic Engineering and the Rationalization of Public Bioethical Debate* (Chicago and London: University of Chicago Press, 2002).

Evans, J. H., *The History and Future of Bioethics: A Sociological View* (Oxford: Oxford University Press, 2012).

Ferriman, A., 'Bristol Inquiry Appoints Doctor to its Panel', *British Medical Journal*, 318 (1999) p. 987.

Fisher, E., *Risk Regulation and Administrative Constitutionalism* (Oxford: Hart Publishing, 2007).

Fissell, M., 'Innocent and Honourable Bribes: Medical Manners in Eighteenth Century Britain', in R. Baker, D. Porter and R. Porter (eds), *The Codification of Medical Morality: Historical and Philosophical Studies of the Formalization of Western Medical Morality in the Eighteenth and Nineteenth Centuries* (Dordrecht: Kluwer Academic Publishers, 1993) pp. 19–47.

Fletcher, J., *Morals and Medicine* (London: Victor Gollancz, 1955).

Fletcher, J., *Situation Ethics: The New Morality* (London: SCM Press, 1966).

Fontaine, P., 'Blood, Politics and Social Science: Richard Titmuss and the Institute of Economic Affairs', *Isis*, Vol. 93 (2002) pp. 401–34.

Foot, P., 'Moral Arguments', *Mind*, Vol. 67 (1958) pp. 502–13.

Foot, P., 'Moral Beliefs', *Proceedings of the Royal Aristotelian Society*, Vol. 59 (1968–69) pp. 83–104.

Foot, P., 'The Problem of Abortion and the Doctrine of Double Effect', in P. Foot, *Virtues and Vices* (Oxford: Oxford University Press, 2002) pp. 19–31.

Foucault, M., *The Will to Knowledge. The History of Sexuality: Volume One* (Harmondsworth and New York: Penguin, 1998).

Foucault, M., *Society Must be Defended: Lectures at the Collège de France, 1975–1976* (Harmondsworth and New York: Penguin, 2004).

Foucault, M., *The Birth of Biopolitics: Lectures at the Collège de France, 1978–1979* (Basingstoke: Palgrave Macmillan, 2008).

Fowler, N., *Ministers Decide: A Memoir of the Thatcher Years* (London: Chapmans, 1991).

Fox, R., and Swazey J., *Observing Bioethics* (Oxford: Oxford University Press, 2008).

French, R. D., *Antivivisection and Medical Science in Victorian Society* (Princeton, NJ, and London: Princeton University Press, 1975).

Fukuyama, F., *Our Posthuman Future: Consequences of the Biotechnology Revolution* (London: Profile Books, 2002).

Gaines, A. D., and Juengst, E. T., 'Origin Myths in Bioethics: Constructing Sources, Motives and Reason in Bioethic(s)', *Culture, Medicine and Psychiatry*, Vol. 32 (2008) pp. 303–27.

Gaw, A., 'Searching for Pappworth', *South African Medical Journal*, Vol. 101, no. 9 (2011) p. 608.

Gillie, A., and Weatherhead, L. D., 'The Institute of Religion and Medicine', *Journal of the Royal College of General Practitioners*, Vol. 15 (1968) p. 286.

Gillon, R., 'Medical Ethics and Medical Education', *Journal of Medical Ethics*, Vol. 7 (1981) pp. 171–2.

Gillon, R., 'Medicine and the Media', *British Medical Journal*, Vol. 286 (1983) p. 715.

Gillon, R., 'Britain: The Public Gets Involved', *Hastings Center Report*, Vol. 14 (December 1984) pp. 16–17.

Gillon, R., 'In Britain, the Debate After the Warnock Report', *Hastings Center Report*, Vol. 17, no. 3 (1987) pp. 16–18.

Glover, J., *Causing Death and Saving Lives* (Harmondsworth: Penguin, 1977).

Glover, J., and Scott-Taggart, M. J., 'It Makes No Difference Whether or

Not I Do It', *Proceedings of the Aristotelian Society*, Vol. 49 (1975) pp. 171–90.

Goold, I., 'Regulating Reproduction in the United Kingdom: Doctor's Voices, 1978–1985', in I. Goold and C. Kelly (eds), *Lawyers' Medicine: The Legislature, the Courts and Medical Practice, 1760–2000* (Oxford: Hart Publishing, 2009) pp. 167–95.

Gostin, L., 'Honoring Ian McColl Kennedy', *Journal of Contemporary Health Law and Policy*, Vol. 14 (1997) pp. vi–xi.

Gostin, L., 'Foreword in Honour of a Pioneer of Medical Law: Professor Margaret Brazier OBE QC FMEDSCI', *Medical Law Review*, Vol. 20 (2012) pp. 1–5.

Gray, J. A. M., *Man Against Disease: Preventive Medicine* (Oxford: Oxford University Press, 1979).

Gregory, J., and Miller, S., *Science in Public: Communication, Culture and Credibility* (Cambridge, MA: Perseus Publishing, 1998).

Gross, M. L., and Ravitsky, V., 'Israel: Bioethics in a Jewish-Democratic State', *Cambridge Journal of Healthcare Ethics*, Vol. 14, no. 3 (2003) pp. 247–55.

Gummett, J., *Scientists in Whitehall* (Manchester: Manchester University Press, 1980).

Gustafson, J. M., 'Context Versus Principles: A Misplaced Debate in Christian Ethics', *Harvard Theological Review*, Vol. 58 (1965) pp. 171–202.

Habgood, J. S., 'Medical Ethics – A Christian View', *Journal of Medical Ethics*, Vol. 11 (1985) pp. 12–13.

Haimes, E., 'What can the Social Sciences Contribute to the Study of Ethics? Theoretical, Empirical and Substantive Considerations', *Bioethics*, Vol. 16 (2002) pp. 89–113.

Haldane, J. B. S., *Daedalus, or Science and the Future* (London: Kegan Paul, Trench, Trubner, 1924).

Hampshire, S., Scanlon, T. M., Williams, B., Nagel, T., and Dworkin, R. (eds), *Public and Private Morality* (Cambridge: Cambridge University Press, 1978).

Hare, R. M., 'Universalisability', *Proceedings of the Aristotelian Society*, Vol. 55 (1955) pp. 295–312.

Hare, R. M., *Freedom and Reason* (Oxford: Oxford University Press, 1963).

Hare, R. M., 'Man's Interests', in I. T. Ramsey and R. Porter (eds), *Personality and Science: An Interdisciplinary Discussion* (Edinburgh and London: Churchill Livingstone, 1971) pp. 97–101.

Hare, R. M., 'Rules of War and Moral Reasoning', *Philosophy and Public Affairs*, Vol. 1 (1972) pp. 166–81.

Hare, R. M., 'Abortion and the Golden Rule', *Philosophy and Public Affairs*, Vol. 4 (1975) pp. 201–22.

Hare, R. M., 'Medical Ethics: Can the Moral Philosopher Help?', in S. Spicker and H. T. Engelhardt, Jr (eds), *Philosophical Medical Ethics: Its Nature and Significance* (Dordrecht: Riedel Publishing, 1977) pp. 49–63.

Hare, R. M., 'What's Wrong with Slavery?', *Philosophy and Public Affairs*, Vol. 8 (1978) pp. 103–21.

Hare, R. M., '*In Vitro* Fertilization and the Warnock Report', in R. Chadwick (ed), *Ethics, Reproduction and Genetic Control* (London: Croom Helm, 1987) pp. 71–90.

Harris, J., 'The Marxist Conception of Violence', *Philosophy and Public Affairs*, Vol. 3, no. 2 (1974) pp. 192–220.

Harris, J., 'The Survival Lottery', *Philosophy*, Vol. 50 (1975) pp. 81–7.

Harris, J., 'In Vitro Fertilization: The Ethical Issues', *The Philosophical Quarterly*, Vol. 3 (1983) pp. 202–27.

Harris, J., *The Value of Life: An Introduction to Medical Ethics* (London: Routledge and Kegan Paul, 1985).

Harris, J., 'The Scope and Importance of Bioethics', in J. Harris (ed), *Bioethics* (Oxford: Oxford University Press, 2001) pp. 1–25.

Harrison, B., 'Animals and the State in Nineteenth Century England', *English Historical Review*, Vol. 88 (1973) pp. 786–820.

Harrison, S., and Ahmad, W. I. U., 'Medical Autonomy and the State 1975 to 2002', *Sociology*, Vol. 34, no. 1 (2000) pp. 129–46.

Hazelgrove, J., 'The Old Faith and the New Science: The Nuremberg Code and Human Experimentation Ethics in Britain, 1946–1973', *Social History of Medicine*, Vol. 15 (2002) pp. 109–35.

Hedgecoe, A., 'Critical Bioethics: Beyond the Social Science Critique of Applied Ethics', *Bioethics*, Vol. 18 (2004) pp. 120–43.

Hedgecoe, A., '"A Form of Practical Machinery": The Origins of Research Ethics Committees in the UK', *Medical History*, Vol. 53 (2009) pp. 331–50.

Heley, M. H., 'Mrs Warnock's Brave New World', *Lancet*, Vol. 324 (1984) p. 290.

Hennessy, P., *Whitehall* (London: Fontana, 1990).

Hodgson, H., 'Medical Ethics and Controlled Trials', *British Medical Journal*, Vol. 1 (1963) pp. 1339–40.

Hoffenberg, R., 'A Tale of Two Cultures: Do We Differ?' *Annals of the New York Academy of Sciences*, Vol. 530 (1988) pp. 16–23.

Hogarth, S., and Salter, B., 'Emerging Science and Established Problems – Governing Biomedical Innovation in the UK', *BioNews*, 11 April 2012. Available online at www.bionews.org.uk.

Honey, C., 'Acts and Omissions', *Journal of Medical Ethics*, Vol. 5 (1979) pp. 143–4.

Horton, R., 'Why is Ian Kennedy's Healthcare Commission Damaging NHS Care?', *Lancet*, Vol. 364 (2004) pp. 401–2.

Hunter, D., 'Bioethics – A Discipline Without a Natural Home?', *Journal of Medical Ethics* blog. Available online at http://blogs.bmj.com/medical-ethics

Hunter, M., 'GMC Agrees New Structure', *British Medical Journal*, Vol. 323 (2001) p. 250.

Illich, I., *Limits to Medicine, Medical Nemesis: The Expropriation of Health* (Harmondsworth: Penguin, 1975).

Illich, I., 'The Medicalization of Life', *Journal of Medical Ethics*, Vol. 1 (1975) pp. 73–7.

Illich, I. (ed), *Disabling Professions* (London: Marion Boyars, 1977).

Imber, J., 'Medical Publicity before Bioethics: Nineteenth Century Illustrations of Twentieth Century Dilemmas', in R. De Vries and J. Subedi (eds), *Bioethics and Society: Constructing the Ethical Enterprise* (Upper Saddle River, NJ: Prentice Hall, 1998) pp. 16–38.

Jacobs, N., 'Which Principles, Doctor? The Early Crystallization of Clinical Research Ethics in the Netherlands, 1947–1955', MA thesis, University of Utrecht, 2012.

Jasanoff, S., *Designs on Nature: Science and Democracy in Europe and the United States* (Princeton, NJ, and Oxford: Princeton University Press, 2005).

Jasanoff, S., 'Making the Facts of Life', in S. Jasanoff (ed), *Reframing Rights: Bioconstitutionalism in the Genetic Age* (Cambridge, MA: MIT Press, 2011) pp. 59–85.

Jasanoff, S. (ed), *States of Knowledge: The Co-Production of Science and the Social Order* (London: Routledge, 2004).

Jenkins, D. T., *The Doctor's Profession* (London: SCM Press, 1948).

Jenner, M., and Wallis, P. (eds), *Medicine and the Marketplace in England and its Colonies, c.1450–c.1850* (Basingstoke: Palgrave Macmillan, 2007).

Jewell, M. D., 'Teaching Medical Ethics', *British Medical Journal*, Vol. 289 (1984) pp. 364–5.

Jewson, N., 'Medical Knowledge and the Patronage System in Eighteenth Century England', *Sociology*, Vol. 8 (1974) pp. 369–85.

Jewson, N., 'The Disappearance of the Sick Man from Medical Cosmology', *Sociology*, Vol. 10 (1976) pp. 225–44.

Johnson, A. G., 'Teaching Medical Ethics as a Practical Subject: Observations from Experience', *Journal of Medical Ethics*, Vol. 9 (1983) pp. 5–7.

Johnson, M. H., Franklin, S. B., Cottingham, M., and Hopwood, N., 'Why the MRC Refused Robert Edwards and Patrick Steptoe Support for Research on Human Conception in 1971', *Human Reproduction*, Vol. 25 (2010) pp. 2157–74.

Jones, J. L., *Bad Blood: The Tuskegee Syphilis Study* (New York: The Free Press, 1992).

Jones, J. S. P., and Metcalfe, D. H. H., 'The Teaching of Medical Ethics', *Journal of Medical Ethics*, Vol. 2 (1976) pp. 83–6.

Jonsen, A., *The Birth of Bioethics* (Oxford: Oxford University Press, 1998).

Jonsen, A., 'The Structure of an Ethical Revolution: Paul Ramsey, the Beecher Lectures and the Birth of Bioethics', in P. Ramsey, *The Patient as Person: Explorations in Medical Ethics* (New Haven, CT, and London: Yale University Press, 2nd edn, 2002) pp. xvi–xxix.

Joravsky, D., *The Lysenko Affair* (Chicago: University of Chicago Press, 1986).

Judt, T., *Ill Fares the Land* (London: Allen Lane, 2010).

Katz, J., *Experimentation with Human Beings* (New York: Russell Sage Foundation, 1972).

Katz, J., 'Who is to Keep Guard over the Guardians Themselves?', *Fertility and Sterility*, Vol. 23 (1972) pp. 604–9.

Katz, J., *The Silent World of Doctor and Patient* (New York: The Free Press, 1984).

Katz, J., 'The Consent Principle of the Nuremberg Code: Its Significance Then and Now', in G. J. Annas (ed), *The Nazi Doctors and the Nuremberg Code* (Oxford: Oxford University Press, 1992) pp. 227–40.

Kavanagh, E., *Thatcherism and British Politics: The End of Consensus?* (Oxford: Oxford University Press, 1988).

Kean, H., '"The Smooth Cool Men of Science": The Feminist and Socialist Response to Vivisection', *History Workshop Journal*, Vol. 40 (1995) pp. 16–38.

Kemp, N., *Merciful Release: The History of the British Euthanasia Movement* (Manchester: Manchester University Press, 2002).

Kennedy, I., 'Alive or Dead? The Lawyer's View', *Current Legal Problems*, Vol. 22 (1969) pp. 102–28.

Kennedy, I., 'The Legal Definition of Death', *Harvard Medico-Legal Journal*, Vol. 99 (1972) pp. 36–41.

Kennedy, I., 'The Karen Quinlan Case: Problems and Proposals', *Journal of Medical Ethics*, Vol. 2 (1976) pp. 3–7.

Kennedy, I., 'The Legal Effect of Requests by the Terminally Ill and Aged not to Receive Further Treatment from Doctors', *Criminal Law Review* (1976) pp. 217–32.

Kennedy, I., 'The Mental Health Act: A Model Response that Failed', *World Medicine*, Vol. 37 (1978) pp. 37–8, 75–6, 81–2.

Kennedy, I., 'Response to the Critics', *Journal of Medical Ethics*, Vol. 7 (1981) pp. 202–11.

Kennedy, I., *Unmasking Medicine* (London: Allen and Unwin, 1981).

Kennedy, I., *The Unmasking of Medicine* (London: Paladin, rev. and updated edn, 1983).

Kennedy, I., 'The Patient on the Clapham Omnibus', *Modern Law Review*, Vol. 47, no. 4 (1984) pp. 454–74.

Kennedy, I., 'The Check Out: A Humane Death', in I. Kennedy, *Treat Me Right: Essays in Medical Law and Ethics* (Oxford: Clarendon Press, 1988) pp. 300–14.

Kennedy, I., 'Doctors, Patients and Human Rights', in I. Kennedy, *Treat Me Right: Essays in Medical Law and Ethics* (Oxford: Clarendon Press, 1988) pp. 385–415.

Kennedy, I., 'Emerging Problems in Science, Technology and Medicine', in I. Kennedy, *Treat Me Right: Essays in Medical Law and Ethics* (Oxford: Clarendon Press, 1988) pp. 1–18.

Kennedy, I., 'The Patient on the Clapham Omnibus: Updated and with a new Postscript', in I. Kennedy, *Treat Me Right: Essays in Medical Law and Ethics* (Oxford: Clarendon Press, 1988) pp. 175–213.

Kennedy, I., 'Preface', in I. Kennedy, *Treat Me Right: Essays in Medical Law and Ethics* (Oxford: Clarendon Press, 1988) pp. vii–viii.

Kennedy, I., 'What is a Medical Decision?', in I. Kennedy, *Treat Me Right: Essays in Medical Law and Ethics* (Oxford: Clarendon Press, 1988) pp. 19–31.

Kennedy, I., 'The Role and Relevance of Legislation', in CIBA Foundation, *Medical Scientific Advance: Its Challenge to Society* (London: CIBA Foundation, 1990) pp. 18–23.

Kennedy, I., 'Patients are Experts in their Own Field', *British Medical Journal*, Vol. 326 (2003) p. 1276.

Kennedy, I., 'Setting of Clinical Standards', *Lancet*, Vol. 364 (2004) p. 1399.

Kennedy, I., and Grubb, A., *Medical Law: Text and Materials* (London: Butterworths, 1989).

Kenny, A., *A New History of Western Philosophy* (Oxford: Oxford University Press, 2012).

Kevles, D., *In the Name of Eugenics: Genetics and the Uses of Human Heredity* (Cambridge, MA: Harvard University Press, 2004).

Kilbrandon, C., 'Chairman's Closing Remarks', in G. Wolstenholme (ed), *The Law and Ethics of Transplantation* (London: CIBA Foundation, 1968) pp. 212–16.

Klein, R., *The New Politics of the NHS: From Creation to Reinvention* (Oxford and New York: Radcliffe Publishing, 2010).

Kleinman, A., 'Moral Experience and Ethical Reflection: Can Ethnography Reconcile Them? A Quandary for "The New Bioethics"', *Daedalus*, Vol. 128, no. 4 (1999) pp. 69–99.

Lansbury, C., *The Old Brown Dog: Women, Workers and Vivisection in Edwardian England* (Madison and London: University of Wisconsin Press, 1986).

Lawrence, C., *Medicine in the Making of Modern Britain, 1700–1920* (London: Routledge, 1994).

Lawson, N., *The New Conservatism* (London: Centre for Policy Studies, 1980).

Lederer, S., *Flesh and Blood: Organ Transplantation and Blood Donation in Twentieth-Century America* (Oxford: Oxford University Press, 2008).

Lewis, J., 'A Commons Subcommittee on Medical Ethics?', *Lancet*, Vol. 331 (1988) p. 1005.

Little, S., 'Consumerism in the Doctor–Patient Relationship', *Journal of Medical Ethics*, Vol. 7 (1981) pp. 187–90.

Lock, M., *Twice Dead: Organ Transplants and the Reinvention of Death* (Berkeley and London: University of California Press, 2001).

Lock, S., 'Monitoring Research Ethical Committees', *British Medical Journal*, Vol. 300 (1990) pp. 61–2.

Lock, S., 'Towards a National Bioethics Committee. Wanted: A New Strategic Body to Deal With Broad Issues', *British Medical Journal*, Vol. 300 (1990) pp. 1149–50.

Lock, S., 'Pappworth [formerly Papperovitch], Maurice Henry (1910–1994)', *Oxford Dictionary of National Biography* (Oxford: Oxford University Press, 2004).

Lockwood, M., 'Ethical Dilemmas in Surgery: Some Philosophical Reflections', *Journal of Medical Ethics*, Vol. 6 (1980) pp. 82–4.

Lockwood, M., 'Introduction', in M. Lockwood (ed), *Moral Dilemmas in Modern Medicine* (Oxford: Oxford University Press, 1985) pp. 1–9.

Lockwood, M., 'The Warnock Report: A Philosophical Appraisal', in M. Lockwood (ed), *Moral Dilemmas in Modern Medicine* (Oxford: Oxford University Press, 1985) pp. 155–87.

Lopez, J., 'How Sociology Can Save Bioethics ... Maybe', *Sociology of Health and Illness*, Vol. 26 (2004) pp. 875–96.

Lowe, R., *The Death of Progressive Education: How Teachers Lost Control of the Classroom* (London: Routledge, 2007).

MacDonald, J. B., 'Medical Ethics Today: A Student's Uncertainties', *Lancet*, Vol. 289 (1967) p. 563.

Macer, D. R. J., 'Bioethics in Japan and East Asia', *Turkish Journal of Medical Ethics*, Vol. 9 (2001) pp. 70–7.

MacIntyre, A., 'How Virtues Became Vices: Values, Medicine and Social Context', in H. T. Engelhardt, Jr, and S. Spicker (eds), *Evaluation and Explanation in the Biomedical Sciences* (Dordrecht: Riedel Publishing, 1975) pp. 97–111.

MacIntyre, A., 'Why is the Search for the Foundations of Ethics so Frustrating?', *Hastings Center Report*, Vol. 9, no. 4 (1979) pp. 16–22.

MacIntyre, A., *A Short History of Ethics* (London: Routledge, 2002).

MacIntyre, A., *After Virtue* (Notre Dame: University of Notre Dame Press, 3rd edn, 2007).

Macklin, R., 'The Death of Bioethics (As We Once Knew It)', *Bioethics*, Vol. 24 (2010) pp. 211–17.

Maehle, A. H., 'Medical Ethics and the Law', in M. Jackson (ed), *The Oxford Handbook of the History of Medicine* (Oxford: Oxford University Press, 2011) pp. 543–60.

Marinetto, M., 'Who Wants to be an Active Citizen? The Politics and Practice of Community Involvement', *Sociology*, Vol. 37 (2003) pp. 103–20.

Marwick, A., *The Sixties: Cultural Revolution in Britain, France, Italy and the United States, c. 1958–1974* (Oxford: Oxford University Press, 1998).

Mason, J. K., and Laurie, G. T., *Law and Medical Ethics* (Oxford: Oxford University Press, 8th edn, 2011).

Mayer, A. K., 'A Combative Sense of Duty: Englishness and the Scientists', in C. Lawrence and A. K. Mayer (eds), *Regenerating England: Science, Medicine and Culture in Interwar Britain* (Amsterdam: Rodopi, 2000) pp. 67–107.

Mazer, D. R. J., 'Bioethics in Japan and East Asia', *Turkish Journal of Medical Ethics*, Vol. 9 (2001) pp. 70–7.

McAdams, R., 'Human Guinea Pigs: Maurice Pappworth and the Birth of British Bioethics', MSc Thesis, University of Manchester, 2005.

McCullough, L. B., 'John Gregory's Medical Ethics and Humean Sympathy', in R. Baker, D. Porter and R. Porter (eds), *The Codification of Medical Morality: Historical and Philosophical Studies of the Formalization of Western Medical Morality in the Eighteenth and Nineteenth Centuries* (Dordrecht: Kluwer Academic Publishers, 1993) pp. 145–61.

McCullough, L. B., 'Was Bioethics Founded on Historical and Conceptual Mistakes About Medical Paternalism?', *Bioethics*, Vol. 25 (2011) pp. 66–74.

McGrath, P. J., 'Is and Ought', *Journal of Medical Ethics*, Vol. 1 (1975) pp. 150–1.

McKeown, T., *The Role of Medicine: Dream, Mirage or Nemesis?* (Oxford: Blackwell, 1979).

McLaren, A., 'Where to Draw the Line?', *Journal of the Royal Institution*, Vol. 56 (1984) pp. 101–21.

Midgley, M., *The Owl of Minerva: A Memoir* (London and New York: Routledge, 2005).

Miller, P., and Rose, N., *Governing the Present: Administering Economic, Social and Personal Life* (Oxford: Polity Press, 2008).

Moazam, F., and Aamir, J. M., 'Pakistan and Biomedical Ethics: Report from a Muslim Country', *Cambridge Quarterly of Healthcare Ethics*, Vol. 14, no. 3 (2005) pp. 249–55.

Mold, A., 'Patient Groups and the Construction of the Patient-Consumer in Britain: An Historical Overview', *Journal of Social Policy*, Vol. 39 (2010) pp. 505–21.

Mold, A., 'Making the Patient-Consumer in Margaret Thatcher's Britain', *Historical Journal*, Vol. 54, no. 9 (2011) pp. 509–28.

Montgomery, J., 'Medical Law in the Shadow of Hippocrates', *Medical Law Review*, Vol. 52 (1989) pp. 566–76.

Montgomery, J., 'Time for a Paradigm Shift? Medical Law in Transition', *Current Legal Problems*, Vol. 53 (2000) pp. 363–408.

Montgomery, J., 'Law and the Demoralisation of Medicine', *Legal Studies*, Vol. 26, no. 2 (2006) pp. 185–210.

Montgomery, J., 'The Legitimacy of Medical Law', in S. MacLean (ed), *First Do No Harm: Law, Ethics and Healthcare* (Aldershot: Ashgate, 2006) pp. 1–17.

Montgomery, J., 'Lawyers and the Future of UK Bioethics', *Verdict – The Magazine of the Oxford Law Society* (2012). Available online at http://eprints.soton.ac.uk/341657/.

Moran, M., *The British Regulatory State: High Modernism and Hyper-Innovation* (Oxford: Oxford University Press, 2003).

Moreno, J., *The Body Politic: The Battle for Science in America* (New York: Bellevue Literary Press, 2011).

Morrice, A. G., '"Honour and Interests": Medical Ethics and the British Medical Association', in A. H. Maehle and J. Geyer-Kordesch (eds), *Historical and Philosophical Perspectives on Medical Ethics* (Aldershot: Ashgate, 2002) pp. 11–37.

Mulkay, M., 'Galileo and the Embryos: Religion and Science in Parliamentary Debate over Research on Human Embryos', *Social Studies of Science*, Vol. 25, no. 3 (1995) pp. 499–532.

Mulkay, M., *The Embryo Research Debate: Science and the Politics of Reproduction* (Cambridge: Cambridge University Press, 1998).

Munro, E., 'The Impact of Audit on Social Work Practice', LSE Research Online (2004). Available online at http://eprints.lse.ac.uk/523/1/Audit-SocialWork_05.pdf.

Nathoo, A., *Hearts Exposed: Transplants and the Media in 1960s Britain* (Basingstoke: Palgrave Macmillan, 2009).

Niskanen, W., *Bureaucracy: Servant or Master?* (London: Institute for Economic Affairs, 1973).

O'Neill, O., 'Medical and Scientific Uses of Human Tissue', *Journal of Medical Ethics*, Vol. 22 (1996) pp. 2–7.

O'Neill, O., *Autonomy and Trust in Bioethics* (Cambridge: Cambridge University Press, 2002).

O'Neill, O., *A Question of Trust* (Cambridge: Cambridge University Press, 2002).

Pappworth, M. H., 'Human Guinea Pigs: A Warning', *Twentieth Century*, Vol. 171 (1962–63) pp. 66–75.

Pappworth, M. H., *Human Guinea Pigs: Experiments on Man* (London: Routledge and Kegan Paul, 1967).

Pappworth, M. H., '"Human Guinea Pigs" – a History', *British Medical Journal*, Vol. 301 (1990) pp. 1456–60.

Pellegrino, E., 'Book Review: *Ethics in Medical Progress: With Special Reference to Transplantation*', *Quarterly Review of Biology*, Vol 43, no. 4 (1968) pp. 478–9.

Perkin, H., *The Rise of Professional Society: England since 1800* (London and New York: Routledge, 1990).

Pfeffer, N., *The Stork and the Syringe: A History of Reproductive Medicine* (London: Polity Press, 1993).

Pickstone, J. V., 'The Professionalisation of Medicine in England and Europe: The State, the Market and Industrial Society', *Journal of the Japanese Medical Society*, Vol. 25, no. 4 (1979) pp. 550–21.

Pickstone, J. V., 'Thomas Percival and the Production of Medical Ethics', in R. Baker, D. Porter and R. Porter (eds), *The Codification of Medical Morality: Historical and Philosophical Studies of the Formalization of Western Medical Morality in the Eighteenth and Nineteenth Centuries* (Dordrecht: Kluwer Academic Publishers, 1993) pp. 161–78.

Pidgen, C. R., 'Bertrand Russell: Moral Philosopher or Unsophisticated Moralist?', in N. Griffin (ed), *The Cambridge Companion to Bertrand Russell* (Cambridge: Cambridge University Press, 2003) pp. 475–505.

Porter, R., 'Laymen, Doctors and Medical Knowledge in the Eighteenth Century: The Evidence of the *Gentleman's Magazine*', in R. Porter (ed), *Patients and Practitioners: Lay Perceptions of Pre-Industrial Society* (Cambridge: Cambridge University Press, 1985) pp. 283–315.

Porter, R., 'Before the Fringe: "Quackery" and the Eighteenth Century Medical Market', in R. Cooter (ed), *Studies in the History of Alternative Medicine* (Basingstoke: Macmillan, 1988) pp. 1–27.

Potter, V. R., 'Bioethics, the Science for Survival', *Perspectives in Biology and Medicine*, Vol. 14 (1970) pp. 127–53.

Power, M., *The Audit Society: Rituals of Verification* (Oxford: Oxford University Press, 1997).

Power, M., 'Evaluating the Audit Explosion', *Law and Policy*, Vol. 25 (2005) pp. 185–202.

Price, A. W., 'Richard Mervyn Hare (1919–2002)', *Oxford Dictionary of National Biography* (Oxford: Oxford University Press, 2004).

Price, K., 'The Art of Medicine: Toward a History of Medical Negligence', *Lancet*, Vol. 375 (2010) pp. 192–3.

Pugh, M., *We Danced All Night: A Social History of Britain between the Wars* (London: Vintage, 2009).

Pullan, B., and Abendstern, M., *A History of the University of Manchester, 1973–1990* (Manchester: Manchester University Press, 2004).

Rabinow, P., and Rose, N., 'Introduction: Foucault Today', in M. Foucault, P. Rabinow and N. Rose (eds), *The Essential Foucault* (New York and London: New Press, 2003) pp. vii–xxxiii.

Rabinow, P., and Rose, N., 'Biopower Today', *Biosocieties*, Vol. 1 (2006) pp. 195–217.

Ramsey, I. T., 'A New Prospect in Theological Studies', *Theology*, Vol. 67 (1964) pp. 527–35.

Ramsey, I. T., 'Biology and Personality: Some Philosophical Reflections', in I. T. Ramsey (ed), *Biology and Personality: Frontier Problems in Science, Philosophy and Religion* (Oxford: Basil Blackwell, 1965) pp. 174–95.

Ramsey, I. T., 'The Way Ahead for Christian Thinking', *Point Magazine*, Vol. 3 (1968) pp. 57–62.

Ramsey, I. T., 'Christian Ethics in the 1960s and 1970s', *The Church Quarterly*, Vol. 2 (1970) pp. 221–7.

Ramsey, I. T., and Porter, R. (eds), *Personality and Science: An Interdisciplinary Discussion* (Edinburgh and London: Churchill Livingstone, 1971).

Ramsey, P., 'Lehmann's Contextual Ethics and the Problem of Truth-Telling', *Theology Today*, Vol. 21 (1965) pp. 466–76.

Ramsey, P., *Deeds and Rules in Christian Ethics* (New York: Scribner's, 1967).

Ramsey, P., *The Just War: Force and Political Responsibility* (New York: Scribner's, 1968).

Ramsey, P., *Fabricated Man: The Ethics of Genetic Control* (New Haven, CT, and London: Yale University Press, 1970).

Ramsey, P., *The Patient as Person: Explorations in Medical Ethics* (New Haven, CT: Yale University Press, 1970).

Ramsey, P., 'Shall We Reproduce? The Ethics of In Vitro Fertilization I', *Journal of the American Medical Association*, Vol. 220, no. 10 (1972) pp. 1346–50.

Ramsey, P., *The Ethics of Fetal Research* (New Haven, CT, and London: Yale University Press, 1975).

Ramsey, P., 'Manufacturing Our Offspring: Weighing the Risks', *Hastings Center Report*, Vol. 8 (1978) pp. 7–9.

Ramsey, S., 'UK "Bristol Case" Inquiry Formally Opened', *Lancet*, Vol. 353 (1999) p. 987.

Rattray-Taylor, G., *The Biological Time-Bomb* (London: Thames and Hudson, 1968).

Raven, C., *Science and the Christian Man* (London: SCM Press, 1952).

Raven, C., *Christianity and Science* (London: Lutterworth Press, 1955).

Raven, C., 'The Continuing Vision: A Personal Retrospect', in I. T. Ramsey (ed), *Biology and Personality: Frontier Problems in Science, Philosophy and Religion* (London: Basil Blackwell, 1965) pp. 9–16.

Reich, W., 'Introduction', in W. Reich (ed), *Encyclopedia of Bioethics* (New York: Macmillan, 1978) pp. xv–xxii.

Reich, W., 'The Word "Bioethics": Its Birth and the Legacies of Those Who Shaped It', *Kennedy Institute of Ethics Journal*, Vol. 4, no. 4 (1994) pp. 319–55.

Reich, W., 'The Word "Bioethics": The Struggle Over Its Earliest Meanings', *Kennedy Institute of Ethics Journal*, Vol. 5, no. 1 (1995) pp. 19–34.

Reubi, D., 'The Will to Modernize: A Genealogy of Biomedical Research Ethics in Singapore', *International Political Sociology*, Vol. 4 (2010) pp. 142–58.

Reubi, D., 'The Human Capacity to Reflect and Decide: Bioethics and the Reconfiguration of the Research Subject in the British Biomedical Sciences', *Social Studies of Science*, Vol. 42 (2012) pp. 348–68.

Reverby, S., and Rosner, D., 'Beyond the Great Doctors', in S. Reverby and D. Rosner (eds), *Healthcare in America: Essays in Social History* (Philadelphia: Temple University Press, 1979) pp. 3–16.

Reverby, S. (ed), *Tuskegee's Truths: Rethinking the Tuskegee Syphilis Study* (Chapel Hill, NC: University of North Carolina Press, 2000).

Reynolds, L. A., and Tansey, E. M. (eds), *Medical Ethics Education in Britain, 1963–1993* (London: Wellcome Trust Centre for the History of Medicine, 2007).

Richmond, C., 'Obituary: John D. Havard', *British Medical Journal*, Vol. 340 (2010) p. 1361.

Robinson, J. A. T., *Honest to God* (London: SCM Press, 1963).

Rogers, B., *A. J. Ayer: A Life* (London: Chatto and Windus, 1999).

Rose, N., 'Government, Authority and Expertise in Advanced Liberalism', *Economy and Society*, Vol. 22 (1993) pp. 283–99.

Rose, N., *Powers of Freedom: Reframing Political Thought* (Cambridge: Cambridge University Press, 1999).

Rose, N., *The Politics of Life Itself* (Princeton, NJ, and Oxford: Princeton University Press, 2007).

Rose, N., 'Democracy in the Contemporary Life Sciences', *Biosocieties*, Vol. 7 (2012) pp. 459–72.

Rose, S., and Rose, H., *Science and Society* (Harmondsworth: Penguin, 1969).

Rose, S., and Rose, H., 'The Radicalization of Science', in S. Rose and H. Rose (eds), *The Radicalization of Science* (London: Macmillan, 1976) pp. 1–32.

Rosenberg, C., 'Meanings, Policies and Medicine: On the Bioethical Enterprise and History', *Daedalus*, Vol. 128, no. 4 (1999) pp. 27–47.

Rothman, D., *Strangers at the Bedside: A History of How Law and Bioethics Transformed Medical Decision Making* (New York: Basic Books, 1991).

Rupke, N., 'Pro-Vivisection in England in the 1880s: Arguments and Motives', in N. Rupke (ed), *Vivisection in Historical Perspective* (London and New York: Routledge, 1990) pp. 188–209.

Russell, B., *Sceptical Essays* (London: Routledge, 2002).

Russell, B., *History of Western Philosophy* (London: Routledge, 4th edn, 2008).

Ryder, R. M., *Victims of Science: The Use of Animals in Research* (London: National Antivivisection Society, 1983).

Salter, B., 'Cultural Biopolitics, Bioethics and the Moral Economy: The Case of Human Embryonic Stem Cells and the European Union's Sixth Framework Programme', Working Paper I of the project 'The Global Politics of Human Embryonic Stem Cell Science' (November 2004). Available online at www.york.ac.uk/res/sci/projects/res340250001salter.htm.

Salter, B., *The New Politics of Medicine* (Basingstoke: Palgrave Macmillan, 2004).

Salter B., and Jones, M., 'Biobanks and Bioethics: The Politics of Legitimation', *Journal of European Public Policy*, Vol. 12, no 4 (2005) pp. 710–32.

Salter, B., and Salter, C., 'Bioethics and the Global Moral Economy: The Cultural Politics of Human Embryonic Stem Cell Science', *Science, Technology and Human Values*, Vol. 32 (2007) pp. 554–81.

Sandbrooke, D., *White Heat: A History of Britain in the Swinging Sixties* (London: Little, Brown, 2006).

Santry, C., 'Sir Ian Kennedy Champions "Fearless" NHS Regulator', *Health Service Journal*, 12 November 2009. Available online at www.hsj.co.uk.

Sass, H. M., 'Fritz Jarr's 1927 Concept of Bioethics', *Kennedy Institute of Ethics Journal*, Vol. 17, no. 4 (2008) pp. 279–95.

Saundby, R. M., *Medical Ethics: A Guide to Professional Conduct* (London: Charles Griffin, 1907).

Schamroth, A., 'A Medical Student's Response', *Journal of Medical Ethics*, Vol. 7 (1981) pp. 191–3.

Scorer C. G., and Hill, D., 'Continuing the Debate: The Role of the Medical Ethicist', *Journal of Medical Ethics*, Vol. 4 (1978) p. 157.

Scott, J., 'History-Writing as Critique', in K. Jenkins, S. Morgan and A. Munslow (eds), *Manifestos for History* (London and New York: Routledge, 2007) pp. 19–39.

Scott, P., *The Crisis of the University* (London: Croom Helm, 1984).

Shapiro, D., 'Nuffield Council on Bioethics', *Politics and the Life Sciences*, Vol. 14, no. 2 (1995) pp. 263–6.

Shaw, G. B., *Doctor's Delusions, Crude Criminology and Sham Education* (London: Constable, 1932).

Sheard, S., 'Quacks and Clerks: Historical and Contemporary Perspectives on the Structure and Function of the British Medical Civil Service', *Social Policy and Administration*, Vol. 44 (2010) pp. 193–207.

Shotter, E., 'Self Help in Medical Ethics', *Journal of Medical Ethics*, Vol. 11 (1985) pp. 32–4.

Shotter, E., 'A Retrospective Study and Personal Reflection on the Influence of Medical Groups', in L. A. Reynolds and E. M. Tansey (eds), *Medical Ethics Education in Britain, 1963–1993* (London: Wellcome Trust Centre for the History of Medicine, 2007) pp. 71–118.

Shotter, E. (ed), *Matters of Life and Death* (London: Darton, Longman and Todd, 1970).

Singer, P., 'Moral Experts', *Analysis*, Vol. 32 (1972) pp. 115–17.

Singer, P., *Animal Liberation: Toward an End to Man's Inhumanity to Animals* (London: Cape, 1976).

Singer, P., 'Introduction', in P. Singer (ed), *Applied Ethics* (Oxford: Oxford University Press, 1986) pp. 1–9.

Singer, P., 'All Animals are Equal', in P. Singer and H. Kushe (eds), *Bioethics: An Anthology* (Oxford: Blackwell, 2000) pp. 461–70.

Singer, P., 'R. M. Hare's Achievements in Moral Philosophy', *Utilitas*, Vol. 14, no. 3 (2002) pp. 309–17.

Singer, P., and Wells, D., *The Reproduction Revolution: New Ways of Making Babies* (Oxford: Oxford University Press, 1984).

Smith, R. G., 'The Development of Ethical Guidance for Medical Practitioners by the General Medical Council', *Medical History*, Vol. 37 (1993) pp. 56–67.

Smith, R., 'All Changed, Changed Utterly', *British Medical Journal*, Vol. 316 (1998) pp. 1917–18.

Spencer, S. J. G., 'Human *In Vitro* Fertilization and Embryo Replacement and Transfer', *British Medical Journal*, Vol. 286 (1983) pp. 1822–3.

Squier, S., *Babies in Bottles: Twentieth Century Visions of Reproductive Technology* (New Brunswick, NJ: Rutgers University Press, 1998).

Stacey, M., 'Medical Ethics and Medical Practice: A Social Science View', *Journal of Medical Ethics*, Vol. 11 (1985) pp. 14–18.

Stark, L., *Behind Closed Doors: IRBs and the Making of Ethical Research* (Chicago and London: University of Chicago Press, 2012).

Stevens, M. L., *Bioethics in America: Origins and Cultural Politics* (Baltimore, MD, and London: Johns Hopkins University Press, 2000).

Strathern, M. (ed), *Audit Cultures: Anthropological Studies in Accountability, Ethics and the Academy* (London: Routledge, 2000).

Swales, J. D., 'Thoughts on the Reith Lectures', *Lancet*, Vol. 316 (1980) pp. 1348–50.

Szasz, T., *The Myth of Mental Illness: Foundations of a Theory of Personal Conduct* (London: Paladin, 1972).
Taylor, G. W., 'Letters to the Editor', *British Medical Journal*, Vol. 300 (1990) p. 395.
Testa, G., 'More than Just a Nucleus', in S. Jasanoff (ed), *Reframing Rights: Bioconstitutionalism in the Genetic Age* (Cambridge, MA: MIT Press, 2011) pp. 85–105.
Thomas, M., 'Should the Public Decide?', *Journal of Medical Ethics*, Vol. 7 (1981) pp. 181–2.
Thompson, E. P., *The Making of the English Working Class* (Harmondsworth: Penguin, 2013).
Thompson, I., 'The Implications of Medical Ethics', *Journal of Medical Ethics*, Vol. 2 (1976) pp. 74–82.
Timmermanns, S., and Leiter, V., 'The Redemption of Thalidomide: Standardizing the Risk of Birth Defects', *Social Studies of Science*, Vol. 30 (2000) pp. 86–102.
Toulmin, S., 'The Tyranny of Principles', *The Hastings Center Report*, Vol. 11, no. 6 (1981) pp. 31–9.
Toulmin, S., 'How Medicine Saved the Life of Ethics', *Perspectives in Biology and Medicine*, Vol. 25 (1982) pp. 736–50.
Turner, L., 'Does Bioethics Exist?', *Journal of Medical Ethics*, Vol. 35 (2009) p. 233.
Turney, J., *Frankenstein's Footsteps: Science, Genetics and Popular Culture* (New Haven, CT: Yale University Press, 1998).
Tutton, R., 'Person, Property and Gift: Exploring the Languages of Tissue Donation', in R. Tutton and O. Corrigan (eds), *Genetic Databases: Socio-ethical Issues in the Collection and Use of DNA* (London: Routledge, 2004) pp. 19–39.
Veatch, R. M., *Disrupted Dialogue: Medical Ethics and the Collapse of Physician–Humanist Communication, 1770–1980* (Oxford: Oxford University Press, 2005).
Vidler, A. R., *Scenes from a Clerical Life* (London: Collins, 1977).
Waddington, C., *Science and Ethics* (London: Unwin Brothers, 1942).
Waddington, I., 'Medical Knowledge and the Patronage System in Eighteenth Century England', *Sociology*, Vol. 8 (1974) pp. 369–85.
Waddington, I., 'The Development of Medical Ethics – A Sociological Analysis', *Medical History*, Vol. 19 (1975) pp. 36–51.
Wainwright, S. P., Williams, C., Michael, M., Farsides, B., and Cribb, A., 'Ethical Boundary-Work in the Stem Cell Laboratory', in R. De Vries, L. Turner, K. Orfali and C. L. Bosk (eds), *The View from Here: Bioethics and the Social Sciences* (Oxford: Blackwell, 2007) pp. 67–83.
Walters, L., 'The Birth and Youth of the Kennedy Institute of Ethics', in J. K. Walter and E. P. Klein (eds), *The Story of Bioethics: From Seminal*

Works to Contemporary Explorations (Washington, DC: Georgetown University Press, 2003) pp. 215–31.

Warden, J., 'Cardiac Surgery Inquiry Given Wide Remit', *British Medical Journal*, Vol. 317 (1998) p. 489.

Warden, J., 'High Powered Inquiry into Bristol Deaths', *British Medical Journal*, Vol. 316 (1998) p. 1925.

Warnock, M., *Ethics since 1900* (London: Oxford University Press, 1960).

Warnock, M., *Ethics since 1900* (Oxford: Oxford University Press, 3rd edn, 1978).

Warnock, M., *Education: A Way Ahead* (London: Blackwell, 1979).

Warnock, M., '*In Vitro* Fertilization: The Ethical Issues (II)', *The Philosophical Quarterly*, Vol. 33 (1983) pp. 238–49.

Warnock, M., 'Moral Thinking and Government Policy: The Warnock Committee on Human Embryology', *The Millbank Memorial Fund Quarterly. Health and Society*, Vol. 63, no. 3 (1985) pp. 504–22.

Warnock, M., 'Do Human Cells Have Rights?', *Bioethics*, Vol. 1 (1987) pp. 1–14.

Warnock, M., 'The Good of the Child', *Bioethics*, Vol. 1 (1987) pp. 141–55.

Warnock, M., 'Government Commissions', in U. Bertazzoni, P. Fasella, A. Klepsch and P. Lange (eds), *Human Embryos and Research: Proceedings of the European Bioethics Conference, Mainz 1988* (Frankfurt and New York: Campus Verlag, 1988) pp. 159–68.

Warnock, M., 'A National Ethics Committee', *British Medical Journal*, Vol. 297 (1988) pp. 1626–7.

Warnock, M., 'Embryo Therapy: The Philosopher's Role in Public Debate', in D. R. Bromham, M. E. Dalton and P. J. R. Millican (eds), *Ethics in Reproductive Medicine* (London: Springer Verlag, 1992) pp. 21–31.

Warnock, M., 'Commentary on "Suicide, Euthanasia and the Psychiatrist"', *Philosophy, Psychiatry and Psychology*, Vol. 5 (1998) pp. 127–30.

Warnock, M., 'The Politicisation of Medical Ethics', *Journal of the Royal College of Physicians of London*, Vol. 33, no. 5 (1999) pp. 474–8.

Warnock, M., *People and Places: A Memoir* (London: Duckworth, 2000).

Warnock, M., *Nature and Morality: Recollections of a Philosopher in Public Life* (London: Continuum, 2003).

Warnock, M., *Dishonest to God: On Keeping Religion Out of Politics* (London: Continuum, 2010).

Warnock, M., and MacDonald, E., *Easeful Death: Is There a Case for Assisted Dying?* (Oxford: Oxford University Press, 2009).

Watts, G., 'New Body Aims to Streamline Approval and Regulation of Research in NHS', *British Medical Journal*, Vol. 343 (2011) p. 7950.

Weatherhall, D., 'The Problems as Perceived by Medical Researchers', in CIBA Foundation, *Medical Scientific Advance: Its Challenge to Society* (London: CIBA Foundation, 1990) pp. 13–18.

Weindling, P., 'Human Guinea Pigs and the Ethics of Experimentation: The BMJ's Correspondent at the Nuremberg Medical Trial', *British Journal of Medicine*, Vol. 313 (1996) p. 1467.
Weindling, P., 'The Origins of Informed Consent: The International Scientific Commission on Medical War Crimes, and the Nuremberg Code', *Bulletin of the History of Medicine*, Vol. 75 (2001) pp. 37–71.
Welbourne, R. B., 'A Model for Teaching Medical Ethics', *Journal of Medical Ethics*, Vol. 11 (1985) pp. 29–31.
Werskey, G., 'The Marxist Critique of Science: A History in Three Movements?', *Science as Culture*, Vol. 16 (2007) pp. 397–461.
White, H., 'Afterword: Manifesto Time', in K. Jenkins, S. Morgan and A. Munslow (eds), *Manifestos for History* (London and New York: Routledge, 2007) pp. 220–31.
Whong-Barr, M., 'Clinical Ethics Teaching in Britain: A History of the London Medical Group', *New Review of Bioethics*, Vol. 1, no. 1 (2003) pp. 73–84.
Wilkins, M., 'Science, Technology and Human Values', in W. Fuller (ed), *The Social Impact of Modern Biology* (London: Routledge and Kegan Paul, 1971) pp. 5–10.
Williams, B., *Ethics and the Limits of Philosophy* (London: Fontana, 1985).
Williams, G., *The Sanctity of Life and the Criminal Law* (London: Faber and Faber, 1958).
Williams, S., *Climbing the Bookshelves: The Autobiography* (London: Virago, 2009).
Wilson, D., *Reconfiguring Biological Sciences in the Late Twentieth Century: A Case Study of the University of Manchester* (Manchester: University of Manchester Faculty of Life Sciences, 2008).
Wilson, D., *Tissue Culture in Science and Society: The Public Life of a Biological Technique in Twentieth Century Britain* (Basingstoke: Palgrave Macmillan, 2011).
Wilson, D., and Lancelot, G., 'Making Way for Molecular Biology: Institutionalizing and Managing Reform of Biological Science in a UK University during the 1980s and 1990s', *Studies in the History and Philosophy of the Biological and Biomedical Sciences*, Vol. 39 (2008) pp. 93–108.
Wilson, J., 'Abortion, Reproductive Technology and Euthanasia: Post-Conciliar Responses from within the Roman Catholic Church in England and Wales, 1965–2000', PhD thesis, University of Durham, 2010. Available online at http://etheses.dur.ac.uk/3076.
Wolstenholme, G. (ed), *Man and His Future* (London: J & A Churchill, 1963).
Wolstenholme, G. (ed), *Law and Ethics of Transplantation* (London: CIBA Foundation, 1968).

Wolstenholme G., and Fitzsimmons, D. W. (eds), *The Law and Ethics of AID and Embryo Transfer* (London and New York: Elsevier: 1973).

Woodford, P., *The CIBA Foundation: An Analytical History, 1949–1974* (Amsterdam: Elsevier, 1974).

Yoxen, E., 'Conflicting Concerns: The Political Context of Recent Embryo Research Policy in Britain', in I. Varcoe, M. McNeill and S. Yearley (eds), *The New Reproductive Technologies* (London: Macmillan, 1990) pp. 173–200.

Zallen, D. T., 'Regulating Research: A Tale of Two Technologies', *Technology and Society*, Vol. 11 (1989) pp. 377–86.

Zussman, R., 'The Contributions of Sociology to Medical Ethics', *Hastings Center Report*, Vol. 30 (2000) pp. 7–11.

Index